THE STONE MONKEY

An Alternative, Chinese-Scientific, Reality

THE STONE MONKEY

Bruce Holbrook

An Alternative, Chinese-Scientific, Reality

WILLIAM MORROW AND COMPANY, INC.
New York *1981*

Grateful acknowledgment is extended for permission to quote from or paraphrase parts of the following:
Stephen Dewar, "A Second Nervous System?" *The Canadian*, August 18, 1979.
Discover: The Newsmagazine of Science, "A Haunt of Flies," October 1980.
Diane Francis, "Sex, Cancer, and the Perils of Promiscuity," *MacLean's*, October 6, 1980.
Brian Inglis, "The Epidemic Trigger," *Omni*, November 1980.
Rick Levine, "Cancer Town," *New Times*, August 7, 1978.
Margaret Lock, lecture on East Asian and Western medicine in Japan, given at the Canadian Ethnology Society conference in Montreal, February 1980.
E. Alan Morinis, "Nature and Mind, Culture and Cancer," lecture given at the Canadian Ethnology Society conference in Montreal, February 1980.
Newsweek, Richard Boeth's review, "The End of Reason," of William Barrett's *The Illusion of Technique*, October 1978:98. Copyright © 1978 by Newsweek, Inc. All rights reserved. Reprinted by permission.
Psychology Today, "How Acupuncture Works: A Sophisticated Theory Takes the Mystery Out," June 1973. Reprinted from *Psychology Today Magazine*, © 1973, Ziff-Davis Publishing Company.
Reader's Digest, reprint of Dewar's article, above, November 1979. © 1979 The Reader's Digest Assoc. (Canada) Ltd. Reprinted by permission.

Library of Congress Cataloging in Publication Data

Holbrook, Bruce.
The stone monkey.

"Morrow quill paperbacks."

Bibliography: p.
Includes index.
1. Science, Ancient. 2. Philosophy, Chinese.
3. Science—Philosophy. I. Title.
Q124.95.H64 501 81-11063
ISBN 0-688-00665-5 AACR2
ISBN 0-688-00732-5 (pbk.)

Printed in the United States of America

First Morrow Quill Paperback Edition

1 2 3 4 5 6 7 8 9 10

BOOK DESIGN BY MICHAEL MAUCERI

To Hsia Po-yan before me
and my children after me

PREFACE

The popular Chinese novel, *Hsi Yu-Chi: The Westward Journey*, by Wu Ch'eng-En tells two stories, only one of which is understood by Western Sinologists. The first is an account of a Chinese Buddhist monk, Hsuan-Tsang, who was protected by a monkey of unique origin. He brought back from India certain Buddhist texts which served as a vital transfusion for the decadent Chinese social culture of the period. The other is an esoteric traditional Chinese allegory, referring to phenomena of which our Sinologists are ignorant. It tells how the monkey, born under unique conditions out of a stone, acquired superhuman powers that qualified him to be the guardian of precious, internationally transported knowledge. In decoded form this allegory describes the education of a bright and high-spirited but totally misguided and troublemaking young man, who is transformed into a man of knowledge and responsible action. The various gods and goddesses who restrain, guide and enlighten him represent his human teachers. That he is born of a mere stone and becomes superhuman is Wu Ch'eng-En's way of emphasizing the distance between ignorance and knowledge, between purposeless and humanely purposeful action, and the evolutionary nature of the journey between them. Wu Ch'eng-En's character is called *T'ai-Shen*: The Intense Spirit, which is descriptive of the quite un-supernatural, human, nature of the product of the esoteric traditional Chinese education-discipline in question.

I do not compare myself to the Stone Monkey. I am not made of the same (rare) stuff, nor have I come anywhere near mastering the traditional Chinese discipline, the study of which led me to write this book. But the intended function of this book, and its background, is enough like the decoded content of Wu Ch'eng-En's book to warrant the title I have given it. As I now understand it, my Western education had certain side-effects closely akin to the mental state of that confused and troublemaking monkey prior to his traditional Chinese scientific education. In addition, my traditional Chinese education was of just the same nature as the one that benefited Wu Ch'eng-En's "monkey." Like him, I went to a foreign country, received knowledge there, and returned to my own country with what I believe is a vital transfusion to my own culture. By "my own culture" I mean its system of beliefs and rules, the structured focus of which is science.

As I see it, the Western sciences, from physics to political science, are based on a dead and deadening view of reality and consequently —despite the brilliance, creativity, and good intentions of our best scientists—they transform the world more negatively than positively. These effects, interacting exponentially, are creating a catastrophe at all levels of survival: physical, biological, psychological, social, political, and spiritual. Of course, many share this perception of emergency. Some believe that our sciences must be fundamentally altered; others believe that our sciences in their present form will save us through further discoveries. I am one of the former. Although I believe that our technology has much more to offer us, I also believe that its scientific basis must be radically altered. Likewise, the social-scientific bases for social planning and government. So far, no one has been able to take any action on such opinions because a fundamentally different alternative to Western science has yet to be provided. That is what this book offers.

Our own scientists, even though some are now trying very hard, have not found a way out of their science to a superior alternative because the basic defect of Western science can be perceived only by standing outside it on alternative ground. Our problem requires an *unanticipated solution*, and that in turn requires an unconditionally heretical procedure of discovery. Cultural anthropology, which links us to foreign realities and suspends time, has served as such a procedure.

Over 2500 years ago, the inventors of Chinese science evolved a live and enlivening view of reality quite different from the one which underlies our sciences. Consequently, Chinese science, which has been evolving to this day, enlivens the world where ours deadens it, and it does so in ways of which we know almost nothing. As I see it, it is superior to our own not only in effect but also as a system of knowing. It is a "dragon" in the Chinese sense: a unified whole, that unites. The traditional Chinese scientific worldview is, or at least could guide us toward, the scientific alternative that we need. Such a vital transfusion of knowledge would promote survival at all levels.

This book, then, shows the basic defect in Western science and explains how it generates negative effects; gives my reasons for believing that Chinese science is superior to it; and introduces an alternative, originally Chinese, worldview, science, and, potentially, reality.

Writers of scientific or academic essays often imagine that they "establish" facts, positions, or truths. Although this book has much scientific content, I make no such claims. A writer on reality can only add a voice to the potentially infinite manifold of human voices which together are a constantly changing song through which humans pattern their future. A writer may assert or prove that a statement is true, but this establishes the writer's own position, not truth. If the writer's voice resonates with enough other voices in a new way, then the human song will change fundamentally, and this may involve establishing truths. If it does not, it has served only as a minor form of entertainment. This book is a statement of my own position, which was formed by thirteen years of research, reflection, and experience. Although it speaks with certitude of facts and truths, it claims to establish none, and it makes no predictions. My only claim is to be quite serious about everything I have said in it, and to be happy to participate in the manifold human song—largely because, along with my own voice, I have brought thousands of years of previously silenced ones back into it.

Not only has Western science reached its dead end, but as it does so its strongest and most creative vehicles are looking, some consciously and some without knowing it, toward Chinese science for a way out. Two fascinating books have been written that compare modern physicists' most advanced understandings to what they mistakenly call "Chinese mysticism," showing that for at least 3000 years the Chinese have been ahead of us in general theory about the physical aspect of the universe.* The same could be but has not yet been said of Chinese physics, chemistry, biology, and social science. As for Chinese medicine, its *efficacy* has been recognized but, because it is foreign and has revolutionary implications threatening to our medical scientists, pharmaceutical companies, physicians, and psychotherapists, its *theory* has been for all practical purposes ignored.**

What I have called "Chinese physics, chemistry, biology, and social science," on the one hand, and "Chinese medicine," on the other, are not separate disciplines. There is one basic, universal Chinese scientific theory of which each discipline is a further specification, and Chinese medicine is at the center of and overlaps with all of them. That is, Chinese physics is partly Chinese medicine, Chinese social science is partly Chinese medicine, and so on. This is because each specific science—each aspect of the world, that is—is recognized in traditional Chinese culture as worth understanding for no other reason than to promote human survival and welfare. Survival and welfare are one with health. This is reflected by the Chinese proverb *Liang-hsiang, liang-i*: He who makes a good scholar-official*** makes a good doctor (and vice versa).

The aspect of Chinese medicine which deals not only with the human body but also with the human psyche, human society, and the relations and interactions between humans and the rest of the world ("physics," "chemistry," "biology," and "social science") is called *shang-i* or *kao-i*:*upper medicine* or *high medicine*. This, then, is a traditional Chinese high-medical book written by a Westerner for Westerners, and its patient is Western science, and therefore, because its knowledge is its axis—Western civilization.

When I arrived in China's Taiwan Province in 1973 to study the living tradition of Chinese medicine, I had prepared myself for six years by studying Chinese language and culture and, at Yale University's Graduate School, the methods of cross-cultural comparison of worldviews and foreign systems of knowledge. Through an extraordinary coincidence in Taiwan I became the *following-son* (*t'u-erh*, pro-

* Fritjof Capra's *The Tao of Physics* and Gary Zukav's *The Dancing Wu-Li Masters*.

** This is less the case in Europe; the North American medical monopoly is self-interested to an extraordinary degree.

*** A person who knows at least the basics about everything and is in an official capacity to put his knowledge to use to serve all of society.

nounced *too-er*) of a mainland-born traditional Chinese savant and medical doctor, who, reciprocally, became my *teaching-father* (*shih-fu*, pronounced *shir-fu*). Following-son is the status one must assume in relation to a traditional Chinese savant (*shih*)* in order to be fully educated in any of the esoteric traditional Chinese disciplines, be it medicine, governing, music, architecture or the martial arts.

My six years of Western "high-educational" and "Sino-anthropological" preparation proved irrelevant to almost all that I learned: I encountered "virgin territory." I discovered that Westerners understand very little about Chinese science and are equipped, if "highly educated," *not* to be able to understand it.

A "traditional Chinese savant" refers to a person who has achieved full florescence as a human being, by being tutored in and practicing taxing forms of intellectual, emotional and physical self-cultivation. I found these men to lack the sanctimonious demeanor of the typical holy-man, regarding it as symptomatic of a covert arrogance and an unnatural form of detachment. They also lacked the patronizing disdain for the common people that is typical—though less so in the present than in past generations—of Western scholars. Finally, they avoid attention and refuse to be objects of research, rightly regarding that as an indignity, preferring instead to pretend ignorance. Consequently, Western missionaries and scholars have met these rare vehicles of the best of Chinese culture, often without knowing it, and failed, not treating them as colleagues but as objects of study, to establish a rapport with them. In turn, out of a mixture of ignorance and convenience, they have often mistaken Western-educated Chinese people of little or no traditional higher knowledge for Chinese savants. For those reasons and others mentioned in the text, very little of higher traditional Chinese knowledge has been consciously absorbed by Westerners, whereas a great body of pieces-out-of-context and spurious representations of that knowledge, which is then distorted into Western terms, is generally mistaken for such knowledge.**

* There are many Chinese doctors (*i-shih*) who are not also savants (*shih*). But all Chinese savants can be or are also excellent doctors. Hence the term *i* (medical) + *shih* (savant) = "doctor." The latter are and always have been rare people.

** There are exceptions. Foremost among them is Dr. Manfred Porkert, a German Sinologist who has written a book called *The Theoretical Foundations of Chinese Medicine* (1974, M.I.T.).

When I returned to Yale in 1974 to write my doctoral dissertation, I wanted to transmit what I had learned. Having grown accustomed among the real Chinese intelligentsia to open-mindedness and respect for all who have good intentions, it came as a shock to find that Western "intellectuals," with rare exceptions, weren't interested in a disclosure of the nature of genuine traditional Chinese science or a fundamental challenge. Furthermore I discovered that the prevailing opinion of China-specialists, Sinologists and anthropologists alike, was such as to deny any possibility of such a disclosure.

The content of that regnant opinion is as follows: First, genuine Chinese savants, largely due to the impact of Western science, no longer exist. At best, there exists a dim and distorted memory of traditional Chinese science of which some may claim possession but of which none could establish a legitimate claim.* Second, anything highly intelligent said by a professed living Chinese savant—even one whose status is widely and officially acknowledged—should be attributed not to his traditional education but to his inborn intelligence and, probably, a sly borrowing of certain Western philosophical and scientific ideas. Third, regardless of its source, be it a classical text or a living human, no traditional Chinese (which means purely Chinese) system of thought or practice, such as medicine, can even provisionally be regarded as scientific, and must be designated as "folk-, pseudo-, or proto-science," not "science," and, at its theoretical level, as "dogma" or "doctrine," and not "theory," "fact" or "hypothesis."

In appreciation of this "intellectual climate" among my academic superiors, my doctoral dissertation was designed to produce what might be called "self-calming" in the typical Westernly-scholarly reader. Consequently it says almost nothing of value about Chinese science or medicine. Since its completion, I have devoted much effort to the question of why we Westerners remain ignorant of, even hostile toward, some of the highest human achievements.

As the reader has guessed, this book derives from a "shadow" dissertation I began at Yale, simultaneously with the one that earned me my doctorate. This is not to suggest that anyone at Yale suppressed my "shadow" dissertation. It is to say that for me then, it was plainly

* I have strong evidence to the contrary. For example, the living tradition makes perfectly intelligible the Chinese scientific texts, which to Western Sinologists are at worst unintelligible and at best ambiguous.

wiser to submit a standard dissertation than to submit one that challenged the very grounds on which dissertations are judged.

By the time I had returned to Yale, after a year and a half in Taiwan, in 1974, my understanding of the nature of Chinese science had evolved through two stages. At the first stage, consistent with Western "scholarly" opinion, I saw Chinese science as an "intuitive" system to which strikingly effective techniques, such as acupuncture, are attached. My task would be to translate that intuitive system accurately into our language and analyze it in what I assumed were "objective," Western-scientific, terms. At the second stage, after comparing the roots of the Western and Chinese systems of knowledge, I wanted to show that there are many correspondences between the most advanced aspects of both.

As my respect for Chinese science increased with my understanding of it, I realized how much confidence I had placed in Western science on faith alone, through social conditioning, rather than through scientific reasoning. It was an unsettling experience, but it was rewarded by a liberation of thought and feeling. And I became able, with my new, bicultural, mind to compare the two paradigms without bias or compunction, my curiosity and the joy of the Trip now far more intense than the anxiety I sometimes felt in free-falling away from my own scientific culture. Distanced, I could see it clearly and as a whole. The reader may find that this book leads to a similar experience. (And because that experience is unsettling, I have taken care to provide an alternative to each Western religious, philosophical, physical, biological, medical, social, or political notion which I jeopardize in this book, so that there is always ground to stand on as the reader travels through it.)

Since that unsettling but liberating experience, after much study and a return to Taiwan, I came to understand that Chinese science, whose brightest flower is medicine, is superior to our own. It was through an intellectual and moral struggle, with tremendous pressure to suppress my best judgments so as to "insure my career," that I acquired the perspective taken in this book.

I am no longer concerned whether I have supplied proofs that the living tradition of Chinese science intensively transmitted to me in China is one and the same as in classical texts up to three thousand years old. If there is truth, beauty and great human warmth in that system, it matters little if it is five thousand or five days old. What matters is its potential benefits to those who are now living, and their descendants—including my own: in this book I simply let Chinese science stand, in the present, on its own ground. I hope that my book, which I know is faulted by many gaps in my own understanding, does sufficient justice to Chinese science and the reader.

I don't expect many academics to agree with this view of mine, but a book's value is intimately related to the extent to which the ideas in it are unlike others'. It is partially for that reason that I have written it.

—Bruce Holbrook

ACKNOWLEDGMENTS

Everyone with whom I have come into contact has been and continues to be my teacher and helper, so the following specific acknowledgments are biased.

I thank my children for giving me purpose.

I thank my wife Jeanine Bitton for giving me my children, for her sensitive and astute social-scientific research on my behalf, for her encouragement, and for her excellent suggestions regarding the content of the book.

Above all, I thank my *shih-fu* (teaching-father), Dr. Hsia Po-Yan of Hsü-chou city, Chiang-su Province, China, for his selflessly tireless instruction and encouragement, and for connecting my brain to my heart. The faults that can be found in this book are due to my failure to have perfectly followed him in intellectual clarity and balance of world-attitude.

I thank my undergraduate advisors at Wesleyan University, Professors David MacAllester and Willard Walker, for their love of creativity and their unconditional encouragement and confidence, and for sending me to China while I was yet an undergraduate.

I thank my tutor at Yale, Cornelius Osgood, for having set the very rare example of a full-fledged Western scholar. Four ways he did so were by emphasizing that an anthropologist should care about human beings, not his pet (and therefore inevitably petty) theory; by not assuming that truth is a Western monopoly; by defining culture subjectively and so recognizing that cultural anthropology is a transmission of knowledge among different peoples, not an analytical game; and by observing that one can know with certitude only after achieving "the proper mixture of Western Idealism (all Western philosophy actually being Idealist) and Chinese Materialism." (I agree fully neither with his prescription nor with his classification of Chinese philosophy, but I have greatly benefited from pursuing his suggestions.)

I thank my doctoral dissertation advisor, Professor Harold Scheffler, for setting an example of the best of Western logical-analytical thinking. That I have employed this influence of his to jeopardize the Western scientific paradigm in which the human logical-analytical capacity is, I believe, trapped will, I hope, elicit some appreciation, if not agreement, from him. It should be mentioned that he never denied my positive evaluations of the Chinese scientific paradigm and even agreed that a good case could be made for it. He simply felt, for practical purposes and genuinely in my interest, that such considerations, as well as my more radical differences with Sinologists and Sino-anthropologists, might unduly complicate acceptance of my dissertation. I agreed and still do.

I wish also to thank Dr. Li Yih-Yuan and his colleagues, of the Academia Sinica's Institute of Ethnology, for their broad-mindedness in publishing the heretical results of my initial research in the Republic of China.

Several of my colleagues at the University of Prince Edward Island have provided me with intellectual challenges and information from their areas of specialization. I especially thank Dr. Fr. Allan MacDonald (Sociology and Anthropology) for forcing me, by challenging it, to fully articulate my moral philosophy and for making me realize that a *genuine* Christian can fundamentally disagree without losing respect or warmth; and Professor Satadal Dasgupta (Sociology and Anthropology) for greatly enhancing my social-anthropological understanding, and for picking me up when I was way down. I also thank Professors Ronald Baker (English) and Thomas Spira (History) for their encouragement and advice with regard to approaching publishers.

I am indebted to my student assistant, Scott MacEachern, a graduate of the University of Prince Edward Island, for his extensive research into Western, especially biological, science on my behalf, his astute recognition of data relevant to this effort, and his challenges to my jeopardizations of Western scientific theories. I should add that he is as yet unsure of the extent to which he might agree with points made in this book and therefore should be held responsible for no more than the provenience of some of the data it is based on and for one of the arguments I have used.*

And I thank my typist, Judi Burke MacKinnon, not only for the excellence of her work but for her kind patience in retyping many pages, thought more than once to be final products, as new ways of expressing certain ideas occurred to me.

I owe the opportunity to have acquired the Chinese data for this book to the confidence and generosity of several agencies who supported eleven years of my research and/or writing. In order, they are Wesleyan University; The National Science Foundation; the Educational Testing Service; the Yale Concilium on International and Area Studies and its Director, Joseph Goldsen; Fulbright-Hays; the National Institutes of Health (1973–75); the Canada Council of the Arts, Social Sciences, and Humanities; and the University of Prince Edward Island (1978–79). Although this product of their support—unlike my dissertation and some traditional academic efforts—differs somewhat from what those agencies normally expect, I hope that they will have perceived and accepted the difference as an effect of a sincere pursuit of truth and humane beauty. Beyond that, understanding that the funds those agencies made available to me originally took the form of United States and Canadian peoples' tax-dollars, I hope that, someday, this book will have benefited those people; at least, that is the intention which underlies it.

* The "intestinal-parasite argument" in Chapter 5.

CONTENTS

FIGURES

PREAMBLE:

THE SEA TURTLE AND THE FROG

. . . a language or a system of a given structure can be somewhat altered from within, but it cannot be *revised structurally* without going *outside* the former system.—COUNT ALFRED KORZYBSKI

Science as we in the West know it is in terminal crisis and must be fundamentally altered in a *truly* revolutionary way for the first time in 2500 years. The crisis is of both the intellect and the heart. At the intellectual level, science is increasingly self-contradictory and inadequate. To borrow an expression from Thomas Kuhn, author of *The Structure of Scientific Revolutions,* our "scientific paradigm" is breaking down. Our initial assumptions about reality are failing to serve their purpose, and there is a growing suspicion that they are false. At the level of the heart, it is becoming clear to both scientists and non-scientists that the *net* effect of science as we know it has been to decrease rather than increase our survival-potential and to impoverish rather than enrich the quality of our lives. For example, the democratic benefit of the electronic reproduction of music is eclipsed by accidents of nuclear fission and a social system which so thoroughly isolates people from each other that most need chemical help to make it through the day, an increasing number switching "life-styles" and self-images in desperation.

The crisis is hard to face directly. As Kuhn has pointed out, whenever an established scientific paradigm is crumbling, our scientists adhere to it until they have created a new, superior one, because if they rejected the old one they would have nothing to cling to.

Kuhn's fascinating and important book is consistent with assumptions which, as I see it, are false. One is that scientists are still scientists, true men of knowledge, when they refuse to confront the chaos into which their paradigm has led them. I agree rather with Confucius, who stated that a true man of knowledge forthrightly admits what he does not know, and never pretends to know in order to fill that gap.

The second assumption is that an alternative, superior paradigm could not already exist outside the European-American tradition with which he exclusively deals. I believe that one does. The required scientific revolution should be a qualitative leap taken with new modes of thought and perception from ours to that one, and the alternative reality it leads to. This would eliminate the fear of chaos resulting from scientific-paradigmatic breakdown and accelerate scientific progress. The paradigm in question is both foreign and of ancient origin, but it survives as a living tradition in an increasingly rare environment, chiefly in Taiwan: the traditional Chinese one. Our scientific "authorities" have called this system of knowledge and action "magical," "mystical," or at best "proto-scientific." But they are trapped and trap all others in their unquestioned belief that they have a monopoly on truth—or the "search" for it. They are blind to Chinese—for that matter, all exotic—science, distorting it into their "objective" terms instead of learning it in its own. So I will evaluate the Western and the traditional Chinese scientific paradigms, with respect to their "scienticity" and their effects on human beings, in the terms of *both* paradigms. The conclusion one must come to, it seems to me at least, is that on both counts actually it is *our* science that is the magic, mysticism, pseudo-science, or proto-science, and traditional Chinese science that is the true science. We have got it backwards and upside-down.

Two important distinctions must be made before the perspective I have taken can be clear. First, there is the distinction between a people's scientific paradigm and a people's science. This book chiefly concerns paradigms, not sciences. The paradigm gives the science intention and direction; it gives it its life-quality and determines the angle at which it approaches and affects Nature. I am attacking our paradigm, not our science. In other words, I am attacking *what* our scientists do and the scientific-paradigmatic conditions which oblige them to do it—not *how well* they do it.

No one could deny the efficacy or the genius of Western science: it has transformed the world more than any other science, and the genius and creativity of its innovators is amazing and admirable. What concerns me is the sector of reality that our scientists, because of the closed and highly selective nature of our paradigm, have never even addressed. What's more, that sector is *most* of reality: objectively viewed, the scope of our science is one-dimensional and microscopic, although it is thought to be exhaustive and "universal." And the inevitable result of this incredibly intense but also peculiarly *narrow* scientific activity is a lowering of human potential for survival and of the quality of life, specifically through *side-effects.*

Second, there is a distinction between a scientific paradigm and the scientists who are directed by it. Our scientific paradigm directs our science in ways of which they are quite unaware and of which, under normal conditions, they could not be aware even if they wished. Consequently, with the exception of the *architects* of our scientific paradigm, who, under abnormal conditions, were aware of this direction and its effects, and with the exception of "scholars" who know about alternative paradigms for more humane science but have suppressed them, I place no blame on our scientists or scholars for the side-effects which they constantly produce. On the contrary, I regard them as victims of these effects. This book is written with affection for them, as much as for those on whom they unintentionally bring side-effects.

Like the thirsty horse sensing the stream, we are already turning toward the Chinese paradigm although we do not know it yet. At the outskirts of our sciences, in avant-garde physics and ecology, our most progressive and brilliant thinkers, in a healthy reaction to the inadequacy of our present paradigm, are producing some elements of the Chinese paradigm—elements whose revolutionary implications have yet to be fully recognized. They fundamentally undermine our scientific paradigm, on one hand, and primitively foreshadow traditional Chinese science, on the other.

As the Korzybski quote that opens this chapter states, one cannot fundamentally revise one's scientific paradigm without standing outside of it. But to stand requires ground. To innovatively leap into space is to orbit around the object leaped from and, ultimately, to crash-land where one started, because one begins with materials from that object. The alternative ground required by *genuine* innovation is a scientific paradigm basically different from one's own. A complete alternative scientific paradigm has never been considered by Westerners. Isolated theories and techniques have been picked up, such as the notion of a life-process-governing energy (in Chinese, *ch'i*—adopted as an hypothesis by Soviet scientists, not ours) and, much earlier, gunpowder, but never has a whole system been provided. As long as there are only fragments of an alternative, there are holes to fall through. One then naturally stands on the grounds of one's own paradigm, rather than fall. For example the present attempt to account for the efficacy of acupuncture is actually less scientific than the unknown Chinese one which grounds the Chinese theory. Never has Western science really been in a position to appreciate another science or to look at itself. This book provides that opportunity. My "half-Chinese" explanation of Chinese science makes possible for the first time an objective (exterior) perspective on our own science and reality.

To explain and understand a truly different scientific paradigm requires new modes of knowing. In terms of my generation, one must be prepared to take a Trip between two realities. If two paradigms are really fundamentally different, as ours and the Chinese are, neither can be clearly understood in terms of the other. What actually goes on when one stands in two realities is a test of the two with reference to a third factor: their relative capacities to make sense of the world and, where the world *is* partly one or the other, the extent to which they commute beauty to the world.

Let me open the way toward such comparison. Traditional Chinese science has persisted as a full-fledged and fully tested system probably twice as long as the 2500 years during which science as we know it has uncertainly evolved to its present impasses. The idea that a superior alternative exists is doubtless harder for many to accept than is the reality of those impasses. This is not for any *scientific* reason. It is because almost all of us have been conditioned to believe that modern science, the Western fruit of classical Greek scientific philosophy, is qualitatively supreme among all systems of knowing. All others are either "primitive" and false or are unsuccessful attempts to attain our level. In short, we have been taught to believe that our culture is the flower of all human endeavor and the goal of the rest of the world, our science being the essence of that flower. When "proof" of this is required, a hydrogen bomb is exploded so as to produce a "flower" which everyone can see. Therefore, were anything better than our science to exist, it would be our creation. Only then could it represent the next step up the evolutionary ladder, a step always taken by Europeans, Americans, or those they have taught to think and act like them. For the sake of objectivity (balance between two subjects), then, I will make every effort to neutralize that conditioning. Accordingly I portray Western science and its social sources and effects very harshly, and portray Chinese science and its social sources and effects very favorably. Needless to say, this could lead to distortion if certain other conditions aren't met; so they are.

First, those portraits are not based on selections of features consistent with a predetermined point I want to make. Rather, they emerge directly as descriptions of what I genuinely understand to be the *basic features* of each paradigm. Second, there is the understanding that neither system is perfectly homogeneous: there is good (or validity) in the bad (or invalidity) of one and bad in the good of the other. Nothing can change that, which means that is natural—beyond human design. What counts is whether or not one system is *basically* good and the other *basically* bad: that is where human intelligence and intention, or "free will," *do* come in. This book, dealing with the basic features of each paradigm, deals with human intelligence and intention. Unlike Nature (that which cannot be changed), they may be evaluated with productive results.

Any people's science must be valid to an extent. Nicely complementing Benjamin Whorf's Theory of Cultural Relativity—the anthropological discovery of the arbitrary influence of language and paradigm (or worldview) on a people's reality, the biologist and brillant originator of Systems-Theory, Ludwig von Bertalanffy, wrote:

> Any organism, man included, is not a mere spectator, looking at the world scene and hence free to adopt spectacles, however distorting, such as the whims . . . of language have put on his metaphorical nose. Rather he is a reactor and actor in the drama. . . . (P)erception must allow the animal to find its way in the world. This would be impossible if the categories of experience, such as space, time, causality, were entirely deceptive.

In other words, *any* human language and worldview, or scientific paradigm, must to an extent reflect the nature of the interaction between humans and the rest of the world, and, therefore, be valid to some extent. The validity of a scientific paradigm, then, has to do with the accuracy of its reflection. All paradigms have something in common, and are good to an extent, because they are all produced by humans; they differ in the extent to which they accurately reflect reality, and thus, to the extent to which they permit humans to harmoniously interact with Nature and survive—preferably, with florescent dignity.

An important implication of von Bertalanffy's undeniable observation is that a people's scientific concepts and paradigm must be aligned with their *perceptions* if those people are to survive. Accordingly, I will devote much attention to the extent to which Western science, in contrast with the Chinese, *is deliberately misaligned with human perception;* to the fact that Western science and our general culture, which it configures, drastically decreases human perception; and to the matter, mastered by the architects of the Chinese paradigm, of heightening perception—*seeing* (*kuan*).

My claim that traditional Chinese science is more valid than ours will have been immediately rejected as outrageous by some, have surprised and raised skeptical interest in others, and have struck a few others as something they had at the back of their own minds. Although this book provides enough reasons and facts, I think, to diminish the strongest skepticism, it cannot open a closed mind. Those whose minds are open, however, are invited to consider the reasons and evidence for my claims.

Let me further open the way to comparing the two paradigms. It may be immediately recognized that anyone who uses Western science to judge Chinese science stands on very thin ground. Although it has been around in basic form for 2500 years, actually European-American science has actively evolved for only about 300. This is because of the discontinuous and repetitive history of European-and-American culture. The evolution of Western science from its root (classical Greek) form was arrested by the Roman conquest; the Romans did little of scientific significance with the Greek thought they absorbed, and what they absorbed was transmitted only fragmentarily to the Teutonic barbarians to whom Rome fell; they, investing in a religious worldview, failed to adopt it as a basic mode of thought. Greek science and philosophy then enjoyed a Renaissance toward the end of the Middle Ages, chiefly due to Arabic transmissions to Europe. Slowly, it became established, about 300 years ago, as the scientific paradigm underlying what we know as science today. In short, our science has been a mere flash in the historical pan. What has been portrayed as the White Miracle of rapidly evolving intelligibility is really an unnaturally rapid and probably short-lived aberration. Is this solid ground to stand on to judge alternative systems of knowing?

In contrast, traditional Chinese science, or high medicine, has endured for up to 5000 years within a great civilization. The extraordinary duration and continuity of that civilization is a testament to the fact that Chinese science *has* advanced human survival, and quality of life. Had it not, it would long ago have been altered by Chinese thinkers and protested against by the people under its canopy. There are, of course, some hypothetical objections to that observation. Each boils down to the unpleasant and quite unjustifiable assumption held to the light and decried by the exceptional Sinologist Pierre Ryckmans, author of *Chinese Shadows*, that the Chinese, because of racial characteristics, are relatively incapable, as compared to Westerners, of free inquiry or democratic protest—unless, of course, they are contemporary Chinese who fostered or have "benefited" from Communism, which, after all, is a *Western* invention.*

It should be clear, then, that what actually is outrageous is not the claim that traditional Chinese science is superior to ours, but the fact that throughout the 300 years during which they have been relatively aware of China, Westerners have unexceptionally failed to entertain the possibility of that being the case.

The stance I have taken is radically heretical, but it directly follows from some precedents of the highest respectability. In 1946, the philosopher of science F.S.C. Northrop published a book called *The Meeting of East and West*, in which he demonstrated that Oriental (Indian and Chinese) science is empirical, that is, factual in the scientific sense, to an extent that makes the empiricism of Western science (on which it prides itself) look quite weak in comparison. In 1947, the great nuclear physicist, Neils Bohr, who with Heisenberg is the chief architect of quantum-theory, was knighted for his achievements and chose as the symbol for his coat-of-arms the *T'ai-Chi* Diagram, the symbol of the Chinese scientific mode of thought. He did so out of a recognition of the consistency between what he saw as the "mystical" high Chinese worldview and the most advanced and revolutionary findings of modern physics, and out of admiration for that ancient mode of thought, to which he had directed his serious attention since visiting China in 1937. As an expression and extension of Bohr's rec-

* As to the proposition that the Chinese Communist Revolution constituted a popular protest against the traditional system, and, by implication, the social aspects of Chinese science, the present effort cannot contain the facts and common-sensical arguments which show it to be wholly false. Suffice it to say that without Western manipulation it wouldn't have happened, and to wait for a Chinese Solzhenitsyn to tell how the Communists used *traditional* ideas to (temporarily) capture and gull the hearts of enough people to get their job done.

ognition, in 1975 appeared the physicist Fritjof Capra's book *The Tao of Physics,* in which he demonstrates in detail how high Chinese and Hindu thought essentially correspond to the view of the universe that modern physics has led to. Then in 1979 appeared Gary Zukav's *The Dancing Wu-Li Masters,* which makes the same point. Meanwhile, since 1956, Joseph Needham has been producing an encyclopedia, *Science and Civilization in China,* to show that traditional Chinese science is empirical, ingenious, and rich, and that it parallels Western science in several respects—one notable respect being a "wave," as opposed to "particle," physics that correlates with the "wave" half of our physics.*

Northrop's book, the only one that treats Chinese *science,* as opposed to Chinese "mysticism," as a system of knowledge that in some respects *challenges* our own, has been conveniently ignored (though widely read) as an oddity, despite its scientific-philosophical rigor and the absence of any rational reasons for doing so. But Bohr's, Capra's and Zukav's recognitions signal that the reality of Chinese science will not be denied for much longer, and Northrop's book, though outdated, inevitably will have its day. This is indicated, for example, by the growing number of young physicists who take high Hindu and Chinese thought seriously and the growing number of European (and to a much lesser extent, American) medical scientists who are heretically taking Chinese medical theory seriously. The principal force among the latter has been Dr. Manfred Porkert, author of *The Theoretical Foundations of Chinese Medicine* (1974). A brilliant German Sinologist and medical scientist who briefly studied under the Chinese physician I am apprenticed to, Dr. Porkert has demonstrated under Western clinical conditions the power of Chinese diagnostic technique, and the theory behind it; and European medical scientists are beginning to take the Chinese theory seriously, in its own terms.

Although recognition of the power of Chinese science has been inexcusably slow in light of the fact that Bohr's recognition occurred more than forty years ago, nevertheless in this century the general view of Chinese science has changed radically. At the turn of the cen-

* Although it is Needham who also established the understanding, here regarded as a grave misunderstanding, that Chinese science is "*proto*science," a system of knowledge that never fully flowered as ours did; nevertheless he more than anyone else has awakened Westerners to the existence of traditional Chinese science. He was the first to compare it and its philosophical roots with its Western correlates on the basis of extensive Chinese data.

tury, it was denied that Chinese science existed. Then Chinese science was declared "protoscientific," and then, perhaps comparable in some respects to our own. My belief that Chinese science is genuine science whereas ours is not, is the inevitable, last step in this progression. All that has prevented other Westerners who have explored Chinese science from making a statement this forceful and heretical is that none have had the opportunity to directly absorb and be absorbed by the Chinese scientific tradition: the traditional teaching-father:following-son relationship is absolutely essential to this. Only in this relationship is traditional Chinese knowledge transmitted whole and at full intensity. In turn, the chance that a Westerner, particularly a "higher-educated" one, would genuinely accept and be accepted into this relationship—one requirement of which is a deep, culture-transcending, personal *rapport*—is as low as could be. All that I will have to say, then, is simply the result of having become able to understand Chinese science in its own terms, as traditionally communicated in the living tradition, instead of in the inevitably distorting translational terms of Western science and of the Western worldview of which our science is a part. It is not that my predecessors in the tradition of open-mindedly investigating Chinese science in relation to Western science have misjudged or lacked in scholarship but that their exposure to Chinese science has been indirect and therefore insufficient.

To turn, from the few who have genuinely tried to touch Chinese scientific reality, to our Knowledge-Specialists as a group, their deliberate ignorance and tacit belief in the inferiority of foreign systems of knowledge is a trait that could only be true of a benighted civilization. Yet it is true of the minds behind ours—more true of ours than any other in history. So powerfully ingrained is the assumption of Western superiority that even the best-intentioned and brightest are unconsciously subject to it. What other than this could explain the fact that, despite his own understanding that the Chinese produced a worldview consistent with his theory of the physical aspect of the universe, Bohr himself tacitly assumed that the Chinese never produced *sciences* consistent with that worldview? We are willing, like Bohr and Capra, to grant Orientals the ability to have "intuitively" recognized the basic nature of the universe, but it is unthinkable that any people other than ourselves could have produced equally powerful *science,* or have used *scientific* methods to arrive at such understandings.

In "Three Masks of the Tao," a paper published in 1979 by the Teilhard Centre for the Future of Man, Joseph Needham, in his maturity and after decades of reflection on the matter, affirms Capra's understanding and asks the question that naturally follows from it:

> Essentially, Capra's argument is that modern sub-atomic physics has made it quite clear that reality completely transcends all ordinary language, and that this was seen intuitively by the Taoist thinkers of ancient China. In the sub-atomic world, the concepts of space and time, the idea of separable material objects, and the usual understanding of cause and effect, have all lost their meaning. Mass and energy are inter-convertible, radiation is "not exactly" waves and "not exactly" particles. Time does not uniformly flow, changes always include the observer in an essential way, and no precise prediction is possible. Polar opposites are complementary rather than antagonistic, particles are both destructible and indestructible, matter both continuous and discontinuous, and objects are relational events rather than substances, spontaneous dynamic patterns in a perpetual dance. Reality is beyond existence and non-existence. Existence itself is a statistical conception. These ideas are similar, he finds, to the thoughts of Taoist philosophers and those forms of Buddhism most influenced by Taoism. The only unanswered question is how it came about that the ancient and mediaeval thinkers of China and Japan came to conclusions so close to those we have now arrived at with a great deal of trouble, building gigantic cyclotrons and following laboriously the traces of hadrons, electrons, photons and the like in bubble-chambers.

This book solves that "mystery": I know how those conclusions were arrived at, because that process of discovery is necessarily recapitulated by every student of Chinese science. I will show that that process is *scientific* in the strictest sense, and I will show the reader how to use his own senses and intelligence to independently recapitulate that process for him- or herself.

Needham, having pointed at the horizon and the galaxies of non-Western knowledge, then withdraws into his study in full agreement with the great majority of Western scholars: "Also, of course, it was one thing to guess it, and quite another thing to prove it; without modern science that could not have been done."*

* Elsewhere, however, Dr. Needham has communicated a somewhat different understanding. To the Taiwanese professor Wu Cha-Jen he stated that Chinese

Our culturally distinctive negative attitude toward foreign systems of knowledge, like any cultural trait that persists, has a function: to perpetuate the cognitive monopoly of our intellectual elites, by obstructing the reception of foreign data which would reveal, with the bright light of comparison, that that cognitive monopoly is an historical aberration with minimal potential as a tool of human survival. The proponents of this negative attitude toward foreign systems of knowledge have in the past been our missionaries and, more recently, a majority of our scholars abroad. Among them the most salient have been our conventional medical scientists, who, of course, have acceded to the sacred status that our men of God once enjoyed. They exhibit the same sacredized closed-mindedness as did their predecessors, single-mindedly propagating their own way, which they assert is *the* true one, and systematically subverting all others, under the guise of benevolent zeal. They study "health-care-delivery" in foreign societies on the unquestioned assumption that they can evaluate, in Western-medical terms, any foreign system of medicine without having bothered to learn it.

Twenty-five hundred years ago, such "men of knowledge" were accurately portrayed with reference to less offensive but as intellectually diminutive types by the Chinese savant Chuang-Tzu, when he spoke of the frog who presumed to invite a sea turtle into his dilapidated well to teach him the truth about water. The sea turtle, unable to fit into the well, declined. Similarly, the true intelligentsia of China has for centuries declined the zealous invitation of the representatives of our Western mission, quietly transmitting their own knowledge and technology, generation after generation, outside the sphere of Western awareness, knowing full well that the day might come when it would find willing receivers in the West. I invite the reader to journey from the well of Western science, in which all our past scientific and social "revolutions" are contained as ripples, to the alternative reality of genuine, Chinese, science to examine my reasons for saying it is both intellectually superior to and more humane than ours.

When something is taken to an unhealthy extreme its opposite

---

social as well as physical science is empirical, that Western science is not in fact empirical, and that Westerners understand much less than do Chinese about machines—*The Chinese Times* [Taiwan]: 19 August 1978. I suspect, and hope, that Dr. Needham, whom I greatly admire and to whom I owe my initial interest in Chinese science, is on the brink of adopting a position similar to my own.

comes into existence to fill the vacuum and restore balance. Such an opposite is cultural anthropology. It arose to reintegrate our fractured sciences and to reconnect us with the rest of human reality. Where our worldview and academic system are highly specialized into disciplines that have increasingly little to do with each other, the scope of cultural anthropology is all human activities, including philosophies, religions, and scientific disciplines. Where sociology, its predecessor, is occupied chiefly with European and American society, cultural anthropology is occupied chiefly with foreign societies. Most important, where history, sociology, and psychology select, describe, and interpret foreign human behaviors in Western terms, anthropology seeks to portray them in their own, foreign, terms, free of the distortion of Western selection and interpretation. Toward this end, a cultural anthropologist tries to absorb the culture of the foreign people he or she studies by becoming a member of the society that is the vehicle of that culture. An anthropologist submits to a total foreign education in order to become able to see and do as do the people in question. In this way, for the first time in Western history an intellectual discipline has sought to get Western consciousness outside of its own narrow bounds and to break the monotony of the lone Western intellectual voice by letting the myriad voices from the rest of the human galaxy be heard.

Necessarily breaking away from its Western-centric philosophical-historical-sociological roots in slow stages of evolution, cultural anthropology has become increasingly open-minded over its one-hundred-year history, and has influenced psychology, sociology, history, and even philosophy in its open-minded and wholistic direction. At first, it was strictly understood that all foreign knowledge is evolutionarily behind ours, a curious object to be recorded, analyzed, and explained in Western-scientific terms, starting with the "earliest" forms of it. To promote this perspective, at first, primitive foreign societies were the sole objects of investigation. But that backfired. The extraordinary Benjamin Whorf, on the grounds of fieldwork among Zuni Indians, realized that thought and probably even sensed "reality" is inseparable from and intermixed with a people's language. Hence "reality" varies from one people to another, and until we recognize this variation we remain trapped, insofar as our scientific understanding goes, by our language, worldview and "reality." A result is often mistaking for "logical necessities" and "facts" mere quirks of our language and worldview which, moreover, have no necessary relation to the rest of the world. Whorf found that Zuni better expressed modern physical reality than did modern physical terminology. So he wrote:

> We shall no longer be able to see a few recent dialects of the Indo-European family and the rationalizing techniques elaborated from their patterns, as the apex of the evolution of the human mind, nor their present wide spread as due to anything but a few events of history—events that could be called fortunate only from the parochial view of the favored parties [who, thirty years later, when our science has obviously backfired, include only our scientists and the industrialists and bankers they work for, not also the people they were supposed to serve]. They, and our own thought processes with them, can no longer be envisioned as spanning the gamut of reason and knowledge but only as one constellation in a galactic expanse. (pp. 215–18)

Then it was discovered that even primitive culture may include systems of knowledge that are more sophisticated than their Western correlates. In the 1950's, Harold Conklin of Yale University discovered that botanical classification among Philippine tribespeople is more elaborate than Western botany's, and "ethnoscience," the study of foreign sciences not as mere curiosities but as alternative systems of knowledge, then became "legitimate." Simultaneously, Needham's *Science and Civilization in China* began to appear, volume after volume, and cultural anthropology widened its scope to include civilizations—with ours as an object as well as a source of understanding. With Whorf's objectivity, ethnoscientists' findings, and Needham's revelations, the floodgates of foreign scientific knowledge were opened.*

A complete leap out of the Western worldview into a foreign one was finally made in the late 1960's by the anthropologist Carlos Castaneda, apprentice to the Yaqui sorcerer, Don Juan. Having made it

* More recently, some cultural anthropologists, led by Magoroh Maruyama, introduced knowledge, chiefly sociological knowledge, from foreign societies into Futuristics, examining foreign traditional social structures, organizations, and values toward making our future superior to our present (see Maruyama, 1978). And, on Whorf's line, Maruyama has recently (1980) investigated the effects of cultural processes of "pattern formation" on "the development of science-theory types," making it plain at the outset that he does not assume that Western science has a monopoly on truth and that he understands that Western science is conditioned by nonscientific cultural variables.

clear that the worldview and reality of the Yaqui sorcerer is simply not intelligible in Western terms, Castaneda opened the Western mind to the reality of the self-contained, intellectually and emotionally restrictive nature of the Western worldview, and the fact that Western science cannot explain all human or natural phenomena better than can any "folk"-science. Castaneda was rejected by the majority of anthropologists as an ingenious charlatan, on the grounds that his description of Yaqui sorcery does not tally with established knowledge of Yaqui culture. It never occurred to his detractors to suppose that he had made contact with an intellectual elite of sorts who, like Western anthropologists, have conscious knowledge of the popular culture of their own people and are therefore not bound by it, but free to innovate. How trustworthy Castaneda is, I don't know: I do not have the pleasure of his acquaintance. That is not the point. The point is that these "scholars" regarded their narrow and easily jeopardized grounds sufficient for dismissing Castaneda because they found the recognition of *alternative realities* intolerable. It changed them from Objective Analysts into naïve explorers of the human reality-galaxy; it threatened them with having to pry themselves loose from their own scientific paradigm in order to understand another. To directly confront an alternative reality, without the armor of one's status as an official Knowledge-Specialist and the insulation of the Objective Perspective, was frightening, because they knew, deep down inside, that without that armor and insulation they have little to guarantee the survival of the system of knowledge they identify with. Castaneda may or may not be trustworthy, but his critics are *definitely un*trustworthy.

Despite primitive defense mechanisms such as theirs, Castaneda's books and other anthropologists' exposures to foreign men of knowledge are having their inevitable, mind-opening effect. The smugly closed Western paradigm must open, because no unhealthy extreme can persist if a people is to survive. And people, through society, powerfully seek to survive. The more powerfully our Knowledge-Specialists narrow our options and misguide us, the more powerfully will alternative knowledge be sought.

I should immediately emphasize that the present effort, and what it describes, diametrically differs from Castaneda's, and is much farther away than his from our scientific reality in an important respect. According to my understanding, genuine science concerns what usually happens, first, and what exceptionally happens, second: it seeks general understanding and, thereby, harmony with Nature. What Castaneda describes is a half-science, for its object is only bizarre and extraordinary phenomena—teleportation, or the impression of it, and the like. Indeed, such phenomena are precisely classified in Chinese scientific terms as *huan*, which means "bizarre and dangerous phenomena resulting from a dangerous, hasty, stressed interaction (*chi*) between humans, on one hand, and the environment, on the other." In turn, whereas Chinese science is unlike Castaneda's Yaqui sorcery, there is nothing more like it than Western physics. If ever there was a bizarre and dangerous phenomenon resulting from a dangerous, hasty, stressed interaction between man and his environment, it was the splitting of the atom. There are plenty of mutants and bone-cancer victims to testify to that, in case anyone has been terminally leached of human instinct by "scientific" propaganda so as to doubt it. I will show that Western science specializes in the extraordinary, bizarre, and dangerous in many other ways, as well. Indeed, it is founded, both in philosophy and in experiment, on an extraordinary, bizarre, and dangerous attitude by which the common people and Nature are treated with the haste and stress merited only by an enemy—just as is Nature in the discipline Castaneda (courageously) endured. (The electric light and the cassette recorder, for example, are, to be sure, relatively gentle and humanly beneficial phenomena, but they are exceptions, and, as I'll show, such exceptions are hard to come by if one thinks things through.)

The following chapters contain not only a scientific comparative test of the Western and Chinese scientific paradigms, but a running analysis of the social-historical effects of the Western one. This reveals patterns in Western society of the most fundamental and widest-ranging variety, which are invisible except from a perspective outside that reality. They are repeated (for example) from the level of classical Greek religious and philosophical assumptions, to the level of the side-effects of Western medicines, to the level of international politics, to the level of the Generation Gap, to the level of behaviors in twentieth-century Discotheques. Being achieved through the Chinese scientific paradigm, the consistency of this analysis will serve as part of the proof of the validity and universalistic power of Chinese science. It is more than an analysis: it is what is called in Chinese a "high-medical diagnosis." And the argument for adopting the Chinese scientific paradigm is what is called in Chinese a "high-medical prescription."

PART ONE:

WAN-CHIN YAO-CHEN:

Diagnosis Worth Ten-Thousand in Gold

CHAPTER

1

The Terminal and Prerevolutionary Condition of Western Science

For the scientist who has lived by his faith in the power of reason, the story ends like a bad dream.—DR. ROBERT JASTROW, Director of NASA's Goddard Institute for Space Studies

THE DRAGON STIRS

Only recently has the dead and asymmetrical crystal of Western science begun to visibly crack. What we call "science" has always been riddled with self-contradictions and yawning explanatory gaps, but now this has become an obvious, even flagrant, fact. From each facet of our scientific paradigm come signs of that paradigm's demise. For example, as Capra has explained, the classical distinctions between "matter," "energy," "space," and "mind" are virtually eliminated by modern physical theory and experimental evidence. This undermines our scientific paradigm very close to its root. Physics is becoming psychology, denying the reality of what its very name refers to: *physis*: substance, matter. David Bohm, Professor of Physics at the University of London, forthrightly confronting the finding that wholes do not have parts, has courageously called for "a new instrument of thought."* Geoffrey Chew, Chairman of Physics at Berkeley, has said:

> Our current struggle with [advanced physics' findings] may thus be only a foretaste of a completely new form of human intellectual endeavor, one that will not only lie outside physics but will not even be describable as "scientific."**

Dr. Jastrow, quoted above, represents many desperate astronomer-physicists who are, as it were, holding church meetings (in fact, some actually are doing so) in reaction to their discovery that they have imprisoned themselves in a mini-universe:

> For the scientist who has lived by his faith in the power of reason, the story ends like a bad dream. He has scaled the mountains of ignorance; he is about to conquer the highest peak; as he pulls himself over the final rock, he is greeted by a band of theologians who have been sitting there for centuries.

He refers to the recently discovered "background radiation" which is taken as evidence of a "Big Bang" origin of the universe: a "Cosmic Egg" of energy-matter is supposed to have instantaneously "come into existence" exploding, to evolve into the galaxies, suns, and planets of today. Hence, prior to the "Big Bang" there were no physical laws and the cause of the Bang—and the whole universe—was supernatural. Jastrow purports that we have discovered religion through scientific investigation, but actually, as I'll show, our scientific investigation is *based on* religion—pure mysticism. It was condemned from its start 2500 years ago to leap into space and end up exactly where it leaped from. We needn't rush back into the arms of our holy men. (Nor is the prospect of then rerunning that old film of religious-to-scientific "evolution" appealing.) Not only is the alternative, Chinese, paradigm consistent with all the relevant data and superbly logical; it gets one out of the religious/scientific double-bind we have been trapped in for 2500 years.

* Capra, *op. cit.*
** *Ibid.*

Where physics is our "hardest" science, the science that all our others imitate, the tremor of its collision with reality forecasts an earthquake right down the middle of the world of science as we know it.

To turn to psychology-psychiatry, it has broken down into a rapidly increasing number of opposed "schools," psychiatrists increasingly doubt their own already discouraging rule that one third of patients are cured, and psychiatric theory is desperately turning toward physiological and chemical solutions, revealing that psychologists and psychiatrists doubt the very existence of the psyche from which their discipline gets its name.

To turn to biology and physical anthropology, the Theory of Evolution (roughly, a biological version of our physicists' tentative "Big Bang" theory of cosmologic evolution) is being supported by mutually contradictory arguments and is therefore being cancelled out. (One argument has it that mutants are selected for survival through competition; the other has it that mutants are selected for their survival by their ability to cooperate.) The sociobiologist Edward Wilson has recently stated that natural selection theory has "severe structural weaknesses."* Without a plausible "natural selection theory," the theory that there has been evolution of one species into another itself is no longer plausible. (I will show why, at least in the extreme form it has taken, it never should have been.) Karl Popper, a famous philosopher of science, has declared work on the Theory of Evolution nonscientific, "metaphysical research."**

Biologists are waging a two-camped war about whether matter is a sufficient condition for life (the Materialist-Reductionist view) or inheres in a specifically biological form of organization which is as basic as the atom (the Idealist, Systems-Theoretical view). Meanwhile, biology is without a definition of life (or certainty that it does not exist to be defined) or a definite definition of living things. It describes the properties of what are generally agreed to be living things (cellular structure, self-reproduction), but neither does it define life nor does it declare that there is no such thing; nor does it circumscribe the sphere of living things, nor does it declare that that sphere cannot be circumscribed. Thus our biologists' central doubts are about what their very science concerns—*bios*: life (or, better, living). And yet our medicine is based on such "biology"!

* Bethel, cited in the *Bibliography.*
** *Ibid.*

Our medical scientists are being outwitted by microorganisms which have mutated to become immune to their antibiotics, and more virulent. They are totally mystified as to how sensation can be registered without spinal-cordal mediation, which is the case when the cord is severed. Yet the nervous system—regardless of what the Chinese say—is what they regard as the regulator of all of the human physiological process. Unlike the other varieties of scientist, despite their impasses, they will not call for "a new instrument of thought." This is because their discipline is distinctively sacredized; it has taken this "right" over from the Church. The best they can do, after decades of now all too obvious failures, is alter their TV appeals for donations for cancer research so that no promise of discovering a cure is held out to the audience. Accordingly, without compunction, Lewis Thomas, well-known biologist-medical-scientist-author (*The Lives of a Cell*) states in his latest book, *The Medusa and the Snail*, "The only solid piece of scientific truth about which I feel totally confident is that we are profoundly ignorant about nature. Indeed, I regard this as the major discovery of the last one hundred years of biology. . . . It is this sudden confrontation with the depth and scope of ignorance that represents the most significant contribution of twentieth-century science to the human intellect."

Then, as one might expect from the head of the Sloan-Kettering Cancer Center, he cites and shrugs off acupuncture as an example of one of the "magics" that desperate people are flocking to after Western medicine fails them. Meanwhile, some competent traditional Chinese physicians are curing cancer, with traditional, non-surgical therapy. That they are believed to be doing so through "magic" must follow, by some special logic, from the facts that (as I'll explain) our cancer researchers haven't a clue as to how they do it, and that the theory behind these cures is not a Western medical one.

Our anthropologists (and sociologists) remain locked in the 2500-year-old battle between cultural relativism and determinism which is a transform of the older theological double-bind between free will and God's total causality which plagued the Babylonians. On one side are the Cultural Relativists who propose that all cultural rules, such as monogamy, are arbitrary, that is, purely a matter of choice: good and bad are culture-relative matters of definition, to solve the problem of crime make what are called "crimes" legal, and so on. On the other side are the Sociobiologists who believe that all human behavior is genetically determined: we are mere hosts for genes (read: mere instruments of the Almighty). These are the extremists' views, to be sure, but neither a moderate form of one or the other nor a combination of the two says anything about what human nature is—yet the name of the discipline is anthropology: the study of humans. An indication of what our anthropologists have "oversighted," and why, follows, to be elaborated on toward the end of the book.

Now that our politicians' and scholars' respect for brute totalitarian force has made Marxism respectable, Western social scientists (psychologists, sociologists, political scientists, and anthropologists) occupy two major camps, the non-Marxist and Marxist, the first of which virtually ignores and the second of which seeks to destroy the very foundation of humane society and the basic, *moral,* feature of human nature: empathy, whose only possible general source is healthy families. As these two camps continue to demonstrate their mutual falsity by demonstrating each other's weaknesses, the credibility of both is rapidly eroding, just as is the credibility of the anti-familial and therefore Capitalist and Marxist social realities which they have failed to appropriately challenge. (Many anthropologists privately share this view with me; I hope they will come out of the closet.)

As naturally follows from the breakdown of our scientific paradigm, our scholars, scientists, and curious and reflective people in general are rapidly approaching the point where they will be obliged to take foreign sciences seriously—despite the lingering official dogma that the world of truth is the exclusive property of white-Western scientists and those they have "enlightened." The seeds of such a general attitude have already been sown. (I ignore the many popular books written by Oriental people who are recognized as savants only by their Western "disciples," or books written by such "disciples," although the currency of such books does symptomize a genuine need that our science does not satisfy.)

Count Alfred Korzybski, an innovative genius whose *Science and Sanity* (1932) proposes an alternative scientific paradigm that correlates with the Chinese one in several respects, quoted the Confucian *Great Learning* in his introduction. He did so out of a recognition that the healthy knowledge-action relation between the individual and society is one of simultaneous feedback whose *relatively* independent variable is the individual's mastery of his or her self. This is the opposite of what is generally conceded by our social scientists, who study systems without studying people and believe that human welfare is made possible by social systems devoid of human factors—unless they be greed, lust, or fear. Jacob Needleman, a philosopher, has shown in his *A Sense of the Cosmos* that the ancient Egyptians knew that the earth orbits the sun, is not the center of the universe. Philip Slater, in his widely appreciated *Earthwalk*, correctly recognized correlations between his humanistic and ecological messages and certain Oriental statements of fact. E.F. Schumacher, whose *Small Is Beautiful* has become a guide for many well-intentioned people, called his just and realistic solution to our industrial and economic problems and their negative social effects, "Buddhist economics." Many of our medical scientists as well as a growing number of people disillusioned with or victimized by Western medicine, already take seriously the aspect of traditional Chinese medicine that we call "acupuncture." The word is finally out, albeit quietly, among our medical scientists that "how it works" cannot be explained in terms of existing Western medical theory. The implication, of course, is that the Chinese theory that it is based on is valid and that our own medical-scientific paradigm is inadequate.

A list of modern scientific discoveries that have been established facts for centuries in China would make Mr. Needleman's Egyptian case look trifling (although I suspect much more could be said, along this line, of Egyptian science, were the data available). For example, *(The Book of) the Savant of Huai-Nan*, which is 2000 years old, explains in unequivocal terms that space *(yu)* and time-and-motion *(chou)* do not exist in themselves, but are interdependent aspects of a whole, space-time-motion. It was only through Einstein's "intuitive leap" to this truth that modern science gained possession of it.

Einstein, of course, was an outsider to the modern scientific tradition: a clerk. I do not want to suggest that Chinese scientists had independently arrived at the complex of pre-Einsteinian theories (although they did arrive at much of it) before making essentially the same contribution that Einstein did. Rather, I mean to say something much more important: that the recognition of general facts such as the preceding one has been a systematically developed science-art in China for thousands of years, and that recognizing such a fact naturally follows from the foundation of the Chinese scientific paradigm.

One could add to our list the fact that for thousands of years Chinese scientists had understood that matter is constituted by and emits, and celestial bodies exchange, wave-particles (*feng-ch'i*)—entities that are simultaneously waves and particles; that light (*kuang*), electricity (*tien*), and magnetism (*tz'u*) are energetic (*neng*) manifestations of these wave-particles; that there is a highly unstable, extreme form of wave-particle emission (what we call "radioactivity"); that gravity (*ti-li*) is a form of energy related to all others, and that there is a single force which underlies all others. These findings are reflected in the debates among modern physicists. They find that light appears simultaneously to be both waves and particles; they wonder if atoms are orbited by particles (electrons) or waves, or, paradoxically, something that is both; and they tentatively posit that "the" Four Forces (electromagnetism, a weak subatomic force, the nucleus-binding force, and gravity) are aspects of a Unified Field of force. They have also concluded that all types of atoms are to some extent unstable, that is, to put it in traditional Chinese scientific terms, that all matter emits *feng-ch'i* (wave-particles). One might also list the recognition of pathogenic microorganisms (germs), described in Sun Sze-Mo's medical treatise, *Ch'ien-Chin Yao-Fang*, of the eighth century A.D., noting that our "germ-theory of disease," the foundation of modern medicine, originated quite recently, with Pasteur. One might add that Shao Yung, in the eleventh century A.D., explained that the sun is not stationary, and has a trajectory, whereas only in this century, with Hubble's discovery that the universe is expanding, did our physicist-astronomers realize essentially the same thing.

One might add that it is standard Chinese-scientific knowledge that space is "curved"—that two "parallel" lines will meet—whereas Einstein, and subsequent mathematical deductions, gave us this recognition only in this century.

As in the first, space-time-motion, example, the intellectual and technological routes to those Chinese findings differed from those taken in the West, but they were neither "intuitive" nor "mystical."

A list such as the preceding might inspire some respect for or skeptical interest in Chinese science, but would be misleading if it were taken to indicate that Chinese science is valid. This is because it suggests that the validity of Chinese science is to be established by showing correlations between it and our science, whereas actually, our science cannot be taken as a measuring stick for the Chinese. This would be to use the frog's knowledge of the water in his dilapidated well to provide the sea turtle with the elements of intercontinental navigation.

One of the reasons Western science cannot measure Chinese science is that, unlike it, it is not unified science based on universal theory. That it has measured many physical, biological, and social phenomena and expresses the results numerically does not constitute *theoretical* unity, as is claimed, but mere *notational* unity. Our scientific paradigm cannot be a unit not only because its theoretical foundations preclude that, but also because it lacks *Humane purpose*. In fact, according to our scientists, it has no purpose at all, because its standard justification, "knowledge is sought for knowledge's sake," is circular, totally meaningless. Research that leads to truth, to universal, unexcepted, logically consistent, empirically grounded theory, must be conditioned by a goal, and that goal, inexorably, is *to serve human beings, as aspects of the whole of Nature.* Without that goal, the selection of what phenomena to investigate, and how to do so, is quite arbitrary, haphazard; and where there is a potential infinity of phenomena and means of investigating them, a goal that *naturally* conditions and unifies the selection is essential. Without such a goal, there arise millions of blind alleys ending in isolated "academic disciplines," theories, "facts," and "side-affecting" techniques. Under such conditions there is no chance of arriving at an integrated, wholistic system of knowledge based on universal principles that may be specified equally in social scientific, biological, chemical, and physical terms—a system such as the traditional Chinese one.

Actually, aside from its legitimate function of fascinating discovery, in which many of our scientists are blamelessly and delightfully involved, our science does have a purpose; our scientists do not wish to admit what it is. It is to discover a phenomenon or technique that will permit merchants to make more money, or, governments to coercively or manipulatively maintain power, and that will commute material wealth, fame, or both, to the researcher. Since the phenomena and techniques that befit such purpose are narrowly selected and one-sidedly conceived, such research cannot lead to universal principles, and therefore cannot be genuinely scientific. Genuine science must be unified by universal principles derived from unrestricted data. In addition, such research as ours is most likely to bear fruit if its topic is *new*. Where all phenomena imaginable to the researcher within the confines of his worldview have already been touched upon, researchers are obliged to seek exceptional, infrequently occurring, or arbitrarily produced data. As the competition for such data increases, the scopes of research projects increasingly narrow, as well. Hence, our scientists operate as far away as possible from the sphere of universal principles, just as they are as alienated as possible from each other. As Donald

Holden, a publishing editor, recently wrote in *The New York Times* (Spring, 1979), "Dissertationese ... is the language that professors use to disguise self-contempt with pomposity. The scholar, who often spends his professional life counting the cats in Zanzibar, secretly knows that what he has to say is unimportant." And as a Princeton physicist recently (1979) reported to *Newsweek*, "We have no idea what the fellow next door is doing."

All told, for our science to be a legitimate measuring stick for traditional Chinese science, it would have to be fundamentally altered not only at the intellectual, but also at the *moral*, level. As I will show, such an alteration would make it the essential equivalent of traditional Chinese science. It would then have only itself to measure. The only measuring stick for any science is unexcepted logical consistency, empirical validity, the inclusion of a truly universal level that unites all its aspects, and suitability to Humane applications. And the only measuring stick for scientists is that they genuinely and courageously strive for all the preceding. Or if their predecessors have already achieved it, that they maintain and transmit it, extending and improving its techniques when possible in adaptive response to changing natural and social conditions.

THE CITIZEN'S ARREST OF DR. SIDE-EFFECT

Recognition of the intellectual inadequacy of our own scientific paradigm, and the possible adequacy of foreign ones, is rapidly spreading among all sorts of intellectually inclined people. Much more important than this is the popular reaction to our scientific institution. Mistrust of scientists and anxiety about the technological applications of their sciences now characterize the attitude of the majority. In their eyes, our scientists, their research, and the kind of technology they make possible are a social problem of major proportions. This perception, since it is natural and since it is that of the rightful ultimate judges of science, the people whom it should serve, is dead accurate. The rising negative reaction to science symptomizes the facts that our scientists are dangerously alienated from humanity and that the costs of perpetuating the scientific institution in its present form far outweigh the benefits.*

Accordingly, our mass media incessantly reflect the negative physical, emotional, and social "side-effects" of this amok machine of an institution. Ivan Illich has written a popular book, *Medical Nemesis*, that discloses the arrogant self-interest of our medical professionals and the resulting alienation of our medicine from actual human needs and sensitivities. Popular newspapers repeatedly run articles on expensive scientific research that is directed toward obvious or socially irrelevant conclusions and which, therefore, increasingly antagonizes the taxpayers who indirectly support it. A recent popular movie (1978), in which, with insectine logic, a computer—a contemporary version of Dr. Frankenstein's monster/Dr. Strangelove's doomsday machine—rapes a woman (a monster-machine violates human body and psyche), aptly symbolizes the net effect of our science and technology, as reflected in popular consciousness.

Now that people see as untenable the so-called higher standard of living which used to "justify" its negative effects, science's *net* effect is being taken into account. It is worthwhile to sample the record behind this recognition that science's "*side*"-effects are its principal ones. Of course, such lists have been compiled; what follows shows something more. First, increasingly our science attempts to cure diseases that it causes and causes diseases with its "cures"; second, this pattern is continuous from the physical level of air and water to the level of society and mind; third, the reason for this spiral of disease-causing "cures" is one and the same at each level.

I begin with the popular impression of science's "side"-effects. For every diverting announcement of a scientific wonder that is flashed through mass media—"Venus has been unveiled! You are looking at computer reconstructions made from radio-photographic information

* As a matter of fact, the sociologists Etzioni and Nunn have established that in 1966 only 56 percent of the public had a great deal of confidence in those who run science. For 1971 through 1973 the average percentage became only 35 percent. See Etzioni and Nunn, cited in the *Bibliography*. Doubtless, with the subsequent nuclear accident of Three Mile Island, epidemic oil spills, and the increasing awareness of the connection between modern science and cancer, it is now less than that. (The modern scientific-technological pollution of all aspects of our environment is a principal cause of cancer.) The public's verdict has already been rendered.

on Venus' surface"—there are several announcements and pictures that amount to a declaration of all-sided scientific-technological war against Nature and all of humanity. At the level of social science, it is an implicit exposé of the cold, nationless-people-less mercantile and power-tripping mentality that our sciences now chiefly serve and, therefore, share: "*The ozone layer, on which all life depends for protection against solar radiation, is being thinned by aerosol sprays . . .—The supertanker Kurdistan is breaking up off the shore of Newfoundland . . .—a train carrying volatile and poisonous gases has derailed in Florida; people within a 300-mile radius are being evacuated . . .—Yet another primary school built of radioactive slag has been discovered; parents are bringing their children to be monitored for radioactivity . . .—The government estimates that there are hundreds of unmarked chemical dumps like Love Canal's throughout the U.S.; the rate of birth defects in such areas is extremely high; the possibility of a clean-up appears remote . . .—Current estimates of deaths from taking birth-control pills range from 300 to 1000 women annually . . .—The rate of leukemia within a 200-mile radius of the 50's A-bomb tests is abnormally high; suits for millions are being filed against the government . . .—Government sources have disclosed that several other nuclear reactors built by that company do not meet minimum safety-standards . . .—And now, our special report on the dangers of over-the-counter drugs . . .—Recent lab tests confirm that red dye, a coloring used in breakfast cereals and children's vitamins, causes cancer . . . —Can cholera be transported by jet plane sewage? Our correspondent in Centerville, Iowa, where a two-ton block of frozen airplane sewage recently demolished the home of the Everett Urquarts, has the story . . .—Residents are warned to stay home today: the pollution level is unacceptable . . .—Radioactive waste containers off the U.S. coast, say government experts, seem to be decomposing; scientists are searching for a more corrosion-resistant metal . . .—But if the reactor is cooled according to the steam-releasing plan, an unexpected wind could spread radioactive fallout over much of New England . . .—The Senate subcommittee has determined that most doctors don't check histories of exposure before prescribing X-rays; children and pregnant women are most susceptible to genetic damage . . .—Ecologists have pointed out that these chemical poisons circulate throughout the food-chain and concentrate in humans . . . mothers milk may no longer be safe for infants . . .—It is hard to predict where the fragments of Space-Lab will fall.*"

"It's not science itself but its *use* that is the problem," many would argue. Rather, it is *both* science itself and its use. Viewed only as a system of knowledge, our science is intrinsically useful chiefly in destructive ways. Briefly put, this is because it is based on a view of the universe that takes into account only one part of it at a time, in isolation—just as does the theory that the problem is not science but its *use*—or the inevitable counterpart to that theory, that it is not its use, but *science* itself. In standard Chinese terms, this is called "Using one part to obscure the whole." Such "science" starts with only one side and when applied it "side-affects" all other sides.

To recognize that pattern, let us examine some of the disease-causing cures of our science, in the context of science's loveless relation to the people who support it. In 1973 the first successful recombinant DNA experiment was performed (a monster was created using chemico-cellular vivisection involving genes). The result was bacteria that are immune to streptomycin—potentially, permanent "strept throat." Despite massive protests by citizens in the areas of such experiments, they continue. Disdaining popular reaction, a researcher at Harvard, according to the popular press, has said that a dangerous mutant bacterium could not even be produced deliberately (would not survive outside the lab). If that is true, if he is so sure, then why has recombinant DNA research for germ warfare been outlawed by federal authorities, and why are researchers decontaminated when they leave the labs?* Allegedly, other researchers who have publicly protested this re-research have lost their jobs. The true motivation for this research is implied by another Harvard researcher: "For the first time, biologists have a chance to get rich, so there is strong peer pressure to go along." These mutant bacteria have immense commercial potential: they may be programmed to manufacture medicines and to produce fuels. As to the manufacture of fuel, it must be social-scientifically weighed against the risk of playing with Nature to produce mutants; the side-effects, including epidemic cancer, of the industrialized society that it fuels; psychological stress against people who are being forced to live in an increasingly doubtful and dangerous environment; and the fact that the hyper-industrialized nature of our society requires seductively or brutally colonizing Third World peoples so that they will supply us

* Our intelligence agencies have concluded that an epidemic of rare and deadly lung-anthrax which struck Soviet citizens in the area of a secret installation must be due to just such experimental research.

with raw materials, interfering with their own cultural evolutions and, in the end, incurring the hate of these peoples. In short, there are side-effects at all levels—physical, psychical,* socio-political, and socio-defensive—which are blithely ignored, if even recognized, by our so-called scientists. That is because our science is Fragmental—pays attention to only a fragment of any actual whole—and ill-intentioned (intention which is not good is bad, be it implicit or explicit). Our scientists are so alienated from us that they no longer think anything of taking so-called "calculated risks," which, moreover, they callously announce to the people whose health is risked. For example, our mass media have coolly reported that the bacteria *E. Coli* used in such experimental enterprise are produced in vat-sized quantities and that "the effect of such amounts on workers is not known." (*E. Coli* is an agent of venereal disease.)

One of the enormous examples of such risking, which, because of its enormity, has received very little exposure, is now specified to ascertain the cold and irresponsible, at its worst, insanely detached and lethally inhuman, attitude which predominates in our scientific community. During World War II, two options were presented by eminent scientists to President Roosevelt: atomically bomb Germany and Japan or possibly be enslaved. There was a third, top-secret option: the possibility of triggering a nuclear catastrophe. In 1959 the third option received brief public attention in an article by Pearl S. Buck about an interview with the Nobel Prize winner Arthur Compton, who took the lead in deciding to explode the bomb on the following basis: "If, after calculation, it were proved that the chances were more than approximately three to one million that the earth would be vaporized by the atomic explosion, he would not proceed with the project. Calculation proved the figures *slightly* less—and the project continued."**

Don't breathe a sigh of relief: the matter has not ended. As the physicist H.C. Dudley observes in his *The Morality of Nuclear Planning???*, "But now the bombs are a thousand times as powerful. Does this lower the odds to 3000 per one million or properly 3 in 1000?" I can tell you that our probabilist-physicists cannot answer that ques-

* "Psychical" should be distinguished from "psychic," which has been appropriated to refer to telepathy and the like. "Psychological" is improper here because it refers to knowing about the psyche not the psyche itself.

** Dudley, cited below.

tion: they agree neither about how an actual event influences probabilities about such an event, nor about what the relation between probability and actuality is. What's more, Dudley points out, the original calculations were made without the present understanding that there is an all-pervading subatomic ether (called *ch'i* in Chinese), constituting more mass than was thought, with which atomic explosions interact in a still not understood manner. Dudley then explains that the nuclear chain reaction has five components and requires a high-pressure hydrogen-rich atmosphere, and points out that:

> The seas also provide a hydrogen-rich environment. Containment pressure measured in tons per square inch exists under millions of square miles of water. The explosion of a conventional fission or fusion device would provide the 100 million degree temperature necessary to initiate reactions (1) thru (5) . . . [there is a] U.S. submarine at the bottom of the South Atlantic, crew long dead, but H bomb rocket heads intact. A Russian sub met the same fate more recently in the Pacific. Same results.

They also left out the "human factor" and the probabilities of "vehicle error," didn't they.

Again: Fragmental, one-sided science guided by bad, all-life-risking intention. In 1945 Fermi made bets with his colleagues on whether the bomb would ignite the atmosphere, and if so, whether it would destroy the entire world, or only New Mexico.* Today, the probability of vaporizing the entire planet just as Darth Vader did in *Star Wars* is doubtless higher than it was when the odds were first calculated. What's more, the "risk-probabilities" of our physicists are based on the assumption that all the parameters of nuclear reaction are known. But our quantum physicists are almost continually discovering new things about atomic and subatomic interactions. They have reached a theoretical impasse so basic that some have even called for "a new instrument of thought."

I turn to an example of the Fragmental treatment of cancer. *New Times*, in August 1978, reported the case of eight-year-old Jimmy Cleffi of Rutherford, New Jersey, who had contracted a rare variety of leukemia. After a year of chemotherapy, his condition dramatically worsened.

> . . . the doctors took Jimmy to a treatment room to remove a

* *Ibid.*

blood clot that had formed in his rectum. He hemorrhaged during the procedure, bleeding heavily from the mouth. "Don't be upset by the way he looks," a doctor warned Vivian [his mother] before Jimmy was returned to his room. "He's swallowed quite a lot of blood." Jimmy's face and smock, in fact, were smeared with red. His voice was raspy and faint. Vivian could see him suck in his nostrils as he gasped for air.

By 6 P.M., the doctors decided that Jimmy needed a tracheotomy to breathe. As they wheeled him out of the room, Vivian heard him scream, "Mom, stop them. Don't let them take me." He returned two hours later, a tube sticking out from his throat, his face expressionless, his body limp.

. . . [Four days later] another set of doctors entered the room to take more X-rays. After fifteen months of doctors, Vivian Cleffi had finally had enough.

"Get out of here," she screamed at them. "Get the hell out of this room. Can't you leave him alone?"

. . . Nine-year-old Jimmy, who was curled up in bed, began to suck his thumb, then looked at his father in embarrassment. "Don't worry, I'll cover you so nobody can see," his father said, leaning over the bed and gently hugging the child. "Tomorrow I'll take you home."

At 2 A.M. . . . Jimmy Cleffi died.

. . . New Jersey, the most densely populated and industrialized state in the country, and one of the most polluted, was found to be the nation's cancer capital. . . . [Among its polluters] Hoffman-La Roche, Inc., a major pharmaceutical firm in Nutley that manufactures, among other products, three drugs used to treat cancer, put its total [chemical emissions into the atmosphere] at 1,800 tons, including 216 tons of acetone, 22 tons of benzene, 56 tons of chloroform, and 14.5 tons of toluene. All these chemicals are considered suspect carcinogens by the National Health Institute for Occupational Safety and Health.

The doctors who treated Jimmy Cleffi undoubtedly did their best. That their best was a nightmarish tragedy forecasting more of the same is not their fault but the fault of a scientific paradigm that is Fragmental. The fact aside that an alternative medical system, the Chinese one, serves to cure non-terminal cancer victims, the genuine, full-fledged medical objective, of course, is to *prevent* cancer. What is called for in instances such as Jimmy Cleffi's is societal, or "high," medicine which precludes (or reverses) the modern-scientific form of industrialization, including manufacture of cancer-treating drugs which has cancer-causing side-effects. Diagnosing and (unsuccessfully) treating only the body of a cancer victim, our doctors are glorified veterinarians hypnotically fixated on a tiny fragment in the virulent living whole of chemical industry, microwave pollution, overurbanization, combustion-engine traffic-concentrations, the hungry mercantile spirit behind all that, the psychical agony of the patient and his or her family, and the anxiety and nightmare visions of the parents of possible Jimmy Cleffis of the future, to which our industrialists and "social engineers" are immune. (Many are "concerned," but this concern, obviously, is only as deep as the action taken on it. Rather than the Fragmental "lowerings of risk-factors," fundamental, Complete solutions are plainly required, not by "The Economy," but by *human victims* such as Jimmy Cleffi.)

Let us turn to Western medicines themselves. Antibiotics, the mainstay of Western medication, are used with only certain kinds of bacteria in mind. The bacteria are killed off, but the body, not having been given the chance to cope with them, cannot, as a consequence, even develop normal resistance to them. In addition, because "good" bacteria and "bad" bacteria are biologically more similar than different, antibiotics evacuate the good with the bad with the common result of severe diarrhea and loss of vital minerals, and insidious, sometimes lethal, metasystemic infections after surgery, from the antibiotic lowering of body heat and evacuation of good bacteria that contribute to normal resistance to infection. (Thus, staphylococcus infections, among the hardest to cure, are known as "hospital infections.") In the end, the patient is usually cured of both the initial and doctor-caused diseases, *but a worse disease has been created*: mutant bacteria and subnormally resistant patients. So, the general abuse of the over-powerful penicillins and mycins has produced mutant, penicillin-and-mycin-resistant bacteria, on one hand, and a futilely dependent population with subnormal natural resistance to even pre-mutant bacteria, on the other—the most defenseless population in human history. For quite some time medical "scientists" have produced these supergerms, then in a spiraling race with Nature repeatedly invented more powerful antibiotics to create super-super-germs. Who do you think will win the race? The scientists who have conveniently forgotten that they themselves are a minor (and ill-fitting) component of

Nature and that their medicines are made of natural materials? Actually, for decades DNA experiments have been conducted *outside* laboratories, with us as the mutants' hosts.

Dr. Margaret Lock of McGill University reported at the 1980 Canadian Ethnology Association Meetings in Montreal that a renaissance of traditional (Chinese) medicine is taking place in Japan, whose people have had the chance to extensively sample Western medicine. It is now popularly said that "Traditional medicine is closer to food, Western medicine is closer to poison."* The same is true in Taiwan: in 1978 over 7000 young people qualified to take the final medical examinations in traditional Chinese medicine, despite Taiwan's open-armed trial of Western medicine. And in Communist China, despite its controllers' basic objective of destroying all traditional culture, the people's will to benefit from Chinese medicine is so strong that a majority of practitioners now being trained have taken the traditional option.** Let us appreciate how accurate the popular Japanese saying is, and how through polite understatement it denotes the disease-causing character of Western medicine.

The *New England Journal of Medicine* (1979), in the wake of the still not understood Legionnaire's Disease (which is described, with its long-attested cure, in the 2000-year-old *Shang-Han Lun* of Dr. Chang Chung-Ching), reports that new antibiotic-resistant strains of pneumococcus (pneumonia-causing bacteria) have come into existence. They tend to strike patients whose immune systems have been suppressed, by Western medication.

A perfect example of disease-causing cures is the relation between Haloperidol and L-dopa.*** Haloperidol is an "anti-psychotic" drug commonly used for schizophrenics. Its side-effects include jaundice, changes in blood composition including anemia, which can render pa-

* Although both East Asian and Western medicines are classified on a scale ranging from foods to poisons, the majority of the former are classified as foods, whereas all the latter are classified as poisons. The dangers of the latter, especially to pregnant women, are increasingly taken into account. Also see Lock's book, cited in the *Bibliography* (254 *et infra.*)

** However, not as in Taiwan, the societal and psychical aspect of traditional medicine has been eliminated according to strictly enforced Communist objects, so that, where it is the *root* of Chinese medicine, traditional Chinese medicine is dying in Communist China despite the people's will. (The necessary connection among all aspects of traditional medicine will be made clear.)

*** The following data on side-effects, through the example of methylphenidate, are taken from Silverman and Simon, cited in the *Bibliography*.

tients susceptible to the just cited new strains of pneumococcus, heart attack, muscle spasms, convulsions, difficult swallowing, paranoia, bizarre dreams, depression, euphoria, insomnia, red skin, rash, sensitivity to light, breast enlargement, false positive pregnancy tests, sexual impotence, loss of facial color, and bronchial spasms.

Of special interest here is Haloperidol's side-effect of the symptoms of Parkinson's disease, which may be briefly described as neural chaos. Reciprocally, L-dopa, used to treat Parkinson's disease, has the side-effect of the symptoms of schizophrenia, which Haloperidol is used to treat.

The following are the possible side-effects of the most widely used antibiotics, a majority of which they all share: fatal colitis, stomach pain, nausea, diarrhea, vomiting, pain on swallowing, itching, rash, difficulty breathing, yellowing of the whites of the eyes and/or the skin (which implies liver dysfunction), liver damage, liver enlargement, changes in blood composition, oral or rectal infestation with fungal diseases, hairy tongue (a symptom of a very grave disease), retardation of bone growth in children, decrease in kidney function, peeling or spotted skin, fever, and chills. Appropriately, antibiotic literally means anti-*vital*.

All "tranquilizers," the drugs most often prescribed by psychiatrists, have the following possible side-effects: convulsions, tremor, muscle cramps, stomach cramps, vomiting and sweating, drowsiness, confusion, depression, disorientation, headache, slurred speech, dizziness, constipation, nausea, decreased sex drive, retaining of fluids and inability to control urination (kidney dysfunction), irregular menstruation, changes in heart rhythm, insomnia, liver dysfunction, blurred vision, itching, rash, hiccups, and *nervousness*. Overdoses can cause comas and paralysis of the diaphragm (shallow to no breathing).

A widely used sedative has the side-effects of causing nervousness and an inability to sleep. Both it and the other kind of barbital may produce difficult breathing, rashes, diarrhea, anemia, jaundice, dizziness, lethargy, hangover, nausea, vomiting, "general allergy," and, in overdose, death by suffocation.

The complement of tranquilizers and sedatives, amphetamine, which is used on people who are depressed and in diet pills, has the side-effects of birth defects (such as brain dysfunction), heart palpitations, rapid heartbeat, muscle spasms, diarrhea, constipation, stomach upset, loss of sex drive, hallucinations (whereupon an "anti-psychotic" may be used to replace it with the side-effect of disorientation), euphoria (whereupon a sedative may be used to replace it with the side-effect of nervousness), and "psychotic drug-reactions" (whereupon Haloperidol may be used to replace them with the side-effect of euphoria, whereupon a tranquilizer may be used to replace it with the side-effect of drowsiness, whereupon amphetamine may be used to replace it with "psychotic drug-reactions"). Or there is the "psychotic" reaction to the "therapy" of having come full circle with the bonus of a child born with birth defects, who can then be a subject of research seeking to determine to what extent brain dysfunctions are *inherent* or acquired.

When the patient proves "over-sensitive" to amphetamine, methylphenidate is often substituted; its side-effects include all those of amphetamine, plus "psychological, educational, or social disorders." It is worth pointing out that the quality of life in our society, basically the work of our consulted *social* scientists, is such that a significant sector of the population takes stimulants for what they call "pleasure"—ample indication that our society is as socially and psychologically disordered as the medical paradigm according to which such drugs are "safe for human use."

The cough suppressants constantly prescribed for children neutralize the cough response which rids the throat and bronchi of bacteria, thus fostering lung infections, ultimately pneumonia.

Cancer-treating chemotherapy and radiation promote cancer. So does x-ray diagnosis of breast cancer.

The Western so-called medicines just described are those most often prescribed, constitute the majority of our physicians' and psychotherapists' actualized repertoires. In contrast, Chinese medicines that may be used, quite successfully, for any of the diseases for which the Western ones above are prescribed have no side-effects at all.

The will to commit suicide is defined by Chinese doctors logically enough as the gravest disease. As the brilliant sociologist Emile Durkheim explained, it is usually the result of recognizing that, to put it roughly, one is not serving the function of one's social position.* It is hardly a mystery, then, that the suicide rate of Western physicians is much higher than that of the general populace, and that the suicide rate of psychiatrists is twice that of physicians.** Indeed, although the

* This rare, valid, and increasingly misunderstood conclusion of Western sociology refers to what is called *anomie.*

** I am indebted for this datum to E. Alan Morinis, an innovative medical anthropologist, cited below in the section on cancer.

suicide rate of the general population of the United States is eleven per 100,000, that of Western psychiatrists, seventy per 100,000, is more than six times as high and is almost twice as high as that of the general population in the country with the highest (reported) suicide rate in the world, East Germany (43 per 100,000). In the case of our would-be healers, the function of their social position is to heal, but their roles serve to transform illnesses into other illnesses. Since this is a matter of life and death, the healers who recognize it sometimes develop the symptom, suicide, of this scientific-paradigmatic disease. When those whom a medical paradigm kills off are among its most talented healers, that paradigm hardly constitutes ground for judging another, and the fact that it is sick itself, is basically disease-*causing*, is plain.*

On December 2, 1980, over a year after I wrote the preceding paragraph, there appeared on NBC's *Today* show Dr. Martin Lipp, author of *The Bitter Price*, who virtually recapitulated all that I have just said. Addressing himself to the problem of the super-normal rate of suicide and drug abuse among physicians and psychiatrists, he observed that there is a "contagious depression" among our doctors, which is due to their failure to meet their own and their patients' expectations. Referring to a doctor/friend who committed suicide, he explained that our medicine is changing in a way that made it too hard for that man to be the kind of doctor he wanted to be, namely, a doctor who could personally respond to his patient's perceptions of their diseases. He concluded by observing that there is the equivalent to one medical-school class of suicides among doctors per year.

An instructive societal-psychical-physical illustration of the basic flaw in our scientific paradigm and medical sub-paradigm is the matter of "the Pill." Our medical scientists have invented and pushed "birth-control pills" that derange the whole specifically female aspect of human physiology at the root. Deferring common sense to the rela-

* An execption to the rule of feudal loyalty to the medical paradigm and profession has finally occurred: an eminent physician has publicly condemned our medical system. This upright man is Dr. Robert Mendelsohn of Chicago; his book, *Confessions of a Medical Heretic*, focuses on our medicine's destructive effects on families and the epidemic of unnecessary surgery. He reveals that not only patients but also doctors, especially younger ones, want to be liberated from our medical system. He calls them "closet heretics," for they write to him, with the qualification that they fear to go public with this, that they are looking for a new medical system.

tively "scientific" method of experimental testing, having used our women as the guinea pigs, they now inform us, without apology, that the following (tentative) conclusions about the side-effects "have emerged": the pills foster blood clotting, brain hemorrhages, heart attacks, (possibly) cancer of the uterus, and defective births. Where female fertility is based on a blood cycle, are blood and heart illnesses a surprising effect of a "medicine" that suppresses that natural cycle? Where the objective is to prevent the conception of infants, are cancer of the uterus and defective births astounding effects?

There is much more to this. As in the example above, the Complete* picture and actual problem is invisible to our "scientists," who deal only with a fragment of it, taking a disease for a cure and, as it were, prescribing Clearasil for leprosy. Let us "spiral outward" from the symptom to sketch the Complete, societal-psychical-physical picture.

Obviously, where there arises the objective of obstructing a human physiological process of *life*-generation, there is already a *fundamental* illness of some kind. One aspect of it is an alienation of women's psyches from their own bodies, without which women could not consider taking such pills. In addition to that psychical aspect, there are two societal-cultural ones. One, pointed out by George Gilder in his *Sexual Suicide*, is that, with the collaborative help of *Playboy*, and other such magazines, Feminists, and Gay Libbers, sexual pleasure has been disconnected from the ends to which it is the means: the pleasure-bonding of couples and the creation of children for whom there are two permanently and totally devoted adults, one of each sex, of whom those children are physical and spiritual extensions. Sexual pleasure has been disconnected from love of children. The other aspect is the impetus to "free" women of the "burden" of motherhood so that they can achieve "self-realization."

The last, as yet unspecified core of the disease, is a swindle perpetrated by social engineers, who adopted it from Marx-Engels and the Bolsheviks, and modified it to work to the advantage of industrialists and bankers in a Capitalist society. To explain: the majority of working-and-therefore-income-spending women do not "realize themselves" through their employment: like the majority of men, who hate their jobs, they find only boring, low-paying, often degrading work. The female "bonus," as in secretarial work, is to be ogled, and, frequently,

* As opposed to Fragmental.

sexually exploited by one's male bosses. So, as Gilder also points out, "liberation" works to the advantage of, proves to really be self-realization for only a tiny minority of women—those with Ph.D.'s or exceptional talent—who *should* be accommodated but *not* at the expense of *all other* women. And, in fact, the majority of working women are not out there seeking self-realization but money to pay their bills, which, unlike their mothers at their age, are for items such as second cars and automatic dishwashers purchased on credit and justified by the fact that, as working women, they deserve their own wheels and should not have to wash dishes by hand. What, then, is the *real* motivation for and function of this aspect of "sexual liberation"?

As social scientists working for advertising agencies working for industrialists and bankers have carefully determined, for rich consultant fees, women are "better consumers" than men: they make 80 percent or more of a family's purchasing decisions and are more easily motivated than men to spend instead of save. "Therefore," give them *their own* money. As Chemical Bank's ad used to put it, "If you're a woman, your reaction is *Chemical.*" Now, the foremost reason for saving, in any society, is obvious: to provide greater material welfare to one's children than one had oneself—a perfectly legitimate and natural goal. "Therefore," remove the motivation for that goal. "Therefore," children are a burden, on one hand, and independent individuals "one should not *own,*" on the other—an obstacle to chiefly female and secondarily male spending. This spending, since two people working for money have the same real buying power as one member of a couple a generation before, takes the form of credit and produces a nation of debtors who, this time, have no New World to flee to. As my *shih-fu* pointed out to me, these debtors are thus made economically subservient and impotent and have no potential whatever to become competitors in the Capitalist sphere. Thereby is insured a Capitalist elite political domination and economic monopoly—a modest correlate of the one established in Communist countries through basically the same means: suppression and misdirection of human nature and destruction of the social structure of the common people. This is what "the Sexual Revolution" for "the masses" is all about (from the informed perspective of those who really engineered it).

Those who do not believe that human nature exists, or regard devotion to one's children as pre-Socialist, arbitrary cultural conditioning, or doubt that modern society is socially engineered, in the way just outlined, against human nature should be made aware of what astute social scientists recommend when their objective is to produce a theory that *works,* instead of one which demonstrates that they are Progressive, interesting people. The *Public Relations Journal* reported in 1953 that Lymon Bryson, a social anthropologist heading six consultant social scientists from Columbia, made the following statements to New York members of the Public Relations Society of America:

> If you are engineering consent, then I think the social sciences would like to warn you that you should begin with a basic analysis of three levels upon which consent moves. . . . The first level, he said, is human nature. He added that little could really be done here to "manipulate" people. The second level was cultural change, which is where you must operate, he said, if you want to influence people's ideas. The third level . . . is where an impulse is running in a certain direction, and some sort of choice will be made regardless. . . . If you are trying to change their ideas, "you work on the second level," where different "psychological pressures, techniques and devices from those successful on the third level" must be used.
> (Vance Packard, *The Hidden Persuaders,* pp. 189–90)

In other words, when given the proper motivations top social scientists reveal they have *found* that human nature exists. It is the aspect of human behavior that cannot be changed, directly "manipulated." However it can be *misdirected* by manipulating "the next level": cultural change, using techniques consistent with "impulses" at the third.

Here are two actual examples of that strategy applied—spanning 5000 "Western-civilized" years. What social scientists regard, usually privately, as human nature is what everybody (because it is human nature to know it) knows it is: focally, the desire, by virtue of which the sex drive and society are so powerful in humans, to be *socially immortal* through one's own or others' children, by helping them to become valued members of a persisting group. Through them, one's persona survives and one's acts continue after one's death. Setting the basic pattern of misdirection of human nature, religion moved in on the desire for social immortality 5000 years ago by substituting immortality in "Heaven." That was the "cultural change." The "impulse" was the original Hebrews', then the Christians', then the Muslims' desperate political-economic situations, in which the prospect of social immortality (here comes the superior army of the Egyptians, the Romans, the Turks) was minimal. The result was church control over matters normally regarded as *private* or *the people's*—marriage, family, economy, morality, law, government. The contemporary correlate of such manipulation around human nature is to drive a wedge between parents and children and sell either Capitalist products, leaders and indebtedness, or Socialist welfare, leaders and taxation to fill the hole. And the leaders are marketed as what they have replaced in the shuffle: *father* types.

Likewise "self-realization" is a "cultural change," and sexual freedom made possible by the Pill is an "impulse."

Where parental inculcating of prudence and frugality—basically, self-control—in order to maximize the next generation's survival is an obstacle to the objective of political-economic enslavement of "the masses," children must be conditioned to disrespect and rebel against their parents, and parents must be conditioned to defer their children's welfare to their own, short-term pleasure-goals ("Heaven" *here and now*). I have explained one way in which the latter is accomplished. As for the children, to quote Clyde Miller's *The Process of Persuasion*:

> . . . if you expect to be in business for any length of time, think of what it can mean to your firm in profits if you can condition a million or ten million children who will grow up into adults trained to buy your product as soldiers are trained to advance when they hear the trigger words "forward march."
> (Packard, p. 136)

Toward that end, the consulting agency Social Research established the "social-scientifically" desired format for kiddie-shows back in the 1950's:

> In general the show utilizes suppressed hostilities to make fun of adults or depict adults in an unattractive light. The "bad" characters (Chief Thunderthud, Mr. Bluster, Mr. X) are all adults. They are depicted as either frighteningly powerful or silly. . . . The good guys were all found to be young men in their twenties organized as a group with very strong team loyalty. . . . "To children," the report explained, "adults are a 'ruling class' against which they cannot successfully revolt."
> (Packard, p. 140)

The average American child spends more time watching TV than doing anything else. After that comes school. Read Packard to find out how as early as the 1950's a firm supplying "educational materials" to schools built subliminal "product-demands" into them, and how were made to engineer the human environment itself toward the same end. Then consider the more generally relevant facts that our schools, now centralized, no longer represent our communities, but social-scientifically determined decisions about what social ideas and "peer-structured environments" best condition our children for "membership in our society," and that an increasingly popular sport is to beat up and/or rape teacher.

The disease of which the physical side-effects of the Pill are a minor symptom, then, is a social structure and value system in which Capitalist desires are allowed to take top priority, where women are economically exploited, the value of children is demeaned, the natural love of children is denied, sexual pleasure is divorced from love for children, conception is treated as though it were cancer, children are taught contempt for their parents, and our men and women are encouraged to regard the vagina as a mere tunnel of amusement—a Love Canal, the pleasures of which outweigh the risk of bearing defective children. To speak for the majority, not a privileged few: women are made schizophrenically "Selves" with objectives diametrically opposed to their own bodies and long-term interests, and husbands who love their wives and therefore can feel jealousy, aware of the degradation of their wives in working environments and a weakening of shared purpose with them, feel degraded themselves and become drunkards or violent or cheat on their wives to express their frustrations, thus raising the divorce rate and the popular demand for birth-control pills. (Of course, the rationale for and the availability of birth-control pills *directly* encourage an increased rate of marital and extra-marital intercourse and thus raise the probability of undesired pregnancies, and fatherless or motherless children, and so on. The "cure" fosters even the sub-illness it is addressed to!)

The disease in question, then, is caused by the architects of our social culture, and is focally their *lack of love for children.** To focus on

* Retrace the argument and that will emerge as the focal variable. As for the venal controllers of our society and the social scientists who work for them, they love not even their *own* children, for if they did they would think ahead to the day when the conned will wise up and take their revenge for the social, psychical, and (even) physical rape they have endured, directly or indirectly, at their hands.

the Pill and its effects is to take a stone in the Great Pyramid for Mount Everest.

Such a Complete, "high-medical," societal-psychical-physical diagnosis of an illness is unthinkable in the West. It is unsettlingly comprehensive. Where does the fear and consequent suspicion come from —our habit of seeing microscopically, fragmentally? We do not even have a societal medicine: anthropology-sociology, its pale substitute, is thoroughly divorced, as a "non-medical" discipline, from psychotherapy and physical medicine; and "social work" deals with *adaptation* to an unchallenged basic system. Obviously, a Complete perspective threatens not only Western "scientists'" fragile ecological niches and research prospects, but also the political-economic machine that our "scientists," some knowingly and some not, are paid to lubricate with disease-causing drugs the rationales for which are deceptive fragments of truth, and Fragmental studies and theories that start nowhere and get nowhere, unless they move toward the increasingly "acceptable" Marxist version of the disease in question. To diagnose this situation is unsettling to those whose interests the situation serves. To substitute a Virus A2 for a Virus A1 is "good work"; it preserves the paradigm. "We need more daycare centers and childless women trained to run them," says a proponent of cure Virus A2. "Of course we social engineers and such 'child-care-professionals' must, unlike mothers, be paid with money, not loyalty and love, for our services. So social welfare taxes will be tripled and taken from the salaries of the male-chauvinist-Capitalist-pig husbands of potential 'child-care professionals' who were after all out to obstruct their wives' self-realization. It's all for the sake of social justice, and the kids will never know the difference: 'Human Nature' is merely the cultural prejudices of a certain stage of political-economic evolution . . ."

Given the stubborn adherence to the Western paradigm and the Capitalist-or-Socialist interests behind it, the Fragmental structure of the disciplines attached to it, and, most basically, the microscopically one-sided perspective and non-Humane attitude required by our paradigm, the present response to the discovery of the Pill's side-effects is comparable to identifying one mere symptom of an unrecognized disease of huge proportions. The teeny-tiny (impeccably "scientific") response is to develop birth-control pills or injections for *men* as affirms the medical profession's dedication to the democratic ideal of sexual equality. Equality for adults only, of course, not also children, the sexual nature of whose kinship-relation (the sexual-reproductive tie) to their parents is conveniently ignored.

I could have begun the preceding illustration of our paradigm's side-effecting, Fragmental nature with any one of the specific illnesses just discussed or implied: the adolescent VD epidemic, the rising divorce rate, the rising parent-absent child rate, the rising teenage abortion rate, the rising suicide rate, sadistic-crime-against-adults-rate of children, the rising wife- (and husband-) beating rate, the rising rape rate, the rising incidence of depression, the undemocratic structure of modern society, credit-enslavement, or the side-effects of birth-control pills. . . . Each is a not absolutely distinct, integrated aspect of one whole disease, which, because of the Fragmental nature of our paradigm, is not even recognized much less diagnosed or prescribed for (see Figure One.) *Indeed, that paradigm is that disease's basic structure.*

I hasten to add that I have by no means painted the whole disease-picture of which the Pill is a component. For example, I do not mean to imply that our work-seeking women would be happier at home, for the modern home, in which a woman can do very little truly productive work, is a prison. That leads to yet another social-organizational and economic dimension. Nor do I mean to imply that any husband, no matter how humiliated and frustrated he may be by his wife's alienation from him in deference to conditions for receiving welfare money, or her work-environment, and no matter how humiliated and frustrated he may be by his political and economic impotence as a modern father-husband, has any right to beat or psychologically abuse his wife, or any excuse, such as the "psychological compulsion" a psychotherapist might grant him, for doing so. That leads to another dimension: modern social science's debased view of the human potential, which has contaminated our courts and the popular consciousness, as well as the inhumane social-economic-work environment in which both our men and our women are trapped.

Now that the general side-effecting pattern of our Fragmental paradigm and the Complete Chinese high-medical perspective* that reveals it have been illustrated, I more simply describe several further Western disease-causing cures or "scientifically" grounded enterprises,

* Of course, only the Humane and Complete perspective, not the scientific content, of the Chinese paradigm has been illustrated so far.

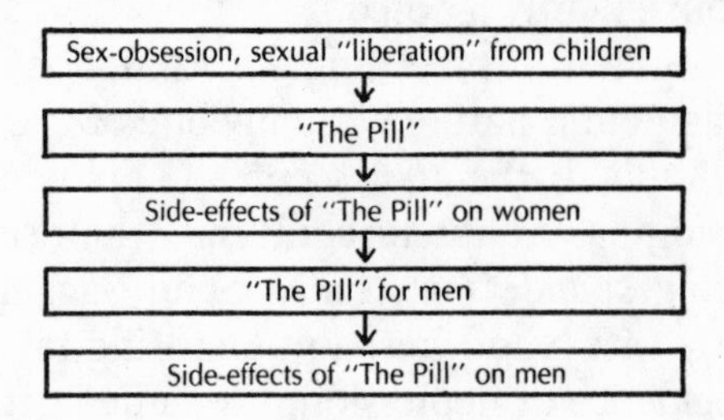

Western, Fragmental, Diagnosis and Disease-Causing "Cure"

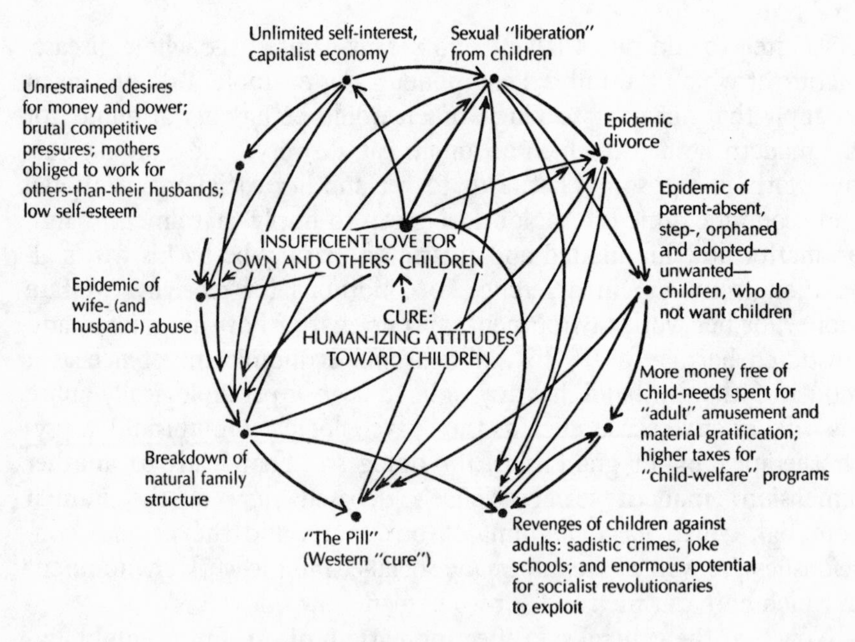

Chinese-style, Complete, Diagnosis and Cure

FIGURE ONE: WESTERN-MEDICAL AND CHINESE HIGH-MEDICAL SOCIAL-PSYCHICAL-PHYSICAL DISEASE-DIAGNOSIS AND TREATMENT and the dangerous-injurious Nature-and-common-people-disdaining attitude behind them.

It has been noticed that cattle raised near electrical power stations are dwarfed. Have our scientists demanded that such facilities be relocated far from human populations? Or that at least household electricity be generated decentrally, and therefore not monopolistically, in gentler form by household solar collectors and windmills paid for with tax monies (which, as things stand, finance research projects such as "The Effects of High Voltage on Cattle: A Projected Twenty-Year Study")?

Chemical wastes dumped decades ago are surfacing, burning, and mutating children. In the United States, sixty-three million tons, five hundred pounds for each American, of hazardous chemical "side-products" are dumped into our environment each year. According to *Time* (19 May 1980), one "expert" says the cost of cleaning up larger dump sites "may be greater than the value of the companies responsible": the side-effect is greater than the benefit *even* in terms of mere money. How many chemists have questioned the health of their discipline or stated forthrightly that there is in fact no way to permanently dispose of chemical poisons? Meanwhile, a year later, the residents of Love Canal have yet to be evacuated.

The first "major" nuclear accident in the United States finally occurred, in Harrisburg, Pennsylvania. As the local population shuddered in an *already* radioactively polluted atmosphere and its pregnant women and children were evacuated, our scientific specialists attempted to distract them by determining if the malfunctions were the result of "human or mechanical errors," as though machines were anything other than human creations, and by patronizingly explaining how they hope to "control the potential meltdown situation." Their message is clear: "We have no intention of putting an end to this obscene threat to human life. Rather, we are out to pacify you with our increasingly suspect line: If Science Doesn't Have the Answer Now, It Will (and Please Forget that We Already Said That It Did)." Rather, as E.F. Schumacher emphasized in his *Small Is Beautiful*, they are firmly committed to condemning you and many future generations to catastrophe of the ugliest kind; they are firmly committed to terminally polluting the planet, because on one hand they don't care about you, and on the other hand they fear and respect the industrialists and energy-monopolists who support their research and pay their salaries.

Chemical herbicides are used with the unconscionable argument that what poisons pests does not affect humans, as residents where they are sprayed vomit and reel with nausea and interminable headaches. Scientists consulting for the government meanwhile disagree on whether herbicides containing Dioxin, associated with skin eruptions, liver damage, cancers, mental problems, miscarriages, and birth defects, is safe and expect to be taken seriously—not to mention Agent Orange in Vietnam.

To turn to the social level (as if the physical and social levels were perfectly distinct), in veterans' hospitals new drugs are tested on patients who do not understand the stakes, often with nightmarish results. According to our media at least one U.S. private clinic for crippled children has put children through sometimes successful physiotherapy and then, to provide interns with surgical experience, amputated their limbs, strongly recalling the sadistic experiments of Hitler's Doctor Mengele. Quite frequently, "doctors" in mental asylums prescribe for patients who bite or refuse to eat the extraction of their teeth; for sexually predatory female patients, hysterectomies. For psychotherapy-resistant patients, until recently, the standard recourse was to vegetable-ize them with shock treatments or lobotomies. Similar atrocities are frequently perpetrated in old-age "homes."

Such atrocities are performed in order to practically maintain institutions in which personnel are greatly outnumbered by patients. But a *doctor* does not obscenely violate the Hippocratic Oath and passively adapt to such conditions, he attempts to alter them for the better. Like the sociologist Erving Goffman, he publicizes the fact that such extreme behaviors almost always arise *after* the patient has been committed and therefore are reactions to being rejected and committed. He recommends that the families of such patients be trained to cope with them as best they can, and encourages their best efforts by lobbying for laws that make such families responsible for such people, making it clear that their choice is between living not perfectly convenient lives and subjecting their relatives to nightmarish atrocities.

Dr. Ewen Cameron, deceased former president of the Quebec, Canadian, and World Psychiatric Associations, embodied the obscene "scientific" attitude that is now being popularly resisted. He administered electroshocks twenty times as intense as those used today to schizophrenic patients who never knew they were experimental subjects, causing many to urinate and defecate uncontrollably and to lose their memories for months to years. All his patients were plunged into mental and emotional chaos as a result. He gave LSD trips under conditions that automatically produced waking nightmares. He put patients into sensory deprivation boxes for up to thirty-five days (where normal people find a half hour intolerable) and blithely noted that "No favorable results were obtained." His objective was to prove that "the therapeutic process may go forward without a strong patient-therapist relationship"—to make psychotherapy inhuman; and, accordingly, his experiments were designed to destroy his patients' humanity, through torture. Cameron, according to the Canadian journal *Saturday Night,** "still excites the respect and protectiveness of his profession."

A rapidly increasing number of female patients claim that their "analysts" sexually exploited them, and 19 percent of 500 psychiatrists polled in 1979 said that they approved of doctor-patient sexual relationships under some circumstances, which could not include that the patient be master of her own wits, or she wouldn't be a patient.

Our weapons, like our antibiotics, are so obscenely out of human scale that, with some help from governments as alienated from human reality as the scientists who work for them, they defeat their original purpose. Just as antibiotics are creating ultimately irresistable sources of disease and weaken the human beings that medicine was originally intended to defend, *our* atomic weapons and *their* atomic weapons, *our* germs of warfare and *their* germs of warfare, are instilling abject obedience through fear into the *entirety* of humanity. They have made possible the transformation of arms from tools for the defense of freedom and dignity to tools of anti-popular social control. The messages are unstated but clear. From the Euro-American governments to their subjects: "You don't like the fact that we use your tax monies and unnecessarily high prices for domestic goods to support Communist economies and, thereby, Communist polities and their suppression of freedom? Well, would you like us to jeopardize 'detente' and risk nuclear war? mutant Black Plague?" The message from Communist governments to their subjects: "You don't like our overriding concern with arms and its attendant requirement that you double your labor for food coupons? You don't like our plans to exploit

* See Valery Ross, cited in the *Bibliography*.

your labor to provide goods to the Capitalists that can be sold at good profit margins? Well, would you like us to lower our guard and provoke an Imperialist nuclear attack? engineered epidemics?"

It is worth comparing the socially alienated scientists who developed guns, airplanes, guided missiles, the atomic bomb, and impersonal, therefore potentially genocidal, war to Leonardo da Vinci, who is, ironically, one of their heroes. Da Vinci designed a machine that would automatically destroy a ship, and thus remove human beings from the ultimately pacifying reality of battle. Reflecting on the nightmarish implications of war without face-to-face responsibility and of weapons that are more powerful than men, he kept the designs to himself.

Lastly I submit to the reader's consideration the socio-cultural side-effect of *aesthetic* and *symbolic* pollution, which, because it is not material and because it is "too close to the home" of those who control our reality, has received very little attention from the well-intentioned social critics who have put so much of the immediately preceding data into print. What I have called "aesthetic pollution" translates into Chinese high-medical terms as "the perversion of *li*," which means natural and beautiful form, and the use of the *hai-chun-chih ma*: the horse that injures the people, which I will explain. Our natural environment is transformed by science and technology toward the objectives of the industrialists, energy-monopolists, and bankers who run our society. Consequently the sounds, shapes, colors, surfaces, odors, and tastes which are produced are designed to attract attention at minimal cost or to hypnotically lower consciousness to frame, as it were, what is designed to attract attention. Beyond that, our environment simply is made ugly, by factories and the *cheapness* of low-cost/high-profit materials. I address myself at first simply to the *quality* of this environment, not its symbolic content and deliberate effects on consciousness and emotion.

Where the quality of any manufacture is basically configurated by the intentions of its manufacturers and their consultants, our manufactured environment has the quality of manipulative intent, selfishness, cold social-scientific detachment, and venal disdain, all forms of hate. That hate can be heard in the Muzak of malls; seen in the spare and characterless, cheaply produced, shapes of modern clothing and buildings; seen in the cheap, untrue colors of food packages and human packages (clothes, cars, and houses); felt in the lifeless surfaces of plastic clothes, upholstery, and Teflon frying pans, and in the alien vibrations of fluorescent light; tasted in coffee whiteners, flour-stuffed hot dogs, and apples grown for color instead of taste; and smelled in chemicals that strip the plastic surrogate-wax surface from the kitchen tile, in the aluminum vapors from the heating vanes on the cheaper-than-iron copper pipes of our plumbing, in the unhealthy body odors produced by the interaction of petroleum-based fabrics and skin, and in the lipsticked kiss.

Even the *human* environment is aesthetically polluted. It is also *symbolically* polluted. To again use a central, sexual, example: at first there was sexploitation imagery: a barrage of cosmetically exaggerated images of young women at their physically most attractive age—when they are most likely to produce healthy children. The purpose was to sell to men attracted to and women identifying with such images products with which they are associated. They are an image from the private, familial, sphere (images associated with sexual intercourse) relocated *out*side it in the public one, "turned-*out*," as pimps put it with reference to girls converted to prostitution. Feminists and female churchgoers continue to object to this, with little understanding of the negative nature of its side-effect, which, as Wilson Bryan Key has observed in his book *Subliminal Seduction*, is to imprint in the minds and hearts of husbands and wives a distaste for women or selves who are not sexually beautiful, be it by genetic accident or aging. With less (but increasing) impact the same has been done with male images.

But there is an even more evil transform of this social-scientific "cure" of consumer apathy which has gone entirely unnoticed. Replacing those sexually beautiful images, step by step, are sexually ugly ones which are excitedly and constantly presented by the mass media as though they were sexually irresistible. These are not images of ordinary looking people or of people who are not at their sexual prime, but images of a very certain kind: of people who specifically look sexually sick or sexually infertile though of fertile age, and whose faces, voices, and gestures are blank, dull, and lifeless, like Hanna-Barbera cartoons. In short, images of physically and psychologically *sexless* and/or *unhealthy* people. A casual tracing of the successors of Patti Page, Marilyn Monroe, Dianne Webber, Suzy Parker, Joe Williams, Gregory Peck, and Burt Lancaster, all beautiful, leads to several examples of human images which, like polyester clothes, are debased, ersatz surrogates for the real thing. Trendy disco-images are one of the foci.

If you want to know for sure that this newer concentration of images, mixed with the "more classical" one, is an effect of deliberate, social-scientific design, read Packard's *Hidden Persuaders*, in which he documents the then (1950's) preliminary use in such social-engineering of data on and experiments with *masochism* (which can be exacerbated by buying products associated with the indicated images) and the deliberate use of psychologically ill-looking faces in ads to provoke masochistic responses. Then reflect for a moment on how well the popular identification with such images fits the image of the "desirable" worker-consumer: someone who is subnormally interested in even his or her own children (by association with the act that produces children: sexless, infertile, impotent), someone who has very low expectations of him- or herself, someone who cannot deserve to buy products of *good* quality: someone who is a nothing, even sexually, at the animal root, and therefore feels that he or she has no right whatever to question or protest, much less attempt to change, the totally side-effecting social, psychological, and physical environment our science and technology have given us.

Such weak and weakening, false and falsifying human images—that is, the *use* of them as models to be generally emulated—are technically called "horses who injure the people" in Chinese high-medical terminology. The expression alludes to a similar but quickly cured manipulation of the popular consciousness and self-image 2000 years ago.

Toward establishing Ku Hai, a son of Ch'in-Shih Huang-Ti, as successor to the throne of the hated Ch'in, China's only one-generation-short dynasty, Chao Kao, eunuch and Secretary of State, brought before the scholar-officials of the court a stag, calling it a perfect example of a horse. Those who called it a stag were executed, and those who agreed with Chao Kao were retained under Ku Hai, for only confounders of the people's common sense and instincts were desired during this regime.* During a popular rebellion which established the beloved Han dynasty, Ku Hai committed suicide and Chao Kao was assassinated. Thus ended the only totalitarian dynasty in Chinese his-

* Strictly speaking, I have made two references here. One is to Chao Kao's deer-as-horse. The other is to a story in Chuang-Tzu's *Nan-Hua Ching* (ca. fifth century B.C.), in which a young horseherd explains to an emperor that good governing is no different from breeding horses. "A horse that does not conform to horse-standards can be used to ruin the multitude (of horses)." A good ruler knows the difference between an inspiring human example and a defective one, and he knows that examples held up by central authorities are followed by the common people.

tory until Communism, which Mao Tse-Tung was amused to compare to it, picturing himself as its founder, Ch'in-Shih Huang-Ti, reincarnate. Secretary of State Chao was the correlate not only of a modern Chinese Communist state propagandist but also of a modern social scientist highly paid to consult advertising agencies employed by industries in Capitalist countries. The difference is that where the Ch'in ruination of popular consciousness employed coercive techniques, the Capitalist one employs seductive ones and the Communist one employs both. So "horses that injure the people" are with us again.

Ultimately, our society is controlled, through its worldview, by such "deer-as-horses": social scientists who are not social scientists, economists who are not economists, and so on, but instruments of an overly Capitalistic, pseudo-Democratic social system, or of its virtual twin—the even more unnatural, disorganized, and common-people-disdaining Marxist one. In the latter, the power interests of industrialists and politicians are united and further inflamed by being co-located in one and the same group of people; and police coercion, government propaganda, and spying on the common people replace our advertising seduction, government propaganda, and advertising consultants' "masked" socio-psychological experiments on the common people.

All our scientists—physical, biological, social—are raised, educated, and conditioned in the general mental-emotional-physical environment just described. Therefore they are aesthetically and symbolically polluted unless they are deconditioned and "washed," which they are not. Rather it is pretended that mere entrance into The Academy automatically instills what is called "Objectivity"—an attitude which, in light of the fact that most of the side-effects just listed are directly related to "Objectivity," is highly suspect. It follows that their emotions and ideas partake of the quality of that environment: of lying and half-truths, of detachment from people, of confusion, of low self-expectation. They are as much the victims of our scientific paradigm as are the common people. The paradigm and the polluted social environment, via such scientists-victims of both, reinforce each other in a vicious cycle designed to perpetuate the system as a whole. As Korzybski, who recognized the relation between science and sanity, pointed out:

> It is psychiatrically known in many instances false knowledge, particularly about ourselves, breeds maladjustments often of a serious character, just because it is based fundamentally on self-

> deception. In the meantime we react and act *"as if"* our half-truths or false knowledge were "all there is to be known." Thus we are bound to be bewildered, confused, obsessed with fears, etc., because of mistakes due to our mis-evaluations, when we orient ourselves by verbal structures which do not fit the facts.

A scientific paradigm which fragments reality into disoriented little so-called parts and deals with each so as to side-effect every other is such a "verbal structure." Dr. Side-Effect is ill.

As Chinese doctors say, and as every Western doctor knows, the healthiest people are those who most strongly resist a disease and who, therefore, develop the most definite symptoms of it. Accordingly, at the psychosocial level of Western side-effects one finds a concentration of our most intelligent and sensitive young people not in our universities but in quasi-religious cults. Such cults sometimes genuinely seek a way out of our side-effecting society. But, in fact, because there are no known truly alternative grounds, they quickly develop exaggerated symptoms of the society they seek to escape and alter. (To leap into space is to orbit the object leaped from.)

For example, the Beats of the 50's and early 60's sought to create communistic noncompetitive communities and, as Paul Goodman in his book *Growing Up Absurd* pointed out, normal families. They produced, instead, a viciously individualistic milieu in which prestige, in the form of Hipness, was competed for at an intensity that makes full professors and Scarsdale ad executives look modest. The Hippies inherited this syndrome, competing for each others' girl friends to emasculate each other and for each others' boyfriends to defeminize each other, under the guise of "lovingness." Erotically assisted by Levi-Strauss' packaging, in one generation the Beat Movement was reabsorbed into mainstream society as college kids and bank clerks cruising malls for designer jeans.

A more esoteric example: there is a loosely organized international cult whose members believe that the present scientific transformation of the natural environment is being directed by extra-terrestrial aliens out to take over the planet: their energy-input is at the radioactive end of the electromagnetic spectrum and they breathe CO_2 (goes the pitch), so our scientists are programmed for nuclear catastrophe and an aerosol-and-solar-energy-beaming-satellites-thinned ozone. The solution? Outsmart them by evolving into bodiless beings immune to any form of biological wiseacreing. One means is to refrain from sexual reproduction: wise to the biological con-game of incarnating sex, the true self is liberated to persist in noncorporeal form. Weaker spirits have the option of collaboration with the enemy: being cryogenically frozen, biologically altered for radiation and CO_2, and defrosted after the biospheric conversion.

Here is a fantastic attempt to escape our past and present scientific reality which reflects that reality with ironic and chilling accuracy. Since classical Greece our Men of Knowledge have, in effect, defined themselves as disembodied beings: *detached* minds, *pure* intellects cultural-evolutionarily superior to those identified with their "animal bodies" and the mere human nature which, with its emotions of sexual and kinship-love, is attached to those bodies. The cult is in this respect classical. A modern, materialized transform of this insanity is the dream of a significant percentage of our so-called biologists: cybernetic superman, a disembodied mind transferred to a feelingless and (if mechanics never strike) immortal machine fit for interplanetary and intergalactic travel. The cult's antisexual means to noncorporeal immortality of course reflects Christian celibacy-for-purity-and-high-ranking-Heavenly-reward, and its prototype, Hindu sexual-asceticism-for-getting-off-the-reproductive-Wheel-of-Life. It also reflects the SexLib movement toward self-realization unburdened by children.

The cult's option for collaborators directly recalls the hottest item in "legitimate" scientific research and planning today: genetic engineering. The cult's ethical options are the analog of the present biological debate about whether life is basically material or basically ideal, and the main ethical "problem" of modern biology: whether to evolve naturally, go for the option of cyborgs, or, as a compromise, to work with what we have through genetic engineering to create superior (better-adapted) human beings.

As a matter of fact, a biological alteration of the human race is in progress, as predicted and planned by elite scientists of the highest rank. In 1967 I attended a public relations lecture given by a representative of one of the elite multi-national think tanks, the Institute for the Future, in Middletown, Connecticut. Directed to reassure the locals of the Institute's good intentions, a spokesman informed us that as industrial pollution increases, *as planned*, the sector of the human race most strongly reactive to it (the healthiest sector) will be "naturally" selected for extinction (remember Jimmy Cleffi), and the sector most weakly reactive to it (the least healthy sector) will monopolize the gene-pool and multiply. The quality of this richly paid government consultant's own attitude about this was expressed in two chalked curves representing each sector over a period of "evolutionary" time. Our social planners are industrio-biologically engineering the human race to become disease-non-immune—indeed, from the genuinely human perspective, to become a disease themselves. Add to their plans the datum that in radioactively polluted sectors of the planet it is non-mammalian, insect and reptile, life-forms which best adapt,* and it becomes clear that said cult does in a certain respect benefit its members in a way that "higher education" does not: its members see what is going on more clearly than do our Men of Knowledge. Dramatically responding to their vision, they also *care* more.

Our Think-Tank-People too accurately recall the extra-terrestrial aliens that cult posits are manipulating our biosphere. Not for nothing does "science fiction" literally mean "causing to come into existence through the use of science." With visions of elite scientists in fallout shelters developing genetically altered insectized human bodies, their eyes insanely switching from radioactive green to radioactive violet as they inhale and exhale through CO_2-converters as aliens land and the rest of humanity fries (or prematurely defrosts) above ground, the members of that cult see our scientists clearly, whereas our scientists cannot even see themselves, much less the *anti-human* nature of the reality they are constructing.

Like revolutionary Western scientific attempts, cultic attempts to generate a new paradigm—being without an alternative to land on—blast off, orbit the world escaped, and crash-land where blasted off from.

What, then, is the defect in our navigational plan?

The cult, Western science, and the Church all share the view that evil in the world is basically the effect of an extra-human variable. For the Church, it is Satan and an error made by Adam and Eve which is genetically inherited as sin and therefore not susceptible to *human* remedy. For science, it is the God of Progress and Evolution, according to whose word, interpreted by men such as that representative of the Institute for the Future, "evil" is a mere concept registered by the

* An all-female, Lesbian, species of lizard has been discovered in the American Southwest. The connection between their survival area and atomic-bomb-testing has not been made. Rather, interest in the biological potential of humans for homosexual reproduction and race has been elicited by their discovery.

"evolutionarily" defunct, reactionary sector of human society. For the cult, it is extra-terrestrial aliens and their human collaborators. The effect of this common understanding, deliberately precluded by the Confucian aspect of the Chinese paradigm, is to relieve humans of *full* responsibility for their own potentially good or evil actions so that the human cause of all our problems is never addressed and, when a fragment of it is, human intention is not *fully* directed toward remedying the problem. Holy men doubtfully pray for salvation, scientists proudly declare that the truth can never be fully known about anything, then demand more research money from a choking populace.

It is time Dr. Side-Effect be citizen-arrested. Are the people to be stressed and socially-psychically-physically mutated to the natural limit before serious consideration of existing alternative scientific paradigms is undertaken? Are our scientists and academicians, given their past record, to be trusted to construct an intellectually valid and Humane paradigm on their own? It is not the professional prerogative but the professional *duty* of our Knowledge-Specialists to seriously consider all alternative systems. Those who are "values-free"of such moral constraints remain subject to a pragmatic one. They can continue in their historically aberrant tradition, ignoring the Humane and more intelligent alternative held out to them from the past and look forward to losing their positions as a consequence of popular disaffection with their research and its applications.

SEEING THE DRAGON IN THE DARK

To appreciate substantiation of Chinese science requires a willingness to confront the fundamental errors which structure present scientific reality and to explore a scientific universe which ours intersects but never touches. From my own experience, I know that even the most willing may be defeated at the outset by a disorienting darkness generated by what anthropologists call "ethnocentrism," the term for assuming without examining the evidence that the knowledge, values, and ways particular to one's own society are superior to all others'. This usually unconscious source of ignorance, being an implicit requirement for full membership in our society, thrives on the seeker's unconscious fear of being rejected by the rest of his society. Its efficacy as a way to keep people ignorant and trapped in their own cultural reality lies in the fact that its darkness increases as the distance between the seeker and the foreign knowledge decreases. The closer one comes to alternative knowledge, the more possible it becomes that one will take it seriously and, therefore, be rejected by one's own society for having done so. Thus, when alternative knowledge is right before the seeker's eyes it is nevertheless quite invisible.

This phenomenon is well illustrated by the reaction to Chinese medicine of almost all Western doctors or medical scholars whom I have met or read. Obliged by the inadequacies of their own discipline, they have begun to take interest in the Chinese one. However, even though they have seen (or believe reports) that it works, they declare that it is either false and its cures accidental or "psychosomatic," or valid but without *scientific* basis. In the latter case they conclude that it would be interesting for *Western* medical scientists to figure out how it *actually* works. It is unthinkable to such men that there could be a valid medical theory that is not their own, or valid medical therapy that is not based on, or consistent with, their own. To so think would imply something intolerable to them: that their own science is limited, perhaps inadequate, perhaps fundamentally false. There is no *reason* behind this attitude; simply, it is ethnocentrism, adhered to by social consensus.

To illustrate, in one case: before the eyes of a Western physician, my teacher in Taiwan used acupuncture to treat a child whose legs had been paralyzed by a Western doctor's ill-placed hypodermic needle. All attempts by Western doctors to cure the child of this not infrequent Western-medically caused disease had failed. The child began to be able to move his legs immediately after the first acupuncture treatment. Unable to comprehend this phenomenon and unwilling to ask the Chinese doctor how his treatment had worked, the Western physician declared the Chinese one to be "a magician," and quickly left, his commitment to remain within the confines of his own discipline much stronger than any desire to cure paralyzed children. (Like twenty-five others treated by the Chinese doctor for this affliction, after several treatments the child was completely cured.)

In another case, during an epidemic of a rapidly debilitating form of what we call cerebral meningitis which results in paralysis of the neck and eyes and strikes preadolescents in particular, my four-year-old daughter was struck with a sudden extreme fever, which, given the epidemic and her exposure to its agent (mosquitoes near chickens), was a strong diagnostic of that disease. The Chinese doctor gave her herbal medicine specifically designed to cure that form of meningitis, and the fever quickly subsided. She regained her normal vigor several hours later. Soon afterward I told a Western doctor about this, and he firmly declared that she could not have contracted the then-epidemic disease because it is incurable (because Western doctors cannot cure it). He was unable to otherwise account for her symptoms and their cure and had no interest in doing so.

Such deliberate ignorance and feudalistic loyalty to the Western medical profession is reinforced by a double standard of scientific evaluation, evident foremost in our medical scientists' reactions to Chinese (and all other) organic medicines. On one hand, a Chinese medicine is declared effective, ineffective, or dangerous in view of the identities of the chemicals isolated from it by Western analysis, and in view of what the analysts know about those chemicals. On the other hand, almost daily these same medical scientists excitedly announce discoveries of new medical chemicals, and new medical-chemical effects of existing ones, including, more often than not, their hazardous side-effects. If it is Western medicine, one is open-minded about its properties, "scientifically" skeptical and subject to refine or fundamentally alter one's understanding of it. If it is Chinese medicine, one has Absolute knowledge of all its medical-chemical properties and can evaluate it with final authority. What makes all this doubly amusing is the groundless, implicit assumption that the medical-chemical theory behind Chinese medicine, if any, must be a proto-scientific version of ours; it never occurs to our authorities on this matter that the Chinese might have alternative medical-chemical theory as sophisticated or more sophisticated than our own, so that our evaluative methods might be irrelevant or even false.

It is the sum of mild to extreme, naïve to vicious, forms of the illustrated ethnocentric attitudes and behaviors that has prevented Western scientists from learning Chinese science. It is this ethnocentrism that accounts for the fact that for forty years after Neils Bohr recognized the parallel between Chinese scientific philosophy and the revolutionary findings of quantum mechanics, not one Western scientist (including himself) stood up to say, "It looks as though the Chinese had recognized the essential nature of the universe at least 2500 years ago; all Chinese science being based on that recognition, it may very well be 2500 years ahead of our own and its power and applications may take forms that we are not equipped to recognize." The logical and natural thing to have done in 1937 was to designate Chinese science as a national research priority. Under normal, not ethnocentrically benighted, conditions, this is what would have happened.

What the scholars who have promoted the belief that Chinese medicine is "not scientific" are really trying to get at is that Chinese medicine works, yet is not based on Western medical theory, and therefore upsets them. Their solution has been to misleadingly call it "unscientific" instead of "different from ours." They have then asked why Chinese "proto-science" never evolved into "science" (Needham's *Science and Civilization in China* is the archetype), whereas they should have asked why Chinese science is unlike Western science. Their answer to their question is that the Confucian bureaucracy discouraged intellectual and empirical inquiry. Regardless of the facts of that matter, since the question is false, there can be no true answer to it.

The answer to the genuine question is that Chinese science is guided by Confucian Humanism, with its central tenet "Of all things in the universe, human beings are most precious." It has therefore always developed knowledge for the sake of human beings, instead of non-human and inevitably dead-ended knowledge for the sake of God and knowledge for the sake of itself, which are the two motivations for Western science, in historical order.

This is worth illustrating. For example, when people have said they can feel pain in a certain area, Chinese medical scientists "concluded" that they feel pain there. When dissection of cadavers failed to reveal nerves in that area, they concluded that there are fine, invisible ones there. A therapeutic technique was then based on theory consistent with this conclusion, and found to work.

In contrast, Western physicians tend to discount their patients' reports of pain for which gross evidence of anatomical-physiological causes is lacking. Either they deny that there is pain or they call it "psychosomatic," meaning either imagined or without a physical basis.

Consistent with such subjectivism, there is a tendency to declare Chinese medical cures "psychosomatic," even when there *is* gross evidence of anatomical-physiological conditions for pain. Thus, both suffering people and foreign theory that disagree with Western medical theory are dispensed with with the same comment, psychosomatic.

Ironically, the relation between the *psyche* and the *soma* (body) is the one to which our medical scientists claim least knowledge, whereas Chinese medical theory accounts for it quite fully.

The Chinese medical scientist is guided by the Confucian Humanist ethic to produce theory which can be applied to heal human beings. The typical Western medical scientist is guided by the alienating and circular principle "knowledge for knowledge's sake" to deny the reality of human pain when it does not fit into his theory. For him, his "knowledge," not the suffering human being, has priority; it is his knowledge, not the patient, that he seeks first and foremost to protect and preserve. When the patient's pain threatens his "knowledge," rather than challenging the validity of his "knowledge," he devotes himself to establishing that the *patient* is invalid.

This situation is extremely dangerous, and, having persisted for three centuries, must be quickly brought to an end. As the Chinese doctor who taught me emphasized: throughout human history, knowledge, like blood, has circulated planet-wide through the social body of human society. This has made it possible for that body to renew and sometimes to improve itself, by producing civilizations in the place of proto-civilizations, and new civilizations in the place of dying ones. Like the human being, no proto- or achieved civilization is an island. Without the vital flow of knowledge, which under normal conditions is unobstructed, there never could have been a civilization, anywhere. There is nothing that any civilization can claim full credit for except, perhaps, styles, as in dress, attitudes, manners, and cuisine. For example, our democratic system (in its original form) derives from the Greek and Iroquois Indian ones. Our very languages (English, French, Russian, German, and so on) are derivatives of an ancient South-Central Asian one. Modern scientific theory (for whatever it is worth) is nothing other than a permutation of Hindu mathematics and early Greek science, which in its turn derives from (decadent forms of) Egyptian, Babylonian, Persian, and Hindu thought. Our predominant religion, Judaeo-Christianity, is a mutation of the one that ranged from Egypt through Mesopotamia in the third millennium B.C. One could go on in this way at great length.

The global flow of knowledge, of which the preceding is a minor aspect, and the mutual exchanges among societies that channel it, are symbolized by the Chinese dragon. The dragon is made up of the parts of many animals, each of which represents a society, and the best of its knowledge. The dragon is the whole of human knowledge, is human knowledge as a self-consistent whole, is wholistic knowledge. It is a unifying principle, a flowing, adaptively flexible being that transcends in vital power that of any individual species. These are the reasons for its majesty in the eyes of those who understand it.

The extreme danger we are in, then, is the danger that a part of a self-regenerating body is in when it is amputated from the whole. It cannot renew itself, it cannot survive alone, it cannot adapt, it dies. It is such an amputated limb that our scholars and scientists insist on calling the true body of human knowledge. It is fitting that it is our exceptionally ethnocentric culture that usually portrays the dragon as an evil being. If our ethnocentrism is St. George, the truth about that story is not that St. George slew the dragon, but that in his own darkness he never found it.

Ethnocentrism, however, like our scientific paradigm, is at its dead end. It follows that something else will be born in its place. To that matter I now turn.

CHAPTER

2

The Western World-Catastrophe

Men do not know how what is at variance agrees with itself.
—HERACLITUS

That view involves both a right and a wrong and this view involves both a right and a wrong: are there two views, or is there actually one?
—CHUANG-TZU

BATTLES FOR WORLDS, WORDS IN FLESH

Traditional China and the West have fundamentally different scientific paradigms, although much of the specific knowledge derived in accord with those paradigms, for example, the visible aspect of human anatomy, is identical. Since scientific paradigms are the fundaments of knowing, it has always been and remains impossible for a Western philosopher or scientist to think like a traditional Chinese one—unless, of course, he has been intensively trained in the traditional manner by a Chinese one.

The entire, almost entirely independent, histories of Western and Chinese science and culture have been determined from the outset by two super-battles, one fought in Greece, the other fought in China. These were battles fought not by armies for territories but by great thinkers for whole worlds. They were battles for *worldviews*, which are systems of knowing understood at a level deeper than is usually intended when scientific paradigms are discussed. Worldviews are systems of ideas which basically condition human thinking and perceiving. They are basic models of and for reality according to which we understand and construct reality. They include ideas of the universe as a whole and ideas of human societies, with their goals, rights and wrongs, truths and falsehoods, beauties and uglinesses, social, political, and economic structures, qualities of life, and institutions of higher learning. So, a scientific or pseudo-scientific paradigm derives from, and is as true or false as, the worldview of the people who use it. Interestingly, the sides taken in the Greek and Chinese battles for worldviews were identical and the battles were fought simultaneously, in and around the fifth century B.C. Opposite sides won. This is to say that had the outcome of the Grecian battle of the fifth century B.C. been different, probably we would now have the essential equivalent of Chinese science, instead of science as we know it.

Once these opposite worldviews were established (actually, in the Chinese case a worldview was *re*-established at the end of a 2500-year sub-cycle), they became real. That is, the world, for the holders of those worldviews, was created according to those worldviews. This sounds mystical, but it is a process that our social scientists have realistically explained quite well. People treat reality as they "see" (perceive) it, and thus transform it, along with themselves, into what they see. This "seeing" is determined by their worldview. Due to this conversion of perception into reality, of worldview into world, of word into flesh; once a worldview has been established and made real, reality proves its validity—gives it a "plausibility structure," as Peter Berger has put it, in his *The Sacred Canopy*. If, for example, it is believed that a struggle between the sexes is natural, the sexes will struggle and "prove" the belief. If a group of people believe they cannot govern themselves in a Humane manner, they *will* not and a church or a manipulative central government will emerge to make that "reality" firmer, claiming superhuman ("divine" or "natural-lawful") power, to improve on "nature." It is obvious, then, that once this conversion of worldview into world has occurred it is virtually inconceivable to people with a given worldview that an alternative worldview and world could exist. For the people within a given world, there is only "the" world.

Under certain conditions, however, "the" world disintegrates, and so does the worldview behind it. It is this complementary aspect of the relation between worldviews and worlds that our social scientists have failed to explain, because they have restricted themselves to their own narrow and exceptionally unstable world, which does not give them sufficient data.

There are three conditions for disintegration. In the first, the worldview is *false*, so that reality can be coerced into accord with it, according to Nature's limitations, only for relatively short-term periods. The result is a series of crises in each of which the worldview is patched up in an attempt to better match reality but then finally disintegrates in a terminal crisis, Nature asserting itself. Here, in a nutshell, is the entire history of Western thought and science, and we are now undergoing that terminal crisis. For example, one specific realization of the one-sidedness of our worldview (exemplified by Dr. Side-Effect) has been an ascription of superiority to men and inferiority to women and a struggle between the sexes that naturally accompanies it. In Egypt, India, Greece, Rome, Europe, and America over the last 2500 years there have been reactions to this, and the worldview has been patched by giving certain male Hindu gods female counterparts, by adding the Virgin Mary to God and Jesus, by making prostitution sacred, by glorifying actresses, and by passing women's-rights laws, for instance. All of these measures are half-baked and designed to leave the original structure intact. Where that worldview and its sex struggle are at their extreme (in Russia, mainland China, and North America, the world's centers of family breakdown), the naturally harmonious relation between men and women is being neutralized, with a consequent dissolution of families that eats away the very core of society. The social catastrophe that is resulting is implicitly disproving the worldview which produced it.

The second condition for disintegration is that the worldview is immature and half-Humane and that its people come into contact with people whose worldview is mature and Humane. The former then follow the example of the latter. In such cases, the breakdown is more a matter of reorganization and new emphases than disintegration and replacement. The assimilation to Chinese civilization, by around the time of Christ, of the two hundred different peoples who now all call themselves "Chinese" is the best example of this. In such cases, the conversion of world and worldviews is voluntary and chiefly or entirely peaceful. The "passive converters" set a desirable example, exhibit superior capacity for self-defense, and—the *upper* class foremost—*marry* the outsiders to replace suspicion with kinship-love. The coerced assimilations of much of the world to the Greek, later, Roman, later, Holy Roman, later, Christian-Capitalist and latest, Communist, worlds are *not* examples of this. The continued refusal of some American Indians to assimilate to the Christian-Capitalist world, the anti-U.S. Muslim rebellion of Iran, and the stubborn, repeated Afghani rebellions against Soviet "liberations," for three examples among many, powerfully demonstrate that the process is coercive and unnatural.

The third condition is that a people's reality is forcibly altered by foreign intervention. This is what is happening to the mainland Chinese. The disintegration is still incomplete and tentative after a century of extreme, systematic, and incessant Western or Western-style politico-cultural manipulation, because their traditional worldview and world, being mature and Humane, naturally persist. First Chinese society was thrown into chaos by British and Japanese invasions. Then it was militarily and economically pressured by Western European and American powers to become like Western Europe and America. At the same time a Communist minority, that is, a group of Westernized Chinese, was given modern Soviet arms, and then it captured modern American arms. In the end, a foreign worldview and reality has been uneasily imposed. Since it is a false one, however, the first condition now applies. And since the Communist worldview is an extreme and exceptionally ugly derivative of the general Western worldview, the Communist Chinese world promises to be shorter-lived than the "evolutionarily prior" Christian-Capitalist one.

As follows from the way a worldview makes the reality a people have constructed "the" (only) world, the majority of scholar-scientists of China were unable to sufficiently apprehend the brute power and predatory character of Western politics, religion, science, and technology when the Western world came into (ultimately forcible) contact with theirs from the seventeenth through nineteenth centuries. Reciprocally, the scholars and scientists of the West, additionally benighted by their ethnocentrism, were unable to even begin to perceive the Humanistic and realistic nature of Chinese politics, religion, science, and technology. Because the consciousnesses of both were trapped inside the worlds that their worldview-establishing ancestors had constructed, the alternative worlds which had come into their spheres were dream-like, not altogether real. Rather than apprehend an alternative reality, each party distorted the other into its own terms without ever really touching it. Our side, moreover, with the help of Western-educated Chinese "scholars" (now succeeded by Chinese-Communist ideologues), took great pains to *selectively* portray the Chinese reality in such a way as to assure itself of its own superiority and manifest destiny to subvert it, while the Chinese side, undergoing a decadent dynastic phase, failed to confront the military, economic, and social-cultural emergency of the West.

Now, in the late twentieth century, 2500 years after the Western and the Chinese worldview battles were decided, both of the worlds created by them are disintegrating. The heat this is generating is melting the crystalline lenses of these alternative worldviews into water, vapor, air, and the alternative realities they have created are becoming simultaneously visible. Consequently, a choice between or a synthesis of the two worldviews (or yet others) is now possible, and the battle for worldviews and worlds has again begun, slowly and imperceptibly, as do all world-changing things.

What, then, are the two sides of this recycling 2500-year-old battle? For short, but appropriately, one may be called the *Polar-Complete* worldview, the other the *Absolute-Fragmental* worldview. Chinese science (with its Humane philosophical foundations) is the essence of the former one; Western science (with its religious-philosophical foundations) the essence of the latter. Interestingly, whereas 2500 years ago it was the Polar-Complete worldview that was challenged by the Absolute-Fragmental one, today, with the latter presently dominating on our planet but disintegrating as it reaches its peak, it is the Absolute-Fragmental one that is being challenged by the Polar-Complete one. We are at an historical juncture of enormous significance for all present and future people, and it is therefore worthwhile to make some effort to understand it.

THE BINARY CON

Here I characterize the intellectual and moral roots—more exactly, the anti-intellectual and immoral roots—of the Western worldview, which was established after the Western Battle for Worlds in the fifth century B.C., and which underlies and configures all of Western science, to this day. Needless to say, since I take the traditional Chinese scientific perspective on this matter, the following will read quite differently from any existing analysis of Western philosophy-science. It will also compare to existing analyses of our "scientific-philosophical evolution" as strikingly simple and somewhat irreverent. Whether or not the reader will find that what follows "clicks," at least he or she will have had the opportunity to be exposed to a non-Western perspective on this matter, and so to see our own paradigm from the outside.

The Absolute-Fragmental worldview, or philosophic-scientific basis of Western science, was made into a closed system, and established in the place of its Polar-Complete opponent, by Aristotle. Its elements had already been set forth, however, by Aristotle's immediate predecessors, foremost Pythagoras, Parmenides, Democritos, and Plato. Further, Thales, a century before, had set this form of thought in motion in a primitive way by making, for the first time, an Absolute-Fragmental assertion about the basic "stuff" of the universe. That is, for the first time it was assumed that basically the universe is a single, homogeneous-monolithic "stuff"—water: the universe is basically water-Matter. It is the choice of *a single* substance as the model for the basic substance of the whole universe, a single substance extracted, as it were, from the mixed and interdependent *whole* of substances, that makes Thales' assertion classically Fragmental. It is conceiving water and the Universal Substance, Matter, as internally *homogeneous*, "unpolluted" by opposite, non-watery, properties, that makes it classically Absolute. It is in this way that subsequent, more "sophisticated" and abstract assertions, below, about the basic nature of the universe, are also Absolute-Fragmental. That is, each assertion at the root of each variant of the Absolute-Fragmental worldview to this day is based on one or more supposedly universal categories, each of which is selected from among a greater set of candidates (and so, is Fragmental), and each of which is defined as a pure, perfectly homogeneous, thing, a thing perfectly distinct from all others (Absolute). Since we take such conceptions for granted, for we have no alternatives to them, it is hard to immediately appreciate their cross-culturally distinctive nature, but it will become clear shortly.

We can and should trace the Absolute-Fragmental worldview further back than Thales, to the twice-as-old religious worldview long current in Mesopotamia and Egypt, or at least, to an Absolute-Fragmental variant of it. In this worldview, of which one expression is Judaism-Christianity-Islam's *Genesis*, the earth and its environs are basically water, just as Thales had it, and in addition, there is a Heaven-Firmament between the subtle waters over earth and the yet subtler super-Heavenly waters, which is the Spirit's abode and the source of light in all senses of that word. Cosmologically, this older view posits two Absolute-Fragments: Spirit-and-light and its opposite, water-Matter, based on a split dimension, "this world," that is, the natural world, and "the other world," that is, the supernatural, "transcendent," world, on which "this" one depends for spirit, light, heat, and life, as well as order, harmony, Humaneness. Thus the West-Asian-Egyptian worldview found in the Old Testament was recapitulated as the root-metaphor* of our "science."

The basic *pattern* (the image-content is of secondary importance) of our Absolute-Fragmental worldview, then, is a separated-out Absolute entity, essentially God and/or Matter, from which all other things somehow derive, plus a division of reality into high and low levels, "that world" and "this world," "science" and "common sense," "sacred" and "secular." As I'll show, it fosters and is upheld by a detachment from humanity, a non-Humaneness called "Objectivity," by which the Knowledge-Specialist identifies himself with the "higher reality."

Although a few admirable attempts have been made, the basic pattern of our worldview has never been fully recognized by Westerners, for three important reasons. First and foremost, our Knowledge-Spe-

* I am indebted to Stephen C. Pepper, author of *World-Hypotheses,* for this concept and the Thalesian example. Unlike him, however, I see no "evolutionary" break with these roots at any point in our "scientific evolution."

cialists have either refused or not had the opportunity to learn alternative worldviews from foreign savants, so that it has been impossible for them to get outside of their own worldview. Second, there are opposed beliefs about what "the" Absolute, original-basic, entity is, so that it appears, to one trapped inside our worldview, that we have all the "logically possible" alternatives covered. In addition, alternatives to our worldview have been obscured by the 2500-year-long argument, internal to our worldview, about what "the" basic entity is. The result of this *conflict*, this conservative escalation into, rather than out of, our worldview, is a variant of our worldview which is *based* on conflict —Marxism. Like the mythical sons of the Greek Titans, it is now rapidly devouring its religious-philosophic parents, just as those who make political use of it are now doing a similar thing at the level of world politics.

Third, our thinkers have agreed, for 2500 years, that philosophy-science and religion are fundamentally different modes of knowledge, so that they have been preoccupied with this alleged difference to a point where the basic nature of our worldview has been quite invisible to them. Like the closely related God-or-Matter split, the religion-or-science split has driven Western minds further and further into our worldview instead of toward its outside. As Arthur Koestler, author of *The Sleepwalkers: A History of Man's Changing Vision of the Universe,* has observed, actually our religious and scientific modes of knowing are often indistinguishable, and support each other. To put it more strongly, objectively viewed these two traditions pretend to respectively specialize in spirituality-mysticism and rationality-science but, actually, neither does either well enough, and, as indicated above, the two are basically identical. They differ chiefly in their practical relations to the human society over which they divide their influences and which they divide.

The functional association of the religious and the philosophic-scientific "halves" of our worldview with each other and with negative social changes is quite evident during periods when the two express their mutual support in a relatively dramatic way. Thales' synthesis of Babylonian-Egyptian religious cosmology with scientific philosophy co-occurred with the transformation of Greece into a centralized, imperialist state representing a "higher reality." About a century later, in his colony in southern Italy, the West Asian, Pythagoras, our first full-fledged Idealist, deliberately combined with Greek "rational" philosophy and originally Hindu "mathematizations" the beliefs and cult of Orpheus, the sacrificed-and-reborn Egyptian-Mesopotamian god-man whom Jesus incarnated five centuries later. In so doing he established, for all thus-far-elapsed Western time, the cross-culturally distinctive and peculiar Western attitude about knowledge, in which the "world of the senses" is distinguished from "reality" as "unreal" and "imperfect." Attending this attitude is the understanding that knowledge itself, rather than service to human beings through knowledge, is the objective of the Knowledge-Specialist. Behind that is a disdain for common sense and a patronizing, breezy, attitude toward the majority of human beings, the common men, women and children, realized as a willingness to sacrifice and injure them for the sake of the Knowledge-Specialist's "higher reality." To this crucial feature I will pay much attention in this and the final chapter.

That non-Humane, and therefore anti-Humane, perspective appears again and again as Western religion and science rediscover their kinship, their common Hindu-Babylonian-Egyptian root. In the first centuries after Christ, our first alchemists, in Alexandria, were Jewish or Gnostic Christian mystics. Their objective was to counterfeit gold, which presupposes a total disdain for the honest labor of common people that gold was to represent. Again, the Westward-riding Muslim savants of the eighth century thoroughly intermixed their religion with Greek rational philosophy, and it was chiefly due to their religiously inspired, sword-enforced influence in North Africa and Europe that the European Christian savants, not to be outdone, thoroughly intermixed their religion with Greek rational philosophy in the Middle Ages, and crusaded against the Muslims. One of the most formidable forces that emerged from this was the Knights Templar, who became the first world bankers of war and foreshadowed the sacredized Christian-Capitalist and then Communist imperialism, subversion of non-Western culture, and war-mongering that distinguishes contemporary history.

Today, a perhaps final mutual back-patting of our religion and science again testifies to the gray detachment from human sensibilities that was made "scientifically legitimate" by Pythagoras 2500 years before. As the Church dies and popular negative reaction to science mounts, as always the Church and Science rush to aid each other, constituting a united, misdirecting front. Our conservative physicists have revived the "Big Bang" theory of cosmic evolution. Many of them, as explained, understand the "pre-evolutionary" conditions theologically (God caused the Bang), and many theologians, in turn, understand that modern physics has scientifically substantiated *Genesis*. Even the religious party to this contemporary version of Pythagoras' synthesis appears to find this intellectual handshaking much more significant than the facts that our physicists use their new knowledge of the electromagnetic spectrum to radioactively pollute our children and that the nations who pay our physicists to specialize in cosmology-without-human-beings support, ultimately with the Atomic Threat, vicious totalitarian regimes in the Caribbean, Africa, and Central and South America. Simultaneously, and with an even colder "scientific" detachment from humanity, the interpreter-priests of Communist Dogma ruthlessly subvert Third World countries, replacing Western-supported totalitarian regimes with their own, and when they succeed, take the results as "scientific" proof of the prophetic accuracy of Marxism, which is a classical synthesis of Western religion and science —not a radical but an extremely conservative variation on our paradigm. How the "higher-realitied" non-Humanism of the Western paradigm directly leads to such negative social effects will be made clearer as the matter is looked at from several Chinese-scientific angles.

Since he was one of the very first architects of our "scientific" worldview and world, Pythagoras may provide us with the introductory illustration of the difference between the Absolute-Fragmental and the Polar-Complete worldviews, and of the arrogant and dubiously sincere character and short-ranged logic of the inventors and perpetrators of the former. Taking numbers and geometric forms to be primary reality, and natural sensed objects to be illusory, secondary reality, Pythagoras was already operating within a microscopic, arbitrarily selected (Fragmental) and subjective confine. Even within that confine, however, he ran into, and refused to confront, the following problem, which inevitably arises in innumerable forms throughout the Western worldview, "scientific" paradigm, and even advanced physical science, and remains not confronted to this day. At the same time, it, in countless forms, demonstrates the scientific validity of the Polar-Complete alternative to our paradigm, for which it is not a problem.

Wishing to advance the notion that the universe consists of numerically expressible Absolutes, that is, discrete, quantifiable Forms, the Pythagoreans were upset by the fact that one of their favorite Forms, the square, had a non-discrete, "formless" aspect: its diagonal, whose measure is an "irrational" number that is impossible to express in terms of the lengths of the square's sides, or any "rational," whole or fractional, number. (An example of an "irrational" number is the square root of 8.)* Even the most Idealized, abstract form, even an idea maximally unlike anything in Nature, then, "refused" to be consistently Absolute-Fragmental. Instead of rectifying the Absolute-Fragmental paradigm, the reaction was to regard "the Formless" as a higher level of reality quite distinct from the level at which "Natural" Philosophy is practiced, and to keep its existence a secret whose disclosure would bring the effects of a standing curse down on the pearl-casting initiate.** Being of lesser intellectual caliber, or simply creatures of Absolute-Fragmental habit, Pythagoras' 2500-year-long train of mathematical-scientific successors have failed even to confront this "formlessness" problem—although, as is characteristic of our scholarship, historians of our science are fond of citing this historical instance of it, and without taking it at all seriously. Twenty-five hundred years later the same game is being played. Just like the Pythagoreans, even our avant-garde physicists attempt, in the face of evidence counterindicating it, to express atomic structure in Absolute numerical terms. Our scientists have always done away with "the formless," the non-Absolute, the non-determinate, handing it over to "God."*** (After all, have we not become the rulers of the world and the monopolizers of Pure Science on the basis of the "not-formless" paradigm bequeathed to us?)

Had the Chinese savants contemporary with Pythagoras been inclined, and they were not, to be narrowly preoccupied with such abstractions, or to see them as Divine encodings of "reality," they might have handled the square-diagonal problem as follows. Not being Fragmentally fixated on one aspect of any problem, they might have *tested* the Pythagorean deduction. If "formlessness" is actually a property only of the diagonal, not also the side, of the square, then a diagonal cannot be drawn first in rationally quantified measure, so that the *sides* have irrational quantified measures. Of course, it can: if the diagonal is, say "4 inches long," then the sides are each the square root of "8 inches" long.**** Hence, the notion that the sides of a square are intrinsically "formed" is one-sided and false, and their "formedness" is

* Diagonal $= \sqrt{\text{side}^2 + \text{side}^2}$. If side = 2, diagonal $= \sqrt{4 + 4} = \sqrt{8}$.

** Koestler, *op. cit.*

*** In his *The Making of a Counter-Culture,* Theodore Roszak makes much the same point, indicting our technocratic science for its aesthetic-emotive poverty.

**** side $= \sqrt{\dfrac{\text{diagonal}^2}{2}}$. If diagonal = 4 inches, side $\sqrt{\dfrac{16}{2}} = \sqrt{8 \text{ inches}}$.

an illusion, perpetrated by the one-sided, Absolute-Fragmental mind.

Of course, whether any "sized" square has irrational sides *or* an irrational diagonal simply depends on what standard of measure *humans choose* in the first place: by another, non-"inch," standard, the square with a rational "4-inch" diagonal can "have" an irrational diagonal and rational sides.* Absolute measure in general, the cornerstone of mathematical Absolute-Fragmentalism, and of our "Scientific Revolution," is an illusion. Rational number is an irrational conception—a subjective projection, mysticism! *Any*-"sized" square has at-once-formed-and-formless sides and diagonal. The "Divine," the Formless, *is one with* the "Earthly," the Formed. They are not at "different levels"; they are aspects of a same thing, ultimately indistinguishable. (Every time a modern physicist successfully applies theory deduced in terms of "irrational numbers," he unwittingly proves this point. For whatever it was worth, we put our astronauts on the moon by using "the Formless.") So, it should have taken no Einstein to have understood, 2500 years ago, that, as he himself later put it, "As far as the laws of mathematics refer to reality, they are not certain; and as far as they are certain, they do not refer to reality."

Western historians of Chinese science are fond of citing the Chinese scientific statement "Earth is square and Heaven is round." They interpret it to mean that this planet is square and that what we call the cosmos is round, and thus indicate that Chinese knowledge is not to be taken seriously. Actually, *Ti*, "Earth," is a metaphor that stands for a *part* of the universe and for what is relatively obvious, immediate, familiar. *T'ien*, "the Heavens," is a metaphor for the *whole* universe, including not only the cosmos but the human conception of it—the whole, Complete—thus, Nature, and for what is relatively non-obvious. Further, it is understood and repeatedly stated in the texts where this proposition is made that "The Heavens and Earth are one"; indeed, the two terms are usually conjoined, as *T'ien-Ti*, to make one term. The implications are as follows, first, that there is a roundness in squareness and a squareness in roundness. More generally, each property is non-Absolute, contains its opposite. Second, where a square is relatively definite, formed, and a circle relatively indefinite, formless; a square cornered, bordered, *sided*, and a circle infinitely, non-sided: a person with short sight and a low, obvious,

*$\sqrt{8\text{ inches}} = 1$ sinch. Therefore this 4-inch-diagonaled square has "rational" 1-sinch sides and an "irrational" $\sqrt{2\text{ sinches}}$ diagonal.

perspective sees only the definitely formed, a person with long sight and a high, non-obvious, perspective sees only the indefinitely formed, and a person of *knowledge*, a savant, who takes the Complete perspective, sees and deals with both simultaneously, as they actually occur. *Jen*, "the human being," is always placed between, in, "Heaven-and-Earth."

According to the Absolute-Fragmental paradigm, one *arbitrarily* separates the formless from the formed, the non-Absolute from the Absolute, the non-quantifiable from the quantifiable, the non-obvious from the obvious, and then totally ignores the formless-non-Absolute-non-quantifiable-non-obvious, handing it over to religious mystics. Or one pretends that it does not exist. And one restricts science to the formed-Absolute-quantifiable-obvious aspects of Natural phenomena. More exactly, one restricts science to the *perspective* from which Natural phenomena can be *partially* described in such terms. Yet this "science" claims to address *entire* natural phenomena, and is "objective," and "rational." If the preceding example is typical of it, actually it is the Absolute-Fragmental paradigm that is *mystical*, highly *sub*jective and *ir*rational. In contrast, according to the Polar-Complete paradigm, there is no Absolutely formed, no Absolutely quantifiable. There is a formless within the formed, a formed within the formless, a non-quantifiable within the quantifiable, a quantifiable within the non-quantifiable, a non-obvious within the obvious, an obvious within the non-obvious. The Polar-Complete paradigm plainly differs from ours in that it is not subjective but objective, not irrational but rational. Our "scientific paradigm" is basically one-sided and deceptive whereas the Chinese scientific paradigm is basically all-sided and truthful.

I now introduce the Chinese, Polar-Complete, paradigm, implicit in the counter-Pythagorean argument, in the primitive way I introduced ours, to give us ground for the other foot to stand on. Where the Western paradigm is based in part on a model of Nature (originally, Thales' water), the Chinese paradigm is directly based on human interactions with Nature, so that where there are Western models of Nature in the Chinese paradigm there are natural phenomena. Among those natural phenomena are some which, because they are widespread in the world, are used with relative frequency as examples of natural phenomena in general. Consequently, whereas the the Western model of Nature's "underlying reality" was originally a separated-out (Fragmental) and, by "rational necessity," homogeneous (Absolute) substance: water (Thales) or air (Aristotle); a central Chinese example was and is an integrated and heterogeneous (Polar-Complete) aspect of the natural whole, which was recognized as *both* a substance and an *act*: vapor (*ch'i*), as water-air shifting among various states.

The Western paradigm begins, then, with fundamental fragmentations (*one* natural phenomenon as model, model *of* instead of Nature itself, subject-without-object) and maximal "rational" input (the universal model "must" be homogeneous), on one hand, and minimal common-sensory input (the senses do not perceive any perfectly homogeneous things "out there"), on the other. The Chinese one begins oppositely to that: non-Fragmentally and common-sensically—empirically.

To further explain: the Greek Physicists took ice and air-vapor to be *basically, essentially* water (*or* basically air *or* even vapor). The subsequent philosopher-scientists, until physics reached its impasse in this century, took the solid, liquid, and gaseous states to be "syntheses," or composites, of the Absolute-Fragments that replaced Thales' water (*or* air *or* vapor): Being, non-Being, Form and Becoming or, analogously, Matter, Space, numerically expressible Form and Energy. (The implicit original model for Energy was the sun, as per Plotinus and, implicitly, Moses.) Our scientific paradigm can still be traced to a water+sun model of Absolute-Fragmental nature: water=matter, sun=energy; matter and energy are each homogeneous and therefore Absolutely different; since they are the "underlying reality" of all, the whole universe is two Absolute-Fragments. As I'll show, post-Einsteinian cosmology retains the same pattern, a hangover, despite its equivalence of energy and matter.

In contrast, the architects of the Chinese paradigm took vapor to be what it appears to the senses to be: at once water-and-air (heterogeneous) which can be relatively watery (what we call "water" and "frozen water") or relatively airy (what we call "air," "gas"). It has non-Absolute extremes, or poles, and is Polar. It is at once a substance and an act, because it continuously changes, cannot be stabilized in one form: *at once* what we call "Matter" and "Energy," a Nominal-Verbal entity—an "event"—as avant-garde physics' terminology now expresses it. Neither air nor vapor nor water is abstracted from Nature as a "basic, essential," Fragment. Each is, to an extent, the other, and the extents continuously vary. As such, it is like the manifold universe of which it is not a Fragment but an integrated aspect—a Polar Complete substance-act, or event. The natural aspect of the universe, then, is theoretically represented as it is at the sensory level and as revolutionizing Western science is beginning to describe it. The relatively subjective mental structure, the Pythagorean square's sides and diagonal, are *at once* formed and formless, both numerically determinate and numerically indeterminate; likewise, the relatively-objective natural phenomena are *at once* energy and matter—energy-matter (*yin-yang*).

Beginning with human sensory interaction with Nature, the Chinese paradigm was and remains thousands of years ahead of ours, which now begins to proto-scientifically imitate it as it self-destructs. Beginning with Absolute-Fragmental theory which, instead of being derived from Nature, was produced from an aspect of the human mind like Athena (Athens, Western culture) out of Zeus' head and *imposed* on Nature, the Western paradigm has reached a dead end within 300 evolutive years, and the empirical evidence against which it has finally collided is comfortably included (along with empirical evidence we have never even considered) by the millennia-old Chinese one.

Needless to say, many more examples of different kinds of phenomena and the Western and Chinese ways they are handled will have to be considered before the general nature of each paradigm can be grasped and evaluated. My purpose at this point is to make clear by example what the basic difference between them is. It is already easy to see, though, that the Polar-Complete paradigm must lead to scientific theories and applications so different from those that ours has led to that one could not infer on the basis of the Western scientific experience what they might be like. By the same token, it is easy to see that Chinese science could be 2500 years ahead of ours along the line it has taken, and that even if that line is *not* more scientifically valid, more humane, or both, than ours, it would be wise for us to learn everything we can about it. It is also easy to see that if we wish to understand anything at all about it we will have to start from scratch, from fundamental differences in worldviews downward to specific scientific applications, because we are simply not equipped to think in Chinese scientific terms. Indeed, we are "equipped," by our educational system, not to. With that, I invite the reader to take a guided tour to the exit from our Absolute-Fragmental labyrinth.

Our sophisticated philosophers of science, who have recognized the common denominators within all of Western philosophy, are fond of saying that "There has been nothing new since Aristotle." Although this is nothing to be proud of, it is true. So the basic repertoire which has produced the many Absolute-Fragmental variations that have been our basic worldview for 2500 years can be fully appreciated simply by examining the three classical variants of Aristotle's time: Materialism, Idealism, and Realism. Using the present, Chinese-scientific, perspective, the examination is fairly simple, for despite the general understanding that one's mind must perform like an adolescent monkey in an undersized jungle gym if it is to recapitulate the thoughts of our original Seekers of Truth, each variant can be simply but nevertheless accurately described. As the Chinese scientific saying goes, what appears to be terribly complex can, and should, be subject to basic, and therefore simple, comprehension. There is a short, Chinese, path to the exit from the labyrinth of Western philosophy, from which most who have entered have yet to return. Let us take it.

It might be said that to portray the whole of our philosophy as a unit could only be an outrageous reduction, but the joke is on those who would say that, those who fancy themselves scholars. It is, in fact, impossible for anyone to represent the works of even a few Great Minds and do each justice, for to do so requires as much exposition as each of those Great Minds required, plus the effort and pages required to relate them to present interests and knowledge. Consequently, over time, "scholarship" increasingly consists of key thoughts, computerizeable bits, expressed by Great Minds. The thoughts of each are reduced to comic book form, as it were, to make it possible to appear to embrace all Great Thoughts expressed through the present. A flat, selective approach is built into Western scholarship: there are simply too many Great Thoughts to handle exhaustively. This is because our Great Minds were never able to agree about anything, never able to make up their minds. Western thought is a cancerously expanding *argument. It is that problem which must be addressed.* And as soon as it is, true scholarship becomes possible. A common denominator running through all Great Thoughts then clearly emerges.

Instead of new Great Thoughts, there are usually inferior derivatives of the original ones, like different versions of a same folktale. Our Great Minds are getting nowhere, have stood still for 2500 years.

Briefly put, there are two reasons for which our thinkers have been unable to reach any conclusions to their 2500-year-long argument. One is that the Absolute-Fragmental structure of their paradigm makes inconclusive argument inescapable. The other is the basic reason for that structure: our paradigm's legitimation and encouragement of a detachment from and suspicious attitude toward the "world" of sense. This world includes the warmths and agonies, the beauties and supposed "imperfections" of the human condition. Socialized according to our paradigm to believe that, God-like, they view the world from above and outside, our thinkers are trapped in another world, a refrigerated sphere of "knowledge for knowledge's sake," dangerously separated from the people whom most of them wish to serve. So, in short, it is because they have been conned into denying themselves access to sensory and living human data, and because they have been conned into denying their natural *emotional* rapport with the data, that our people of knowledge accept and submit to the *unnatural,* Absolute-Fragmental structure of our paradigm which, in turn, obliges them to inconclusively argue for as long as they remain trapped in it. Direct and unrestricted interaction with the data they deny themselves access to results, as I'll show, in an alternative paradigm in which argument is about details, not root concepts. I'll describe these two dimensions of our scientific-paradigmatic prison in some detail, but first it is best to look at their immediate result, the Absolute-Fragmental generator of argument and conflict—the Binary Con.

The terms for the three major variants of our worldview—Materialism, Idealism, and Realism—are taken from the *ontological* content of each. Ontology is the aspect of philosophy which has to do with the basic character of the universe, what exists, and causality, the question of why or how things move and change. A Materialist believes that the universe is basically Matter. That is, everything is substantial, physical, concrete, solid in the sense common to "solid, liquid, gas." Thought, ideas, forms, are fine, subtle, forms of Matter. So are energy, soul, and spirit, if they are thought to exist at all. There are two subvariants. The first, which has not been preserved in our tradition, has it that Matter is a continuous mass and that movement is an illusion. Parmenides held this view. The other variant, preserved through this day, has it that Matter is made up of a (finite or infinite) number of indivisible, tiny, basic units—"atomic particles"—and that their different shapes, sizes, and densities account for movement. Democritos held this view, which pictures the universe as a kind of perpetually motioned atomic machine.* To explain movement causally, all one has to do is analyze and describe what moves. The relation between

* See Copleston, cited in the *Bibliography.*

these physical things is conceived as cause and effect: in their movement-relation, object A is all, Absolutely, cause; object B is all, Absolutely, effect. So has physics remained until quite recently, although with Newton and then Einstein, Energy has increasingly acquired something of an identity of its own, and physics' cited avant-garde is clearly recognizing the fallacy of Absolute causes and effects.

The Materialist variant is both ugly and intellectually inadequate. With respect to its ugliness, it gives us a *dead universe*. With respect to its inadequacy, it posits causes which are not, in turn, accounted for as effects of other causes*—especially human ones, such as the laboratory experimenter (as Capra, after Heisenberg and Einstein, has pointed out). The result, as I'll show in a moment, is that Idealism is needed to fill the gap, that Materialists are "closet Idealists": they imply or posit metaphysical ultimate cause—a God of some kind. "Natural laws" themselves, the Materialist scientist's product, are *ideas* to which causal power is subtly attributed.

Most of our scientists, chiefly our "hard" scientists, but also a great percentage of our social scientists, are Materialists, or schizophrenically combine Materialism with another variant of the Absolute-Fragmental paradigm.** The implication is rather scary. A majority of our scientists believe, or operate as though, human beings were soft-machines, robots created by God or "Accident" or "Laws," whose hearts are mere beef, whose emotions are mere chemical reactions, and whose eyes are not windows of human souls but opaque spheres of flesh. That the nightmarish character of Materialism is ignored by Materialists—and most Idealists as well—exemplifies the non-Humane detachment that is basic to our scientific paradigm. Can truth be established without regard for beauty? Can an ugly worldview have other than ugly effects? Dr. Cameron, the still respected and defended electroshock-and-LSD-torturer mentioned earlier, was a quintessential Materialist, and a good example of how one's worldview becomes one's reality, including oneself: only a soulless robot could have done what he did. The dreamer became the nightmare, the audience the horror-movie. Elite Communism, whose basis is the Marxist Materialist variant (until, like Mr. Brezhnev, one begins to die, and then says things such as "God will forgive us"), also fosters and attracts person-

* The "Big Bang" is the central example at present.

** I am indebted to my tutor, Cornelius Osgood (then at Yale) for calling my attention to this.

alities consistent with the Materialist ontology. True Communists are able to call people "the masses," by analogy to the mass of Matter they believe (or pretend) the universe to exclusively be and to manipulate and overwork people as though they were mere machines, tools toward an End of Higher Reality. Marx, with his open disdain for the life ways of farmers, who were economically and politically independent as compared to labor-selling factory workers, of capitalists and their central-governmental representatives, betrayed this elite Communist personality trait at the outset, but little notice has been taken of it.

Idealism, the second variant, has it that the universe is basically made out of Universal Ideas caused by God, and that Matter is an illusion or a gross, derived form of Idea. In this view, individual concrete things, such as rocks, plants, animals, and people, are imperfect realizations of Forms, or Ideas. For example, Rock, Plant, Animal, People, and, "above" those, the universal Ideas, Form, Matter, Being, Non-Being, and Becoming—all are effects of God's Mind. Before having been conditioned by a univeristy education to reason in the way that leads to these conclusions and allows one to take them seriously, one might respond to Idealism as one of my college friends did. Bash the Idealist's head against a tree until he admits that the non-universal, individual, material tree and his non-universal, individual, material skull and nerves—and everyone else's—are quite real.

My friend's "counter-argument" to Idealism is a genuine one from the perspective of a Humanist, one who is sympathetic to human pain, but it does ignore certain features of Idealism (below) that are superior to those of Materialism. But, more importantly, and as is quite obvious to anyone whose common sense has not been amputated by higher education, Materialism and Idealism are two halves of an implicit whole, each of which, if adhered to, leads to the other as a counter-variant, like the legs of a very pigeon-toed and knock-kneed walker. Here we encounter the central fact about our worldview's so-called evolution. As Koestler puts it, as of the fifth century B.C.:

> The pendulum has been set swinging; its ticking will be heard through the entire course of history, as the bob alternates between the extreme positions of "all is body," "all is mind"; as the emphasis shifts from "substance" to "form," from "structure" to "function," from "atoms" to "patterns," from "corpuscles" to "waves," and back again.

The efforts and experiences of the people of that Asian peninsular colony, Western Europe, and their own giant colony, America, constitute only a minor part of human history, but outside of that (ethnocentric) oversight, Koestler here begins to get at the nature of the Binary Con. Now let us take a neither knock-kneed nor pigeon-toed walk.

Materialism posits one Absolute-Fragment, Matter; Idealism posits opposite Absolute-Fragments, Spirit and Form. As a result, neither system is complete, each has its strengths and weaknesses. But in the end, they are quite similar. Where Materialism, unlike Idealism, "grants independent existence" to Matter, still, like Idealism, it only casually takes the evidence of the senses, "this" world, into account. And where Idealism, unlike Materialism, "grants independent existence" to Spirit and Form, and thus to life and thought, nevertheless, as its defective effects aggravate, it ultimately yields a universe almost as dead—and ruthless—as does Materialism. To this matter I now turn.

By denying basicness to life and thought, Materialism implicitly portrays human beings as passive effects of "more fundamental" Material things (the "Environment" which the Materialist seeks to forcibly control), and suppresses recognition of the human being's spiritual capacities. For the moment these may be described as those that permit human beings to *make connections* between sensed data and between people. To illustrate this making of connections, I observe that this thing and that thing are more like each other than either is like any other things. I give them a common name—rock. I can now talk about, share knowledge about, and generally investigate the nature of rocks, and rocks, henceforth, involve my thoughts about them as much as their own physical nature. Denying the spirituality of science, this ability to make connections, the Materialist view of Man and Nature makes Man relatively stupid, plays down the intrinsic relation between Man's spiritual capacities and the natural order. This "legitimation" of Man's supposed passivity, stupidity, and *plasticity* is one of Materialism's attractions for totalitarian revolutionaries, for it "philosophical-scientifically" "justifies" doing just about anything, including fostering hate and war, in order to attain their goal.

In turn, pendulum-swinging back, Idealism in its philosophical or religious disguise has notions and influences almost as ugly, just as pernicious, and much like those of Materialism. Idealism's ultimate is God, whose nature is deduced from the Idealist ontological assumption that Form is more real that Matter, analogously as Materialism's passive-stupid-plastic soft-machine image of human beings is deduced from the Materialist ontological assumption that Matter is more real than Form. Idealism deduces God by equating logical generality, "height" of category, to primacy of reality-level. "All rocks are things; all vegetables are things; all animals are things; all human beings are things; all inanimate phenomena are things. All things exist. The relatively powerful things, human beings, think. The highest category, therefore, is Existence, or Being, and the highest action is Thinking. Reality is a reflection of these hierarchical categorical relations. Therefore there is God, a Pure Being Who Perfectly Thinks and Absolutely Causes all movement." It follows from this making-real (reification) of abstract-generalizing thought (ideas) that human beings and their senses of pain and comfort, human beings and human, as distinguished from Godly, goodness, are not as "real," or important, as God's; indeed, they are "unreal" and virtually valueless before God's. This dubious train of logic *makes human beings' expectations of themselves lower than they might otherwise be,* and preoccupies the savants, who are responsible for helping along a Humane social order, with God and/or knowledge for its own sake, *so that they are less attentive to humans and society than they might otherwise be.* Demoralized on one hand, the people are deprived of the full benefits of their best minds, on the other.

As Koestler puts it, the religious-Idealist worldview is "socially sterile." What is sterile cannot enliven; it can only destroy. It is this religious-Idealist attitude, disdainful of "this" world and "sterilized" of the natural, "this-wordly" Humanism that is found in the hearts of ordinary men, that has given us rapists who are "saved" and their secular transform, rapists who are psychoanalyzed, instead of punished for the sake of protecting women. It has given us pogroms, genocide, and "holy" wars, and their secular transform, wars "for freedom," that were actually waged for wealth and power, and the wealth and power of a few. It has given us "divine" instead of *human* justice. In what way do "holy" wars differ from the Marxist-Materialist-Communist's "necessary bloody revolution," forcible colonizations, tortures of freedom fighters, and "necessary seedings of chaos" in societies that fail to evolve to the prescribed "prerevolutionary" stage?

An examination of the historical record substantiates this connection between both halves of our worldview and predatory warfare. In contrast with ours, traditional Chinese wars, a realization of a non-Absolute-non-Fragmental worldview, have been relatively infrequent and unexceptionally self-defensive.

Social conflict, war, is simply the historically most dramatic and obvious effect of the Absolute-Fragmental paradigm. More generally, its effect is imbalance, staggering, the side-effect syndrome. So, the list in Chapter One of Dr. Side-Effect's technological crimes is no more or less an Absolute-Fragmental effect than is the Western penchant for social conflict and predatory war.

Our paradigm is a *basic pattern;* its effects are innumerable manifestations of that pattern at all levels. The Idealist one-sidedly chooses Spirit and Form as his basic concepts, and the Materialist one-sidedly chooses Matter as his basic concept, so that Christian-Capitalists and Communists war for human territory. Likewise, at a more specific level, the Western medical scientist, for example, chooses to pay attention only to certain bacteria instead of the whole illness and the whole afflicted individual. Dealing with only one of many sides of the problem, he produces a disease worse than the original problem. Just as the Idealist and the Materialist make their one-sided choices, medicine, for example, is divided into two incomplete halves called psychiatry (soul-iatry) and physical (body-al) medicine. The former treats the patient as though he or she were a disembodied soul (psyche) and the latter treats the patient as though he or she were a soulless body. Our psychiatrists accordingly neutralize their patients' physical responses to their illnesses, "free" their souls, by administering tranquilizers and stimulants. At the extreme, the teeth of the biting patient, the ovaries of the promiscuous patient, the male sexual organs of the patient who is told he is a female psyche trapped in a male body may be surgically removed. Complementarily, our physicians, whose ultimate technique is surgery, learn their art by practicing on corpses, soulless-lifeless bodies, and their powers are at their zenith when they have reduced the patient to the closest thing to a corpse—the totally anaesthetized patient on the operating table. Like our one-eyed Idealist-or-Materialist social-political policymakers, our Idealist (soul-iatrists) and Materialist (body-al) "doctors" are at odds (war) with each other and the people they should be protecting, as though the people were a "Third World territory."

Needless to say, Chinese medicine has alternatives to the preceding practices which are as agreeable as these are unpleasant, as balanced as these are one-sided, as Humane as these are inHumane. In turn, our psychiatrists and physicians are not to be blamed for the inadequacies of their sciences, because they do not know any better and are doing the best they can. The only blame to be placed is on those who have been ethnocentrically ignorant of, suppressive of, alternative scientific paradigms—yet another form of one-sidedness that our paradigm fosters.

In general, the effete (the Idealists) and the brutal (the Materialists) have bequeathed to us a Binary Con: an either-Spiritual-or-Material worldview in which human beings are encouraged to be passive subjects of God or of the Material Environment, and by which our Knowledge-Specialists are freed of the obligation to carefully reflect on the human consequences of their work, "scientifically licensed" to generate one side-effect after another—all told, to *fail*, on the basis of an always defective and fragmental view of the whole. Materialism promotes a cold detachment from humanity and short-term practical measures; Idealism promotes a "hot" detachment from humanity in which human beings are overlooked for Abstract Principles appealing to Academics—ultimately, "God"—and promotes long-term impractical measures. The Materialist, for example, reconditions the juvenile delinquent as though he were a soft machine, *à la Clockwork Orange*. The Idealist prays for him and asks him to sacrifice himself for God or a Principle instead of the reality of other human beings. Anywhere between the *pure* Materialist and *pure* Idealist extremes there is still the deadly detachment from humanity, the lack of direct Humaneness, that guarantees error. InHumaneness, be it hot, cold, cool, or warm, remains inHumaneness.

The Con in the Binary Con is that our worldview-and-"scientific"-paradigm is divided into two supposedly opposite halves which are actually mutually supporting members of a defective, un-whole whole. The history of Western science, and, indeed, of Western social culture, is a Dialectic, a frantic, unbalanced, switching-back-and-forth, from one defective side to the other, an *either*-this-*or*-that pincer-movement that produces continuous catastrophe, emergency, from the scope of individual health to the scope of war and revolution and the ruination of the natural environment. The Binary Con reduces the quality of life at all levels. Trapped by our Absolute-Fragmental reality and therefore unable to conceive of a whole, balanced, stable, life-promoting system in which each side reenforces each other at the same time, the Western mind, terminally inflamed by 2500 years of friction, has produced Marxism, in which the (easily disproven) assumption is made that our Dialectic, our tiresome switching-back-

and-forth from one side of the Con to the other, each side "*negating* itself" in alternation, is a natural, inevitable phenomenon. The Absolute-Fragmental proposition that the basic-original entity in the universe is God, Absolute Spirit, is the seed, and Marxism, an Absolute-Materialist system based on and promoting conflict itself, is the fruit. We need a non-Absolute-non-Fragmental way out of this Con, and that is the reason to take genuine interest in the Polar-Complete paradigm of traditional China.

Because the present objective is to clearly describe, for the first time, the Chinese and Western paradigms from an insider's view of both, I am using terms that are black-and-white, clarity being the foremost prerequisite. I am emphasizing the differences. As I go on, it will become apparent that there are similarities as well. It is crucial to understand the differences, however, because our natural tendency is to identify foreign ideas with our own, by translating them into our own terms. Sometimes the differences that are thus neutralized are subtle, but this makes them no less crucial, because, as the Chinese proverb has it, "A little error in the beginning becomes a great error at the end." For example, and as can be made clear only after the reader has absorbed the next chapter, a subtle difference in the way that Western scientists and Chinese scientists conceive the relation between Energy and Matter leads to Western medical techniques that are barbarous by Chinese standards and Chinese medical techniques that are miraculous by Western standards. The reader who has thus far found my portrait of our own scientific paradigm—and social culture—too hard-edged might keep the preceding considerations in mind as he or she proceeds.

Let me add two Chinese proverbs. "The best medicine has the most bitter taste." "It is the person who wishes you well who tells you things that are hard to take, and it is the person who wishes you ill who tells you only things which please you."

Having identified the two basic sides of the Binary Con and seen some examples of the various specific forms they can take, let us turn to the Aristotelean "synthesis" of the two, which made our worldview and paradigm a self-perpetuating, closed system. The great Aristotelean "synthesis" of Idealism and Materialism to which our best thinkers trace their "scientific" worldview is essentially a way of trapping the thinker between the two, to preclude seeing through the Con to something truthful and Humane. It added two wrongs and called them a right. Aristotle granted reality to Ideas (Forms) and Matter equally, while lopsidedly positing God (*Theos*) as initial cause. He does say that God is inside, not outside, the world-machine, and thus superficially departs somewhat from the Idealist stance, but his distinction between Spirit and Matter remains Absolute, so that the same effect is obtained: God is cause and Ideas and Matter are effects; God is perfect and Ideas and Matter and their knowers, human beings, are imperfect; God is infinite and indeterminate, the world, finite and determinate. This God, then, plainly is not a natural phenomenon, and therefore is an unnatural if not also a supernatural one. The world remains Binary in this view, is seen through two false half-views that pretend to be a true whole. Since our scientists analyze only things that can be quantitatively expressed, or, if they are "non-hard" scientists, self-consistent" Absolute-Fragmental ideas as well, that is, analyze only "the finite and determinate," they can do so as Idealists, Materialists, or Realists, for it makes no difference in the way they see, or refuse to see, the data they work with.

Aristotle linked Idealism and Realism by positing that Matter, the "substratum," becomes Form. Matter and Form are inextricably bound to each other, although they remain Absolutely different. Matter is potential and Form is actual. Thus, a given thing at the potential level is all the Matter of which it is made, and at the actual level is all the Forms that it takes: boulder-rock-sand-dust; seed-sapling-tree-firewood. Things are classified according to their Forms, as specific forms of Being and Becoming, much as in Platonic Idealism, but are understood also to have an equally real, substantial, Material substratum, as in Democritosan Materialism. Change, motion, is the Formal aspect of Energy, which is an effect of Matter becoming Form. Thus, Forms are "final" causes (goals) of, and Matter the condition for, energy. This implies that energy, of which life would be one form, is a function of, is secondary to, derived from, Matter and Form. It is a mere relation between Absolute, dead, things. What animates living things (the *psyche;* or soul), then, is also a relation between Absolute, dead things. So, to use a modern analogy, the whole universe in Aristotle's view is much like a spring-powered clock: Matter is the spring, hidden from view under the dial of the senses; the Forms are the different shapes of the spring at different times and the different positions of the hands; and energy, which takes the form of the soul in living things, is the unwinding, itself. The winder, of course, is God.

In this way, Aristotle preserved both the Materialist view of life and human beings as mechanical and basically Material, and the Idealist view of life and human beings as depending on and being totally inferior to God. The result is as insidious as its derivation is clever. Later I will demonstrate that it is also intellectually (logically and empirically) false; it is already fairly clear that, because of its Idealistic-Materialist *devaluation of human beings*, it is inHumane. It is easy to see, however, that Aristotle did no more than appear to resolve the contradiction between Idealism and Materialism, by introducing a Potential-*vs.*-Actual dimension that "permits" Matter and Form to coexist as "equally real things." Therefore, he departed in no basic way from his antecedents' mode of thought. He simply added another Absolute dichotomy, and made Matter a mystical concept to complement the already mystical notion "God." Had he actually resolved that contradiction, there could hardly have been the many resurgences of Idealism and Materialism, as alternatives to his Realism, that have occurred to this day.

The effect of the Binary Con on the academic community, which in its turn affects the people, has been to perpetrate a 2500-year-long dispute as to which of its variants is true: Materialism, Idealism, or both at the same time minus Materialism's denial of God (Realism). One should find it odd that our "great thinkers" did not realize after one thorough discussion 2500 years ago that each variant must be false, because none is Complete and each leads to the others. As Chuang-Tzu observed during the battle against the short-lived Chinese version of the Absolute-Fragmental worldview, in the fifth century B.C.:

> The one side affirms what the other denies, and vice-versa. . . . That view and this view produce each other. . . . That view involves both a right and a wrong and this view involves both a right and a wrong: are there two views, or is there actually one? They have not found their point of correspondence, the pivot of the Way.*

That "point of correspondence" was precluded by the very feature that defines the Absolute-Fragmental worldview of which each of the three preceding views is a variant: its Absoluteness and requirement of Fragmental one-sidedness—its Binary structure. The universe is basically *only* Matter or is *only* Ideas; X is *only* Spiritual or *only* Material or *only* Ideal. How could people with such Absolute views find a

* After Watson's translation, cited in the *Bibliography*.

point of correspondence? Indeed, could people who ascribe to such views genuinely wish to find one? Each of the three variants has an opposition to the other two built into it. Those who stand for one or the other variant know this from the start and are therefore inclined least of all to seek a point of correspondence. The Absolute-Fragmental, either-or mentality totally permeates Western intellectual activity, even to the level at which fundamental innovation might occur. Indeed, it is impossible to otherwise account for the fact that from the fifth century B.C. to today no fundamental innovation has been made, although there have been partial successes which never took hold, and the three classical variants remain the basis of all higher Western thought.

To be sure, there are many philosophers and scientists who deny this, imagining that contemporary philosophers and scientists have somehow transcended these three variants. This is simply a conceit, typical of all active periods of Western thought, whereby contemporary thought is declared superior to anything that has preceded it, without any serious substantiation for that claim. Meanwhile, science proceeds with increasing detachment from philosophy, taking itself to be "hard-nosed." (Our scientists are increasingly unsophisticated philosophically as our educational standards decline.) Because science ignores established, and does not seek to generate new, philosophy, it remains strictly trapped within the confines of the philosophy that spawned it. For example, despite the fact that avant-garde physics, partly due to its Western-philosophical naïveté, has produced marvelous data that all but spell out the Polar-Complete alternative to our paradigm, it is rapidly converting from classical Materialism to classical Idealism, in precise obedience to the Binary Con. As Capra makes clear, the "new" dogma is "All is mind."

THE DEATH- AND LIFE-FORMULAS AND THE OLD CHARACTERS BEHIND THEM

The basic structures of our Absolute-Fragmental worldview and the Chinese Polar-Complete one, which I have "cartoon-sketched," may be expressed as formulas. These formulas are not my own but those of the architects of each paradigm, who, unlike the successors and victims of the Western world-architects and unlike Western China-specialists, knew very well that these expressions were the formulas for one and the other worldview's structure.

The Absolute-Fragmental, therefore side-effecting, therefore lethal, formula was expressed by Parmenides, and its implications for knowing, reality, and truth were then expressed by Plato. (Those who find logical argument unpleasantly dry will be compensated shortly, by "medical isolation" of the non-logical, *personally*-characteristic sources of each formula, actually the "ultimate causes" of each paradigm.) Parmenides (a Materialist) declared:

THAT WHICH A THING IS, IT IS NOT ALSO NOT.

In other words, reality consists of pure, homogeneous Absolutes which, because they are unique in this way, have nothing in common with each other, and therefore are discrete—Fragments. Reality is a set of Absolute-Fragments. Plato (an Idealist), basing all his thoughts on this formula, then declared:

> Of the world of sense it is true that opposites intermingle. The same object may appear simultaneously as hot and cold, the same experience as pleasant and painful, the same amount as double and half. But this signifies only that sense-experience is not a valid medium of truth. For truth . . . must be rational, i.e., characterized by perfect self-consistency.*

(This view is shared by Idealists, Realists, and Materialists, alike.) This "self-consistency" is a Fragmental, superficial self-consistency: one not of theory but of individual categories. He means by it, simply, Parmenides' formulized homogeneity. It follows from this stance that knowing is basically homogeneous-categorical thinking, that "reality" is non-sensory and highly abstract, and that truth is about that abstract reality, not "the world" of sense, which is regarded as an array of illusory "appearances" masking said "reality."

It is immediately obvious that this is an *anti-empirical* paradigm, for empiricism is based on sense-data; that the universe it posits is Absolute-Fragmental: nothing real has anything in common with anything else that is real; and that it is the path of coldly elitist "whip-carrying *horse*-riders": the world of common (or should we say

* See Wheelwright, cited in the *Bibliography.*

healthy) people, being based on an integrated whole of sense, common or "*horse-*" sense and emotion, is "not real."

To this day, the preceding is what Western philosophy and its offspring, science, most pride themsleves on. The Western savant imagines himself to be looking at—or rather, for—a reality and truths which underlie the common people's pseudo-reality. Ours alone, of all the paradigms of the world, has been clever enough not to be fooled into taking perceptions, common sense, and natural human feelings seriously.*

Having been present at the arrest of Dr. Side-Effect, the reader will quickly make a connection between the Absolute-Fragmental formula and way of knowing and the life-denying character of our paradigm. It is based, simply, on intellection, and on a narrowly selected kind of intellection which gets only to dead ends in science. It excludes all the other human capacities to know and make sense of things. Indeed, being based on a tiny part of a rare (peculiar) variety of human being, it excludes humans as whole beings and in general. It inevitably produces less-than-half-knowledge which when applied inevitably produces side-effects, and it is guided by an attitude that disdains the reality of the great majority of humans. It views reality as an array of Fragments: splits the world into little dead pieces. It is a destructive, lethal little paradigm.

Parmenides was a Materialist; Plato was an Idealist; and the Realist Aristotle shared with both the death-formula. The differences among the three variants of the Binary, Absolute-Fragmental, Con are secondary and, when taken seriously, misleadingly turn into a maze designed to keep bright people permanently trapped.

Accordingly, regardless of philosophical "school" or whether one is a philosopher or a scientist, the Western worldview consists of dead Absolute-Fragments: Being/non-Being, positive charge/negative charge, animate/inanimate, only-male/only-female, ruling class/ruled class, Monarchists/Democrats, Capitalists/Communists, Progressives/Reactionaries. Each member of each Absolute-Fragmental pair, it is dogmatically insisted, is the pure opposite to its partner—or, as in

* Even the great Sinologist Porkert, whose understanding of the Chinese alternative is qualitatively superior to any other Sinologist's, has written to the effect that the Chinese missed out to an extent by not having looked, like Western savants, for a real "substratum" (underlying reality). See Porkert 1975, cited in the *Bibliography*. We will see that that is not the case; rather, they looked for it and found it in terms and in a way much sounder than ours.

Marxism, is also the *potential* (hear Aristotle?) opposite-of-itself, realized by "self-negation." But in truth, *as I'll make much clearer*, each member is non-homogeneous, partly its partner, therefore one with its partner—and always present with its partner. Things which have, or believe they have, nothing in common are *enemies* (and therefore "negate" each other and themselves). But where each thing is one with its partner, and each pair, directly or indirectly, is one with all other pairs, reality, as according to the Chinese worldview, is whole, integrated, Complete—therefore living and healthily harmonious. Things which, even if opposite, have each other in common are interdependent, are *friends* or *lovers*. Let us turn, then, to the Polar-Complete formula. Paraphrasing Lao-Tzu, Parmenides' contemporary, Chuang-Tzu said:

IF THE UNIVERSE POSSESSES THIS,
THEN IT DOES NOT POSSESS NOT-THIS.

In other words, if there is a pure, perfectly homogeneous anything, then its opposite does not exist. There *are* opposites. Being/non-Being, . . . male/female, . . . rulers/ruled. With Chinese statements of truth, because they are brief but extend far, one has to think things through.* Therefore (he said "if"), the universe does *not* have anything purely homogeneous.** Each member of a pair of opposites is partly its opposite. It is chiefly itself and secondarily its opposite, and not on a Potential/Actual split dimension but always and actually.

The Chinese formula is the exact opposite of the Western one. (The Chinese principle is not excepted by this fact: there is an Absolute-Fragmentalness, required by inherent characteristics of the *lower*, categorical, aspect of human thought, in the Chinese paradigm; and there is a Polar-Completeness, due to the built-in self-contradiction of the Western paradigm as a whole, in the Western paradigm.) Being the exact opposite of the Western one, it has the opposite quality and effects: it is lively and enlivening; it makes connections among all things, recognizing their interdependence and harmony; thus it fosters

* Confucius called this "being provided with one angle and seeing the triangle."

** The Sinologist A.C. Graham made the connection between Chuang-Tzu's statement and the non-Absolute existence of things in an extraordinary article, "'Being' in Classical Chinese" (*Asia Major*: #7: 79–112, 1959). I am indebted to Professor Floyd G. Lounsbury of Yale for calling this paper to my attention.

unity, health, and life, whereas the Western paradigm fosters conflict, illness and death.

As 'll show in the next chapter, the Western paradigm also makes far less sense of things than does the Chinese one; but for the moment I wish to make clearer the basic nature of each paradigm, and to identify their actual sources.

There is one more recognition to add. There is a Western *unity* and a Chinese *unity*, which are as different as Western opposites and Chinese opposites, and which, at the level of constructed reality, are the one dead and the other alive.

To the Absolute-Fragmental, categorical, "rational" mind all opposites do, of course, have something in common unless they are (Western) ultimates. For example, male and female have in common, are kinds of, sexuality, which is not an ultimate, and man and woman have in common, are kinds of, human being, which is not an ultimate; and Being and non-Being have nothing in common because they are ultimates. The Western unity, then, is a member of a triad: opposites X and non-X (or X and Y) and their unity—the component in each which is identical between them, which each is a kind of. The triad looks like Figure Two.

Constructed according to that "rational" paradigm, which is nothing more than a small part of a certain kind of mind, our social reality chronically transforms into what, as distinguished from true unity, should be called "the dead neutrality which is inevitably consequent to conflictful opposition." This occurs during what we call "declines and falls of civilization." For example, our males and females, because of the general masculine-dominant, women-suppressing character of our (Hindu-Babylonian-Egyptian-Greek-Euro-American) society, conflict with each other for centuries. Never understanding what they have in common, that they are not Absolute opposites, they exhaust their sexuality through struggle, and then *non*-femaleness and *non*-maleness, a negative unity, often called "equality," is "discovered" and takes the place of opposition. The latest example of this cycle* is the phenomenon of Unisexuality and the popularization of homosexuality: there is just *sex*, not also male and female. (Whereas, in truth, without male and female—*together*—there can be no sex, for sex is an abstraction, a generalization, about male and female.) According to

* Regarding this cycle, see Amaury de Riencourt's scholarly and exhaustive *Sex and Power in History.*

FIGURE TWO: THE WESTERN TRIAD OF UNITY AND OPPOSITION

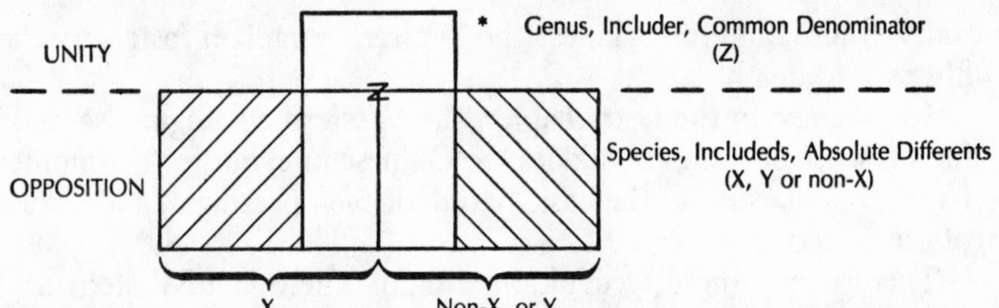

*Note that even the concepts of Unity and Opposition, themselves, cannot be diagrammatically (topologically, spatially) represented Absolute-Fragmentally: X and Y are included by *but also include* Z.

this "unity" there is just the magnet, not also the positive and negative poles!

The "genius" of Marxist Dialectical Materialism is nothing other than a systematization, misrepresented as "Natural Law," of this tiresome, life-obstructing process of conflictful opposition followed by dead neutralization. (Marxism is historically where Unisex last came from). Communism, the *classless* society, is the (imaginary, impossible, and therefore "as yet unattained") "unity" consequent to the exhaustion of the struggle between the Absolute opposites, owners and owned, of which male and female are said to be varieties.

To go to the opposite end of the "scientific" spectrum, another example is the matter/energy opposition. For 2500 years, off and on, they were thought of as the correlates of Being and Becoming, or non-Movement *vs.* Movement (really, as Whorf indicated: Noun *vs.* Verb*). Tired of struggling between them, our physicists were happy when Einstein equated the two. As I will show, they are now unable to distinguish between them (but need to) and are terribly confused about it, really at a dead end, theoretically "de-magnetized."

So, that dead Western unity is not to be mistaken for the Chinese one. The Polar-Complete paradigm is not an Absolute-Fragmental triad but a dyad-unity; it does not have two Absolutely distinct levels, one for opposition and one for unity, so that it staggers from one unhealthy extreme—conflict or lifelessness—to the other. Rather it has one "level" with both near-identity and non-Absolute, Polar opposition within it. It looks like Figure Three.**

* Linguistic categories considerably vary from one culture to another as to what each includes.

** Here the inevitable fact that includer is also included, and reciprocally, is represented: Unity and Opposition, themselves, are also a Polar-Complete pair.

FIGURE THREE: THE CHINESE MONAD-DYAD OF UNITY-OPPOSITION

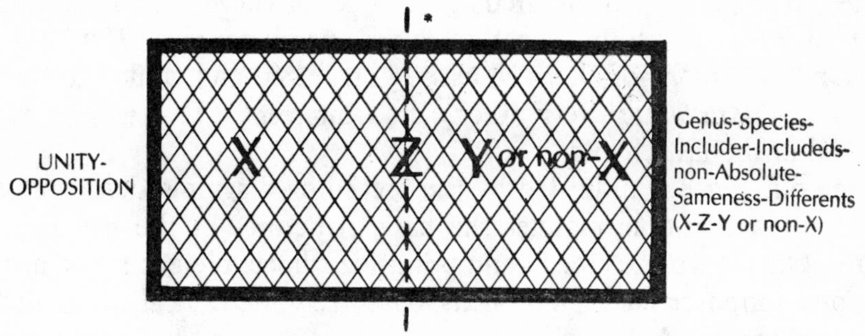

*Here the inevitable fact that includer is also included, and reciprocally, is represented: Unity and Opposition, themselves, are also a Polar-Complete pair.

For example, correlative to the preceding Western ones, the maleness-in-femaleness and the femaleness-in-maleness are taken into account (penis-like clitoris, milk-giving-nipple-like nipples, and so on), and the feminine role in masculine role (for example, nurturance based on protectiveness) and the masculine role in feminine role (for example, protectiveness based on nurturance) are recognized and realized as a true unit: sexual *equality-with-sexual-difference** and the common purpose of a family of procreation and social immortality (survival intergenerationally as a persona and as one's beneficent effects on people, centrally, one's descendants).

Instead of the Western ruled-*vs.*-ruler opposition, Chinese social theory and—except during decadent dynastic phases when the basic social culture is fallen away from—practice is that the ruled, being empowered by the Confucian right to rebel and able, because of their organization into armed leagues of extended families, to do so, also rule; and the ruler (who as emperor obligatorily calls himself, reciprocally, "The Orphan") is also ruled. The Western paradigm gives either the fragmented-and-conflictful society of self-interested opposed social classes (now called "*interest*-groups") with nothing (consciously) in common, or the collapse of that society into a classless "unity" in which the Proletariat, conveniently "represented" by the

* Present space does not permit showing how anti-Chinese "intellectual" and political propaganda about the cruel suppression of Chinese women by Chinese men is based on narrowly selected data about the worst people under exceptional stress or during decadent dynastic phases. Suffice it to say that the Communists have torn down the city gateway erected in honor of mothers and that, as Philip Slater pointed out in his *Earthwalk*, unlike Western women, Chinese women make up one third of the famous scholars, artists, and so on described and honored in the dynastic histories.

State "temporarily" incarnated as a super-privileged, police-armed elite, dictates. Instead of that, the Chinese paradigm gives *T'ai-P'ing*: the Great Harmony of all social classes, organized and guided so that their self-interests are every other class's interests. (This will be explained in the last chapter.)

The difference is indicated by the preceding diagrams that express in a pure and simple manner the basic structures of the two paradigms. Ours looks like a dead and primitive edifice, the geometric artifact of an impoverished mind; theirs looks like something alive—subtle, blending, moving, balanced.

Creating Absolute-Fragmental classes according to the Binary Con, we inevitably react to the extreme confusion and stress by wishing for a classless scientific view or society. We go from unnatural rigid structure to no structure. It is like being ill in a way that opposes each bodily organ to each other one and then wishing for a body which has no organic structure, something less than a mere cell, something worse than ill: dead. *The class-less scientific view*: buried in a potentially infinite morass of Absolute-Fragmental theories increasing in number with each research grant, periodically our scientists react and go to the opposite extreme that the Binary Con has prepared for them: mysticism, usually including thoroughly distorted, Christianish, references to Far Eastern worldview in which the notion "All is One" (Z without X and Y) is central. *The class-less society*: choking in a potentially infinite morass of Absolute-Fragmental social structures and laws increasing with each new social contract and each litigation between husband and wife, child and parent, employee and employer, taxpayer and taxers, community and criminal, periodically our people react and, guided by our Social Engineers, go the opposite extreme our Binary Con has prepared for them: mysticism—class-less society, Heaven-on-earth-cults of one variety or another, be they Christian, Communistic, pseudo-Buddhistic, or science-fictional.

Our heart either beats too fast and arrhythmically or, after it has done so too long, it goes lax and weak, and our limbs, deprived of vital energy, atrophy. This atrophy, this impotence is what is presently recommended to us by our "Progressive" savants, who are now at the fore, because we are tired of conflicting. We have been ripped off, over and over again, by a dead paradigm 2500 years old. There is no masculinity without femininity and no femininity without masculinity. The more feminine, the more masculine, the more masculine, the more feminine. The more powerful the positive pole, the more powerful the negative, and the higher the charge. Long live the difference: life *is* that difference. And that difference is also a sameness. To take the example from Chinese science made famous by Needham: take a magnet and cut it in half. There are now two magnets, one of which has a negative pole where it was positive and one of which has a positive pole where it was negative. The process can be repeated infinitely. The poles are opposites *yet also identical.* Likewise, in sexual intercourse the opposition is ultimately lost. In Chinese it is called "attaining the ultimate Pole"—the equivalence of male and female. And likewise, in good government the ruling/ruled opposition is lost.

The *I Ching*'s* 64 hexagrams (*kua*) include one that is the image of dying, of terminal disease, of catastrophe. It is called *P'i.* Its image is *yang* (positive) and *yin* (negative) going in opposite directions, separating. In Western terms, they separate by thinking of themselves as perfect opposites according to Parmenides' formula. They separate by thinking of themselves as identical according to the Western illusion of negative "unity" built into that formula. We have gone through the conflict-collapse cycle many times and are in the process of a potentially final collapse. Nature only permits so many repetitions of mistakes. It is happening at all levels, from subatomic wave-particle-smashing toward the fragmental unity of the quark, to the struggle among all classes and nations of people which is irresponsibly disguised as an evolution toward fair play, equality, unity. It is an anti-organic, anti-vital death-throe. But death is life's way of renewing itself. I am as optimistic as the illness I am diagnosing is grave."**

Our people of knowledge have been conditioned to simple-mindedly handle only one Absolute-Fragment at a time and to regard the "terribly interesting" results as a justification for such activity. But there are many among them, in this incredibly creative and exceptionally well-intentioned society of ours, who, once they have perceived the trap that their paradigm actually is, could easily perceive and then wish to restore the natural manifold harmony of society—of men and women, parents and children, merchants and consumers, rulers and ruled, farmers, artisans and scholars, heroes and whores.

* This book is the oldest extant Chinese scientific text and contains 64 abstract images of universal reference.

** As Thoreau put it: "In the Spring I burned over a hundred acres till the earth was sere and black, and by midsummer this space was clad in a fresher and more luxuriant green than the surrounding even. Shall man then despair? Is he not a sproutland too, after never so many searings and witherings?"

There are enough who would recognize that our paradigm, by so reducing and cheapening the world, has made it less worth being responsible for. There will always be, in all countries, those who are afraid of the life-energies produced within a manifold harmony, between men and women, between rulers and ruled, between savants and the common people. It is only necessary to recognize that such sensation-fearing types have no right to monopolize consciousness, and that the inconclusive read-outs and chronic breakdowns they call "science" are not genuine science. Such people, also, have a natural place in the manifold harmony, but that place is not at its center.

That brings us to the core of the Stone Monkey's high-medical diagnosis. The Polar-Complete paradigm is based on sense: there *is*, as Plato grudgingly remarked, hot in cold, small in big, and so on. It is based on common sense: it is good and natural, for example, to love one's children and to be proud of being a good mother. It is based on the common people's interests: for example, their desire for social immortality—*human* survival. The Polar-Complete paradigm is also logical and empirical: if the diagonal is "irrational," "formless," then so are the sides. To add one preliminary example: where *real* Being, non-Being, and Becoming are mental abstractions from sensed, *empirical*, reality, which—to grant Plato's observation—has no Absolute-Fragmental properties, it follows that Being is not *purely* Being and, therefore, is in part non-Being; and where no Natural object is perfectly stable, Being is also partly Becoming. Therefore Being, non-Being, and Becoming are non-Absolute-Fragments, each of which is itself but also, to a lesser "extent," the others. Of course, this is precisely the evidence of modern physics: "matter" (Being) is energetic waves (Becoming) at such high frequency and short length that it is only *almost* non-changing (*almost*-Being, *almost*-not-Becoming).

Opposite to the Polar-Complete paradigm, the Absolute-Fragmental one denies sense: all "real" and therefore non-sensual properties are Absolute-Fragmental. It denies common sense: human nature is cultural prejudice subject to evolutionary alteration; it is primitive to cleave to your children. It denies the interests of the people: it's not social immortality that you want, but divine immortality; not social immortality and admiration for your self-mastery and dignity but hot fast sex and/or immediate admiration for your silk disco-shirt and disarming, sun-lamped good looks; not social immortality along an expanding generational line of your spirt and flesh, but immersion in The People's struggle for a Classless Society where no one is distinguished for anything good or bad* and which we will be available, if called upon, to dictate on your behalf.

What is the source of such an odd, anti-sexual, non-common-sensical, anti-popular—unnatural—"scientific" paradigm? What is the basic disease-causing variable? Obviously our paradigm is not, contrary to scientific-philosophical dogma, based on logical necessity. The basic death-formula for it, is a simple *assertion* without prior logical grounds: *that which a thing is, it is not also not.* Nor, as Plato admitted, is there any empirical, basically sensory, evidence for it. There is no logical ground for the Absolute-Fragmental formula—the formula for every "informed" Western act, be it one of thought or nonverbal action, be it social, psychological, biological, or physical.

The "rationale" for that formula and the denial of sensory reality which goes with it are contained in that Bible of our philosophic-scientific elite, Plato's *Theaetetus*, a dialogue in what he called *dialectic* form (it remains contemporary) between Plato's teacher, Socrates, and a student, Theaetetus, whom Socrates seeks to corrupt because he has been exposed to the earlier, Greek, version of the Polar-Complete paradigm.** The *Theaetetus* is a central record of the shady victory of the Absolute-Fragmental paradigm in the Western Battle for Worlds 2500 years ago, a victory imposed only after the chief proponent of the Polar-Complete paradigm, Heraclitus, had died. In as entertaining a manner as I can, in the Appendix, I will expose Socrates' central arguments, of which all his others are variants, to indicate that they are self-contradicting and irrational.

Socrates admonished Theaetetus not to "judge the gods." I would like to do so here. The most sophisticated explication of God is common-sense reality expropriated, relocated in, and monopolized by God as His divine properties. That explication of God, relied on to this day, is Aristotle's. On the Absolute-Fragmental assumption that all changes are a chain of purely-causal causes and purely-caused effects,

* There are state-approved exceptions. But of course. by immutable Dialectical law, your status as Hero of the Proletarian Struggle will be temporary, since, as you may have noticed, the Hero during one phase is the Enemy during the subsequent one, and records tend to get lost as Socialist society dialectically zigzags through opposite phases toward Communism.

** I need not expound it since I will give the better-preserved and fuller-fledged Chinese version. But compare with what I will say Protagoras and Heraclitus in Capra's *The Tao of Physics* on avant-garde physics' proofs of Heraclitus' worldview. Also see the Appendix.

he (logically) concludes that there must be an Ultimate Cause, external to and unlike this natural chain—*Theos*, God. God must, therefore, be un-Caused, totally independent of everything else. He must also be non-Moving, because everything else moves. How could something be a cause without moving? (he asks). Obviously, by being "an object of desire"—in this case, human desire. But, setting a classical precedent of short-range logic, as he fails to reason: God then needs humans and their desire in order to be a cause, and therefore is neither totally independent nor totally causal nor, having a need, unmoved. So, *logically* viewed, God is at once causal and caused, independent and dependent, un-moved and moved—Polar-Completely patterned, just like Pythagoras' square and, as we'll *see*, everything else. Aristotle also endows God with masculinity (cause: activeness) and must then explain "Him" in terms of being caused and desired, pursued—as feminine. Again: "God" is Polar-Completely patterned.

That God, so argued, is nothing other than a projection of the Classical Greek Academician who, typically a pederast, hoped that he was an object of his disciples' sexual desire. Accordingly, our theologians should dissociate themselves from that theological argument. And this would help free them to foster their truly Humane moral code on the basis of human interests and human nature, rather than misdirect the desires of all humanity "upward" on drafts of organ music, as God would not, in fact, desire.

Implicitly, the use of God as a Sacred Container or Sacred Garbage-Disposal-Unit of all the data that do not fit into an Absolute-Fragmental paradigm has been recognized by Mary Douglas, a brilliant cultural anthropologist and former Catholic nun. In her *Purity and Danger* she explains that what is sacred in a people's worldview is what is dirty—that is, what does not fit into a people's classification of natural and social phenomena. Being dirty in this sense, what is sacred has a dangerous (dis-ordering), polluting aspect. One cannot directly deal with it, and if one does deal with it, one must take special ritual precautions—confessing of sins, anointment with oil, or the acquisition of a Ph.D., for example. To adduce my own example, the Polar-Completeness of all sensory data (Plato's intermixed light and dark, wet and dry, and so on) does not fit into our philosophic-scientific paradigm; indeed, strictly understood only a certain variety of thought does. "God," the dangerous Great Mystery to whom modern cosmologists, as documented, are returning, contains the rest. Douglas goes further, to propose that there are two varieties of worldview: "composting" and "non-composting." The composting variety, examples of which she finds in pre-civilizational cultures, confronts the mystery (the chaos—what does not fit into what I have called Absolute-Fragmental worldview), and uses it as fertilizer for the growth of the relatively definite worldview it doesn't fit. For example, an Eskimo shaman retires from society to go crazy, to incarnate chaos, and as a result of having digested it, returns to society enabled to produce its opposite: order, wisdom. Like the shaman, but not schizophrenically, the Chinese savant remains in society and *simultaneously* takes into account the formed (order) and the formless (chaos), which is, as even Plato admitted, present to everyone's consciousness at all times under normal conditions as a big that is also small, the "rational" side of a square that is also "irrational," and so on. Our worldview is of the non-composting variety: the bellies of our savants, unlike those of Eskimo shamans and Chinese savants, lack the size and fortitude required to digest the chaos-that-is-in-all-order.

Be they "religious" or "scientific," Western savants are consequently in an attitude of prayer toward the "stubbornly non-Absolute-Fragmental" aspect of reality, either on their knees in a darkened room or peering into a darkened bubble chamber so as not to have to face a "less-manageable" human and natural-phenomenal reality. From the perspective of "Satan," this state of affairs is more than ideal. Not only are our savants so far removed from reality as to be totally useless to those who provide them with subsistence, but they kneel, eyes closed and ears attuned to "another world," their heads bowed so as to expose the napes of the necks. He is virtually invited to walk up behind and decapitate them: they protect not even themselves. Worse, they have popularized and for centuries institutionalized this prayerful attitude, infecting the general population. Instead of thanking Mother and Father for the food they have sweated to produce on their farm, children have been enjoined to thank "God." What one does not know is his is lost. As a "Satanic" consequence, our children are now enjoined to thank MacDonald's or The Proletarian State, and their mothers tune into *Another World* on NBC.

The actual source of the Death-Formula and the paradigm it generates should now be obvious: the insecurity, fear, and common-people-disdaining Narcissism of the architects of our worldview. The causes of this human *weakness*, in turn, are impossible to specify. They are in the mists of pre-Grecian India, Babylonia, and Egypt, where those Greeks got their ideas, in the specifically Greek cultural conditions for the synthesis of the West-Asian–North-African ideas and attitudes which became our paradigm, and probably in ill-fated marriages, bad diets, and ill-starred genes. What matters is understanding the nature of the character behind the formula, "psychoanalytically."

Insecurity: They demanded that one's concepts of truth and reality be perfectly homogeneous, determinate, have definite, cartoon lines around themselves, be rational-numerical sides of squares, pure Being, homogeneous colors or *no* "real" colors. This is plainly a symptom of intellectual and emotional insecurity. A person who so deeply doubts Nature's gifts of perception and intelligence, and the natural environment itself, can hardly accept being the measure of all things, and is uncomfortable around any other persons who do. He therefore devises a microscopically narrow, logical game and calls it "searching for truth." And he seeks to politically obfuscate and eliminate his more-than-worthy opponents, as Parmenides, Plato, and Aristotle did to Heraclitus. He floods popular consciousness with his game to create a cognitive monopoly and general disdain for alternatives, and collaborates with seekers of undeserved and inordinate power who, like him, are uncomfortable with a common people inspired by men who believe in themselves as humans. ("Human nature is evil or non-existent; it is basically simian and subject to cultural evolution; hence we need a powerful central government guided by Reason," pronounced, in sacredized tones, REE-ZUNN. . . .)

Fear: A character who for no logical or scientific *reason* denies reality to sense and common sense is plainly afraid of sense and common sense, of human warmth and human suffering. In contrast, Sakyamuni Buddha, whose Way took root not in Absolute-Fragmentally caste-ed India (part of the West), but in China, stipulated as the highest accomplishment first entering into a state of total detachment and *then* re-entering, fully human, into human society, to set the pro-survival, florescently dignified example of digesting the human suffering that living necessarily entails. (In the central hall of each Chinese Buddhist temple is mounted a huge plaque which reads Hall of Strong People.) Basically, a character who denies sense is afraid of people: afraid they will discover that he or she does not measure up to average human standards. (Therefore, "man is not the measure.")

Indeed, that fear of discovery is institutionalized in the guise of a "gentlemanly" intellectual standard. All Academic Debates must be at the intellectual level only: nothing can be said about the debaters as persons. (Accordingly, Academicians call themselves not "intelligent *people*" but "Intellectuals.")* When, as occasionally happens, one debater is genuinely outraged at the character of the other hidden in the other's ideas, personal criticisms are leveled; the other shrinks in horror and, with the Intellectual audience in chorus, screams "*ad hominem* attack!"—a ritual Latin phrase, meaning "attack on *my person*!," whose function is to affirm that no Man of Knowledge should be required to take personal responsibility for any of his ideas or their effects on people. "My theories and plans have served to pollute your air and water for the next fifty generations, but it was nothing *personal*, you understand."

Of course, when a debater challenges the dogma of the day, suddenly attacks on the person (his) become legitimate. For example, the misguided genius, Wilhelm Reich, who correctly stated that our natural science never deals with the phenomena it is addressed to, but with experimentally altered transforms of them, was declared insane and incarcerated in a federal penitentiary, where he died. (Now, avant-garde physics agrees with him, but no one, save a few Reich groupies, seems to be able to recall his observation or his fate.) Likewise, anyone who challenges the Soviet social system is declared insane and injected with drugs so agonizing they may make the eyes pop out of their sockets; or, if one is a Soviet Feminist objecting in the late 1970's to the double workload of women, she may be sent to a prison where she is forced to strip in front of leering male guards and, if she objects, "steam-hosed"; or, if one objects to Chinese Communist suppression of the peasant cottage-industry which is the farmer's source of extra cash, one is shot or sent to a work camp to be brainwashed and starved, as was Wei Ching-Sheng in 1979.

In contrast with the "pure and gentlemanly" intellectualism of Western debate, and its modest side-effect of torture and murder of those who threaten its paradigm, traditional China had an elite debating society called the *Wu-Che Ta-Hui*: The Great Association Which Has No Cover. All scientific debates were understood to be directed toward use for the survival and welfare of the people, and therefore to be matters of life and death. Such seriousness is out of place among our game-playing academics for three reasons. One, their inability to conceive of themselves—or, therefore, anyone else—being certain

* There are no such de-humanizing words in Chinese. *Scholar* translates as *one who studied*, *worker* as *working man*, *man* and *woman* as *male human* and *female human*. Even *whore* translates as *prostituted female being*.

about the people's welfare. Two, their tacit knowledge that they do not have the people's welfare sufficiently at heart. Three, the convenience of their own, Humane, system, which permits entrenched Men of Knowledge to suffer no consequences for their bad influences, while those who call for accountability may be psychologically or physically tortured or executed out of harm's way. "The Humane man is not Humane"—Chuang-Tzu.

We, too, sometimes execute thinkers. The difference is that we do so on a basis of academic power, whereas the Chinese did it on the basis of popular will—democratically.

Narcissism: To take a small aspect of one's intellect for the sole medium of truth, to make it such as to be accessible only to a few people sophisticated into intellectual oblivion like oneself, to claim a monopoly on truth, and to impose this small aspect of oneself on the entire universe as its "Essential Structure," are all plainly symptoms of megalomaniac Narcissism.

What, then, is the source of the Chinese Life-Formula and the paradigm based on it? As I will show in the next chapter, full-fledged logic and sensory interaction with Nature are one source. Of interest here is the *character*-source. Obviously it is the opposite of the Western one: self-confidence as a human and confidence in Nature, love of life and common people, and the opposite of the Narcissistic imposition of one's worldview, be it through selfishness or forceful benevolence: harmonizing oneself with Nature and people, so as not to obstruct the natural process—perhaps even to help it along. This empathy with Nature and all people, and its cultivation among intellectual and political leaders as well as their followers, is called *Jen* (pronounced Ren). It is the root concept of Confucius. And it is virtually absent from all Western natural or political philosophy, with one short-lived exception: the imperfect but well-intentioned democracy of late eighteenth century New England, which tragically has become little more than a memory. It is almost absent, because only the uniquely spirited and unbounded circumstances of revolutionary America have had the force to override our paradigm's central distrust of Nature and of the human potential for self-regulation. It is that spirit of independence from top-heavy central intellectual-political control which resonates with the Chinese one, and it is one with a scientific confidence in Nature.

There is a Chinese term not only for the rightly empathetic spirit of genuine science but also for those who have misdirected it: *Te-Chih*

Tzei: Stealers of Virtue: stealers of human self-confidence, of knowledge, of spirit, of empathy, of harmony, of life. The Stealer of Virtue is the character behind Dr. Side-Effect, who, as the contemporary scientist, the suicidal doctor, is one of his many victims.

Socrates, the corrupter of Theaetetus, discloses that character through his metaphor for the "enlightenment" of Theaetetus: taking away and killing his child. Likening the Polar-Complete paradigm of Protagoras and Heraclitus to a child of Theaetetus and comparing himself to a midwife, Socrates says:

> And now that he [the child] is born, we must run around with him, and see whether he is worth rearing, or is only a fart and a sham. Is he to be reared in any case, and not exposed? Or will you bear to see him rejected, and not get into a passion if I take away your first-born?

Here I exercise the same rigor of judgment that Socrates pretended to demand, but I do not compare myself to a midwife nor am I after a child. I am after a thoroughly corrupt 2500-year-old ghost. Actually, he is twice as old. He can be traced back to Babylonia—history's most gods-fearing and, accordingly, a Hell-hole of a pseudo-civilization—and later found speaking through the Old Testament where he terrorizes and dupes Abraham into agreeing to sacrifice his own son, and his humanity, for the sake of God, not, as might justify such sacrifice, for one's people, as in defensive war. He appears again in the Christian dictum: "First, God; second, the family." And finally, in extreme form, in the Communist dictum: "The State replaces the family." As God, Reason, the State, the Economy, or a cult's image of the coming alien masters of our planet, that 2500-year-old ghost's projection of himself as poison into the human heart is one and the same, because love of the divine is misdirected love for humans. As soon as there is divine love, human love is diminished, and the family, economy, and polity are crippled. As soon as there is divine justice, human injustice is guaranteed, for humans can only have *human* criteria of justice. And there arise instead of social harmony, social-"scientific" laws and top-heavy central control.

I should elaborate on this. All peoples whose worldviews contain a version of the notion that there is something "higher" than humans with which they as a distinct group are identified are victims of the architects of their worldviews. There arises, along with "top-heavy" government, a tendency to regard those not of one's group as inferior, a view which, unless it is balanced by humane popular culture, becomes virulent. Recognizing this *common and mutual* victimization neutralizes that tendency. Because of differences in popular cultures, the Jews have understood God to be intelligence and virtue more than a force, whereas Christians, and after them Muslims, have understood it to be more a force, specifically one behind their world domination, just as Marxists do with their version of God, Dialectical Materialism. Accordingly, scapegoating outsiders is a Christian/Muslim/Marxist, not a Jewish, characteristic. It is the tendency to discriminate and its virulent potential, namely torture and genocide, to which my "arrest" of the "5000-year-old ghost" is also addressed. The human world needs common participation in the construction of a basically Human reality, in which humans are the "highest" thing.

As my Chinese doctor put it:

> One must fully realize that one is human, that humans are an aspect of the living whole, before one can reflect about what is true or false, right or wrong. It is because humans exist that we are here asking these questions. They are *humans'* reflections. To forget that disqualifies one from making any reflections at all. As soon as there is even a little misunderstanding about this, one has to invent a God of one kind or another to make up the difference.

I have mentioned but not cited as such admirable attempts by Westerners to escape their paradigm. I now cite what I see as the most forthright and successful among them and then show how it failed. Count Alfred Korzybski, in his *Science and Sanity*, recognizes two periods of human development. The first he calls "infantile" and characterizes as "aristotelean," recognizing its either/or and Absolute this-or-that structure, and the fact that it is based on the "Laws" of Identity, Non-Contradiction, and The Excluded Middle (each of which is a further-specification of Parmenides' death-formula): "Whatever is, is"; "Nothing can both be and not be"; "Everything must either be or not be." These should be regarded as attempts to "legislatively" coerce reality into accord with the Academic "requirements" made of it, much as a small child, after being admonished, pronounces his parent "is bad" with minimal certitude that serves as an implicit request for further enlightenment. The second, adult stage Korzybski calls "non-aristotelean," characterizing it as cognizant of the paradoxical, self-inconsistent, and non-self-existent nature of all things which, therefore, cannot be fragmentally abstracted from the whole or designated as "underlying reality." He then recommends *observation*, an exercise to which the next chapter is devoted, as the therapy required by Western consciousness. Unwittingly, he recapitulates much of the Chinese scientific paradigm and declares it to be at an evolutionary stage beyond our own.

But he misses the most crucial element: the non-intellectual, human-intentional one with which I have concluded this chapter. So fundamental in our paradigm and culture is the divorce of the mind from the heart, of intellection from common sense, of Goodness from human empathy, of intelligence from personal character, that even as daring and revolutionary a thinker as Korzybski cannot see it, and therefore is ruled by it as he attempts to act as a high doctor to Western society. The Confucian root, *Jen*:Human Empathy, is what we must most fix upon and seek to develop if genuine, valid-and-Humane, science is to be produced in this part of the world. Observation without Humane intention will not suffice.

CHAPTER

3

Science in Living Color

Man is the measure of all things.
—PROTAGORAS

Before there can be truth there must be a true man.
—CHUANG-TZU

SCIENCE AND PSEUDO-SCIENCE

I have said that our impression of the traditional Chinese world-view and world has been distorted by our religious and secular "scholars," and then by the Chinese Communist ideologues, who, understanding that these distortions make the Western "scholar" feel secure, thereby perpetuate Western blindness to the Human* richness of traditional Chinese culture while they use our negative view of it to

* In light of the previous chapter, I now use "Human" instead of "Humane" to translate *Jen*.

justify Communist ends. Unjust and sordid social conditions naturally occur at the *end* of each dynasty, and in the tiny areas where colonialist Western powers wounded, and brought forth and concentrated the worst of, Chinese society. These conditions have been taken to represent *all* of traditional Chinese society at *all* times, to justify destroying the kinship-based clan-system of local self-government and imposing a totalitarian but, "happily," originally Western and therefore "promising" political system. Complementarily, ignored is the fact that the lengths of Chinese dynasties usually exceed the lifetimes of entire Western civilizations, each of which also ended with unjust and sordid social conditions. Also ignored are the facts that the longer durations of the Chinese dynasties, and their successions of one another without the "evolutionary" alterings of the social-political paradigm that distinguishes Western civilization phases, plainly imply that the overall Chinese system was more satisfactory to its people than were any of the Western evolutionary-civilization-experiments to theirs. Chinese history is a life-flow; ours is a zigzagging series of violent mutations with short-lived offspring.

In short, the Western view of traditional Chinese society is an Absolute-Fragmentally selective and ethnocentric one whose gross inaccuracy has, therefore, been guaranteed. It is a physician who diagnoses a healthy patient by examining one square inch of his cheek, there finding a pimple and pronouncing the patient to be a leper—whereas the physician has leprosy himself.

Beyond such distortions, lies have been told. An example is the favorite of Chinese Communist ideologues and our Marxist Sinologists,* that Confucius (like our Greek philosopher-heroes and Thomas Jefferson) advocated slavery, whereas, as that superior Sinologist, Yale's Jonathan Spence, has pointed out, Confucius never even used that word; and he left no place for such atrocity in his worldview. In turn, never have we seen fit to give ourselves a taste of our own "medicine," by, say, taking the homosexually prostituted "peg-house boys" of late nineteenth-century San Francisco, and their present successors from there to the Mideast where the "pegging" custom originated,** to exemplify the American institutionalization, inspired by classical Greek Academics, of love for children. One could go on in this nega-

* Who multiply as it becomes more and more necessary to Sinologists' careers to have "by-the-way-just-returned-from-the-mainland."

** See Lloyd, cited in the *Bibliography*.

tive way at great length but this brief, bitter taste of our own medicine should sufficiently make the point that our evaluation of other civilizations as compared to our own has been terribly biased and selectively based. In turn, if the *Complete* pictures were to be painted of Chinese and Western civilization, at all their different times in all their different geographical and social zones, including all their Human florescences and all their atrocities, it would be fairly obvious that the Chinese one has been more humane. That is, it has, on the average, created and sustained a higher social and material standard of living—more human warmth and dignity and more real owning of property, basically, *land*—for its people as a whole, and has better protected its citizens by fostering relatively peaceful relations with neighboring peoples. To be sure, modern civilization's material welfare—aside from the pollution it entails—is the highest in human history. But as has become obvious, such a standard is short-lived, involves substandard material welfare elsewhere, and may lead to wars in other parts of the world. And our psychological and social welfare is a catastrophe. We have been emotionally and socially impoverished. But there is a common denominator, to which I have referred and which I will expand upon at the end of the book: Chinese Human government and Western liberty in its original form. (It is the marriage of Western liberty—and creative spirit—with a more moderate standard of material welfare and traditional Chinese trust in Human self-government which, I believe, we must seek.)

Our distorted view of modern as compared to other civilizations need not be elaborated upon. It is sufficient to recognize the unique and misleading conditions under which we are accustomed to evaluate our own social reality. Our savants have always provided future, Utopian, images of society and propagandized us into believing we are on the way to making them real. When it becomes obvious that Utopian Image 1 will not, in fact, be attained or has backfired into a manifold Side-Effect, they switch to Utopian Image 2, as from Western-style holy-monarchy to Capitalism to Socialism. Consequently, we never *have* the society we have in mind and being deprived of it, yet promised it, instead of recognizing the social reality and irresponsible misdirection and *stress* that we are enduring, we fantasize the Utopian Image that we hope our reality is Evolving into, *confusing* social reality with an unrealistic image-goal which is actually devoid of human characteristics. It is humanly void because it is based, via an impoverished geometric-systematic Social Plan, on our Knowledge-Specialists' disliking for and alienation from what they call "human imperfection" and the inevitable suffering that living involves, which they are too "sensitive" to withstand. Accordingly, when presented with a foreign social reality based on Human-ness and realism—what may be called an *optimal-and-real,* as opposed to an *extremist-and-half-dreamed,* social reality—we tend to find it and the acceptance of it accorded by the people benefiting from it somewhat repugnant.

Our understanding of Chinese science has shared the semi-fantastic quality of most of our "scholars' " understanding of traditional Chinese society. It is because one cannot take seriously the science of a people whose society one does not respect that I have begun this chapter by showing that we do not have grounds to disrespect Chinese society, and I have suggested that on the contrary, the traditional Chinese society succeeded at several things which we have dreamed of but not yet attained. Unfortunately, the messianically one-way (*our*-way) Western view of Chinese knowledge and education, despite even the influence of admiring students of it, has been so severe to this day that the term "Chinese scholar" is more often interpreted as meaning "Western scholar of Chinese culture" than "scholarly Chinese person." Westerners have been persuaded by the majority of their "China specialists" to believe that traditional China was an enslaving, woman-and-child abusing, unjustly stratified society, whereas it was the most democratic large society* ever created and whereas the solidarity and strength of its familial relationships, including an unquestioned equal status of men and women, was its most striking social feature—one that persists, despite Japanese and Western influence, chiefly in Taiwan today.

* Democracy is people having real power and a central government that is sensitive to their needs. Armed and organized in landowning-and-landusing farming communities which constituted potential people's armies, and imbued with the Confucian right to disrespect and rebel against corrupt officials, the traditional Chinese people always had real power. An example of their government's sensitivity to their needs is that it traditionally dissolved nascent landowning corporations because of their potential for abuse of tenant laborers and for holding onto land for (undeserving) generations, and permitted communities to deal with (consequently, rare) child-corrupters in a Human manner—a manner Human, therefore, toward *the children:* execution of the child-corrupter. In the last chapter, I'll expand on this, showing how the Chinese social paradigm is reflected in the attitudes of *Western* people, which account for the many merits of our society despite the influence of our scientific paradigm.

Similarly, we have been persuaded to believe by Western scholars who have never approached the actual sphere of traditional Chinese savants that those savants have for the last 2500 years religiously supported their beliefs by reference to the founders of their schools of thought. Confucius said this, therefore it is true; Lao-Tzu said that, therefore it is true. By thus attributing behaviors to Chinese savants which actually characterize our own religions (and to a lesser extent our sciences) our vehicles of "cultural exchange" have not merely distorted, but diametrically contradicted the truth about traditional Chinese knowing. Along the same line, Chinese folk superstitions and individual efforts that never became a part of the Chinese scientific tradition, and which the Chinese common people have taken seriously only at times of dynastic breakdown (for as Chinese scholar-officials have observed, superstition rears its head only when society falls into disorder) have been taken by Western Sinologists and anthropologists to represent the system of knowledge of Chinese savants. An example is Ko Hung's book on physical immortality, a con perpetrated on the decadent rich by an angry physician, which is taken by our Sinologists to represent the Chinese correlate of European alchemy.* Such portraits of Chinese science are comparable to taking a Louisiana snake cult and the theories of California's psychic healers to represent the religious and medical traditions of Euro-America. Finally, it is generally supposed, after Needham, that China has always been split between two opposed "schools of thought": Taoism and Confucianism (whereas it is our Knowledge-Specialists who have always been split, by the Binary Con).

There are and were, in fact, no "schools" of Chinese thought except during the Battle for Worlds in the sixth and fifth centuries B.C., and, again, during a phase of the decadent court of the Sung dynasty, in the middle of the 2500-year Chinese cycle, one thousand years ago. As my Chinese teachers, like the founders of those supposedly opposed schools, emphasized, where there are opposed schools of thought one or both of them must be false. Taking this scientific principle seriously, the Chinese, within a short period, 2500 years ago, rid themselves of the schools of thought inconsistent with the Taoist-Confucian one that became the Chinese worldview. The capacity to sup-

* Astutely, Nathan Sivin, a specialist in Chinese science, has recently challenged that assumption, even though unaware of these biographical variables.

port opposed schools of thought over long periods of time without embarrassment as to the anti-scientific implications of such circumstances appears to be a distinctively Western trait. As Nathan Sivin, a Sinologist and specialist on Chinese science, has pointed out, in turn, there is no evidence whatever in Chinese writing that such a thing as a "Taoist" or "Taoism" (save in folk religion) ever existed. As Dr. Hsia explained to me to relieve me of the misconceptions taken for "basic Sinology" in the West, Confucius and Lao-Tzu were the originators of the two, mutually consistent and interdependent, aspects of the Chinese paradigm—the Human and the Natural-scientific. I have begun to show how this is so, that is, how the natural-scientific pole of the paradigm produces genuine, Human, social science and how the Human pole of the paradigm produces genuine scientists. Accordingly, *Tao, the Way,* and *Jen, Humanness,* are terms frequently used by both Lao-Tzu and Confucius (and all their successors), and are always the most important concepts.

Relieved of the false Fragmentation of Chinese knowledge into typically Western "opposed schools," we can turn to the question of science, of knowing, itself. The assertions and arguments of Lao-Tzu, Confucius, and their successors are not quoted in the awed and self-hypnotic way that Westerners quote Moses, Jesus, and Mohammed, Pythagoras and Aristotle, and Einstein, Keynes, Freud, and Marx, nor are they dogmatically transmitted from one generation to the next. Rather, they are constantly challenged, for Chinese science, as ours claims to, recognizes that a datum which is an exception to a to-date universal rule may arise. Were it otherwise, as my *shih-fu* emphasized, how could one develop understanding in which one genuinely believes and which one is capable of defending, not from a dry pedant's throat, but from the heart, with a breath that, as Chaung-Tzu put it, "starts not at the back of the mouth* but at the heels." There is no place for "faith" or dogma in Chinese science, or for debates to which scholars playfully attach themselves instead of debates which are *personal actions* of scholars as representatives of all people, serious business. That is one of the major reasons why, where it took Westerners 2000 years to develop the skeptical scientific attitude, the

* This shallow, crack-voiced or nasal, symptom of uncertainty, self-doubt, and fraud is epidemic among contemporary politicians and academics and has become a standard identifying behavioral pattern of the most elite amongst them, who now seek to cultivate it as a kind of badge of authority.

Chinese (re-)established it quite quickly more than 2000 years ago; and why, where most Western scholars continue to regard intellectual discussion as an elite game, and to propose tentative and predictably ill-fated theories without embarrassment, most Chinese scholars have regarded intellectual debate as a matter of other people's lives and deaths.

Confucius and Lao-Tzu are constantly quoted by Chinese savants precisely because their theories and attitudes have been challenged and borne out for 2500 years. Indeed, as I'll now show, the scientific rigor of the traditional Chinese process of knowing far exceeds ours.

Which worldview, if either, has produced genuine science, the Absolute-Fragmental or the Polar-Complete? "Scientific" means "intellectually and morally true," that is, "true" in the intellectual sense that our scientists usually intend, *and* in the sense that humanity is served—made more able to survive, made more comfortable, and cultivated to full flower. (If that is *not* what "science" means, then there is no reason to take interest in science.) E.F. Schumacher stressed in his *Small Is Beautiful* that if two competing theories have equal intellectual validity, but one has good social effects and the other has bad ones, the former should be adopted and the other discarded as a "bad idea." One can go further, to show that a bad idea cannot be as intellectually valid as a good one. If a system of knowledge is intellectually true, *it is also morally true, and conversely,* so one is inevitably dealing with both aspects of science when one focuses on either one. I have shown how the intellectual structure and moral character of a paradigm are really one thing. Here, I focus on one aspect of science, *intellectual* truth and falsehood—what, because of one of our worldview's basic Fragmentations, the separation of mind from heart, we tend to regard as the only kind of truth with which science could be concerned.

The criteria for, the roots of, "intellectual" truth have been stated by both Chinese and Western philosophers of science. However, our philosophers and the scientists under them, unlike their Chinese counterparts, have not practiced what they preach. Consequently our own so-called science can be shown false, and traditional Chinese shown true, according to the very criteria that our most serious scientists claim to respect. These criteria are two: *logical consistency* and *factuality.* Logical consistency is true of theory whose parts are consistent with each other and that is also general. Ultimately, then, a true theory must be universal, able to account for all data, or must be a specialized derivative from such a theory. The Chinese scientific term for logical consistency in all these respects is *t'ung*. My *shih-fu* always emphasized that I must test what he taught me for this quality and, as a genuine seeker of truth, reject *all* of it if I were to find it lacking. I have not found Chinese science logically inconsistent, but have found Western science to be throughout, as I will demonstrate.

That is not to say that every specific Western scientific theory is false. (We have gone much further, and ingeniously, in several specific, especially microscopic, directions than have the Chinese, as in atomic physics and microbiology.) It is to say that the paradigm that underlies each of our specific theories is fundamentally false, and that each of those specific theories, as a result, is from fundamentally false to distorted, a Fragment out of place. For reasons that will be made clear, and as Dr. Porkert generously pointed out to me, falsity in specific Western science is at its lowest degree in sciences of the inanimate and at its highest, almost entirely alienated from reality, in sciences of people. Below, I demonstrate the falsity in our sciences of the inanimate, so as to make it clear at the outset that even the foundation of our science, physics, is mixed with too much sand and not enough concrete.

Factuality, the second criterion of intellectual truth, is achieved when a theory is *empirical*. Such a theory is based on and predicts *phenomena that can be sensed*: seen, heard, smelled, tasted, felt. Most of our scientists do not also regard *ideas* as empirical data. Such thinkers argue that, unlike sense-data, ideas cannot be observed. For example, you and I can both see the moon, a sense-object, but we can only assume that the meaning of a word or sentence communicated between us is one and the same to each of us. The moon is "objective," the idea, "subjective." The problem with this Absolute-Fragmental argument is that it is impossible to Absolutely distinguish sense from idea, perception from conception, objective from subjective. The moon is subject to individual perceptions that cannot be verified, but only assumed, to be the same, so it is not purely "objective." And the visual perception of the moon is intimately mixed with ideas about it: you may "see" craters on its surface whereas I, if I am a northern American Indian, may "see" the face of a wolf. Did you perceive, conceive, or both, the lunar craters? In turn, ideas approach perceptions in that even highly abstract ones have images that are, or are made up of, perceptions; for example, the highly abstract idea "relation" has the following concrete image or something like it: ◯ – – – – ◯

whose form originates in perceptions of connected physical objects. So, common sense shows that, actually, sense-data, perceptions, are not significantly more "objective" than idea-data, and that both types of data are subject to observation—the former, chiefly sensory observation; the latter chiefly mental, or spiritual, observation.

Since the predominant Western opinion is that sense-data are more "objective" than idea-data, however, I will emphasize sense-data in the following demonstration of the factuality, the empirical rigor, of the Chinese as compared with the Western scientific paradigm. Factuality may be basically defined, then, as *faithful reflection of sense-data.* In scientific philosophy, this is sometimes called "epistemic correlation"* —the idea or thought correlates with the sense-datum. *T'ung* is the Chinese term for epistemic correlation as well as for logical consistency, for epistemic correlation and logical consistency are simply two varieties of *consistency*—two varieties which cannot be Absolutely distinguished, because (as Einstein recognized) what is sensed and what is thought are intermixed, not Absolutely distinct.

Taken together, the two criteria of intellectual truth, logical consistency and factuality (*t'ung*), call for a system of knowledge that has logically consistent theory that is factual, facts which are logically consistent with each other, and universal scope. Only then is a system of knowledge genuinely scientific.

Where it is impossible in actual practice to investigate all data in the universe, universality must be practically defined. On one hand, the whole universe insofar as it has been sensed and logical-consistently thought must be accounted for. On the other hand, no new data can be ignored, nor, if new data empirically disconfirm (disagree with) one's theory, can they be set aside for future investigation, to falsely "preserve" one's theory.

It is worth putting the preceding discussion of truth into cross-cultural and historical context. Our students of Chinese culture universally "recognize" the Sung dynasty (tenth through thirteenth centuries A.D.) as the one in which Chinese thought "matured" to become systematic and "proto-scientific," because most published scholars of that period, in which the Absolute-Fragmental worldview was almost established, remind them of themselves. During the Sung, the savants Ch'eng Hao and Ch'eng I attempted unsuccessfully, although their effort bore fruit later on, to restore Chinese science as represented by

* Term originated in Northrop, *op. cit.*

government officials to its usual rigor. The preceding discussion of criteria of truth could all but serve as a translation of passages from the Ch'eng text toward that end, which, in turn, is explicitly based on the philosophy of Lao-Tzu and Confucius (sixth century B.C.).

To complete the historical context, listen to serious Western philosophers' criticisms of the tradition and the present state of Western knowing. In this century our philosophers recognize with increasing frequency that they (and our scientists) have not been and should start being good observers. William Barret has recently written a courageous and admirable book, *The Illusion of Technique,* about the terminal, actually non-empirical, therefore fact-less condition of our way of knowing. He blames its failure on Aristotle's separation of thought from being (an aspect of the overall Binary Con). He then traces this split through Descartes, who "abolished" being and called mathematical logic the essence of truth. (We have already understood that was nothing new: this is Pythagoras again.) Descartes rediscovered this worldview while ill, watching through his window a tree sway in the wind—a window that became the Cartesian coordinate-system. The psychological health of this man was such that he once attempted suicide. Barret continues his tracing of the Thought/Being split through Bertrand Russell and Alfred North Whitehead, who invented a "universal logico-mathematical language" and then found it useless. He then looks to Ludwig Wittgenstein, who submitted himself to the same non-common-sensical and arrogant exercise (one must use "imperfect" non-mathematical, "folk," language to generate this "perfect" mode) and then, with just enough energy left to do so, declared: "Don't think, look!" That is, "logic" alone leads nowhere because it starts nowhere; sensing must be involved.

Richard Boeth, who reviewed Barret's book in *Newsweek* (2 October 1978), concludes that "the primary function of the mind is to *perceive,*" and makes an innovative leap which serves as an excellent Western introduction to this chapter: "And it is the obligation of the philosopher to take his empirical evidence from those who have spent their lives in pure perception." With the qualification that a genuine savant's perception is not "purified" of cognition and he does not spend *all* of his life in perception, but also theorizes and takes responsible, informed actions, Boeth's perception is very "Chinese high-medical." Of course, the "*un*-knowing," "common," people of the West, unlike their Knowledge-Specialists, have never troubled themselves with a split between thought and perception because they have not suffered from scientific-philosophical "education"; hence they say of a sound theory: "That makes *sense.*"

Let us now *scientifically* examine the Chinese and the Western worldviews, their scientific or pseudo-scientific paradigms, or cores, and the actual research practices of the scientists that are their vehicles. Let us responsibly engage in the recycled Battle for Worlds to make its outcome a good one, or, at least, one based on a full-fledged choice.

TRUTH IN BLAZING COLORS . . .

The Chinese scientific paradigm is richly metaphoric, just as was ours in its original, Hindu-Mesopotamian-Egyptian-Hellenic form. One of the central metaphors is *Form is color*: the term *se* means both *color* and *Form.** *Se,* then, encodes the proposition that all forms, what we call "appearances," are like color forms. This metaphor is perhaps the best door to the traditional Chinese scientific (the scientific) universe. In this chapter I will explore color from both the Chinese and the Western perspectives, by way of exemplifying the fundamental difference between the two paradigms, making the Chinese one clear in its own terms, and establishing to what extent each may be scientific. It will then be possible to proceed with relative ease through other physical, then biological, then social phenomena to the level of universalistic (Chinese) science.

Color is a central phenomenon from both the Western and Chinese scientific perspectives. Physics is our "hardest" science and is the implicit model for all our other sciences. It really is the most highly developed of our sciences, and colors are the chief data that our physicists (and chemists) take into account in investigating the nature of "matter," "energy," "space," and "the cosmos." Hence the validity of our physics, and of our science in general, not only depends, but depends to a major extent, on the validity of our understanding of

* I have already introduced the *ch'i* (vapor) metaphor for energy-matter; I will return to it in the next chapter.

colors. From the Western perspective, then, an examination of the Chinese and Western understandings of color is crucial to comparatively evaluating Chinese and Western science.

In turn, from the Chinese perspective, if one is to begin with any single phenomenon to attain scientific understanding, color would be the one. One does *not* begin with a single phenomenon when learning Chinese science. On the contrary. But for the purpose of this comparison it is useful to do so. Color is the best phenomenon to start with because it is the only quality present in all things exterior to the human body: it is the only universal empirical datum. We can sense air's or water's colors, but not their shapes, if any; we can sense a star's colors, but not its heat. Rather, we logically deduce its heat from its colors by analogy to objects whose heat can be sensed, and so on. There are many sensible objects of which we cannot sense heat, shape, sound, odor, tactile (touch) qualities and/or taste, but we can sense color of all of them. The object may have to be thick before its color is perceptible, as in the cases of water, air, and "colorless" crystals, but the colors can be sensed. Hence, it is empirically and logically scientific to take color as the foremost example of form in general. This accounts, no doubt, for the fact that the term for forms (*se*) is also the term for color, instead of for heat, shape, sound, odor, tactile qualities or taste. The empirically universal quality, color, is the basic meaning of the (metaphorically widened) term for form in general.

A fact necessarily involves human sensing of the phenomenon it concerns. To establish the facts about color, then, requires sensing colors, and the resulting theories must be faithful reflections of what is sensed, if they are truly to be facts. As soon as descriptive or explanatory theories about color depart from one's sensings of them, the theories are not factual, not empirically derived; rather, they are subjective mental constructions, mystical statements. To begin with the empirical aspect of the truth about colors, then, let us sense them.

Our visual sensing tells us that they vary continuously: they are not discrete, not absolutely distinct one from the other. On a natural spectrum, as in a rainbow, violet-blue-green-white-yellow-red, there is no "line" between any two adjacent colors, no point at which one ends and another begins. The colors at the ends, violet and red, overlap and inside the range of none of these colors is there an Absolutely specifiable "pure" region of that color. Further, one's perception of a color depends on its color-context. What one might think of as "pure" red when placed next to what one might think of as "pure" green, appears yellowish when placed next to what one might think of as "pure" blue. Likewise, "pure" black becomes gray in its middle and blacker at its boundaries when surrounded by "pure" white, and obversely, "pure" white becomes gray in its middle and whiter at its boundaries when surrounded by "pure" black. Our eyes tell us that colors have no boundaries and are relative to each other—are not Absolute-Fragmental.

Logical extrapolation and empirical testing, based on the preceding data, establishes facts about color which modern scientists suppress and ignore to this day, because they imply that their theories about color (indeed, about all electromagnetic phenomena) are basically false. Ultimately, these facts imply that their entire scientific paradigm must be discarded. First let us fully appreciate the *fact* of the non-Absolute-Fragmental nature of chromatic (hued) colors. If one color is continuous with, not discrete from, its neighbors (red and violet, at the opposite ends of the spectrum, are here included as neighbors), and no color, even at the center of the range called by its name, takes a pure form, then *each color is both itself and its neighbors.* That is, yellow is both yellow and green-and-red, red is both red and yellow-and-violet, and so on. But a color's neighbors are also themselves and their neighbors, so a color is also its *neighbors'* neighbors: yellow is not only yellow and green-and-red, but also blue-and-violet; red is not only red and yellow-and-violet, but also green-and-blue, and so on.

A color is foremost itself, less its neighbors, and yet less its neighbors' neighbors. Thus *a color is even its opposite,* to an extent. It follows that the very *existence* of a given color is relative. A *given color, in itself and taken alone, does not exist; rather, it relatively-exists, more exists than not.* Put otherwise, each color is a product of the relations among all colors, so, rather than there being various colors in the Absolute-Fragmental sense—"just blue, just red," and so on, as "parts" that are "added up" to make the whole of color—there are *relative-colors* each of which is the totality of colors (white) "slanted" in a certain way. Blue, for example, is all the colors, but with relative-blue in greater proportion than the others. No color exists except as a relative aspect of a *whole,* color, and the whole, color, *does not have parts.* If we wish to mean something real when we use the word "colors," then we must redefine the term non-Absolute-Fragmentally. We must understand that when we use a specific color-term we mean not a "piece" of the spectrum, but all colors in a complex relation of relative intensity, relative presence, relative "thereness," that gives that whole a relatively specific character.

Since first writing this, I came across Capra's *The Tao of Physics* and in it found a virtually identical understanding, about another aspect of Nature, directly related to colors, originated by Geoffrey Chew, whom Capra rightly regards as a particularly enlightened avant-garde physicist. This remarkable correlation powerfully substantiates my assertion that as Western physics is forced by Nature, despite its paradigmatic basis, to undermine the paradigm out of which it grew, it is turning precisely toward the Chinese one:

> Newton's universe was constructed from a set of basic entities with certain fundamental properties, which . . . were not amenable to further analysis. [These are basically the Absolute-Fragments, Being, Becoming, Non-Being, Form, Energy, Matter, and their Materialist correlates, the kinds of atom, or basic units of matter.] In one way or another, this notion was implicit in all theories of natural science until the bootstrap hypothesis [of Chew] stated explicitly that the world cannot be understood as an assemblage of entities which cannot be analyzed further. In the new world-view, the universe is seen as a dynamic web of interrelated events. *None of the properties of any part of this web is fundamental; they all follow from the properties of other parts, and the overall consistency of their mutual interrelations determines the structure of the entire web.* . . . The picture of hadrons [relatively massive particles] which emerges from these bootstrap models is often summed up with the provocative phrase, "*every particle consists of all other particles.*" . . . *Rather than "containing" one another, hadrons "involve" one another in the dynamic and probabilistic sense of S-matrix theory, each hadron being a potential "bound state" of all sets of particles which may interact with one another to form the hadron under consideration.* (emphases mine)

(Heraclitus has finally found a follower, in Geoffrey Chew.) If Chew is right, then is it the "world of sense" that is deceptive, or is it the Western scientific paradigm which for 2500 years prevented our physicists from achieving such Polar-Complete understanding, because it is in fact anti-empirical, because it denied, for example, the immediate, sensible reality of colors? The "underlying realities" "penetrated" mathematically and experimentally by our physicists are but a microcosm *of what we have always had before our very eyes,* but never saw, because of the mysticism of the Western, Absolute-Fragmental, paradigm.

The reader has probably already thought of many other phenomena of which the preceding is also obviously true, for example, night and day, and the relation between sides and diagonal denied by Pythagoras, but let us continue to attend only to colors, in order to appreciate, through one exhaustive example, the nature of truly empirical science, and the difference between Western and Chinese science. As we do so, let it be clearly understood that I am making a case for the Chinese paradigm and against ours not only, or even centrally, on the basis of color phenomena. Rather, I am using color phenomena to exemplify the differences between the Western and Chinese scientific approaches, with the understanding that colors will be shown to share their non-Absolute-Fragmental, Polar-Complete, nature with all other things.

All the preceding statements about colors, logically derived from sense-data, are now empirically confirmed, as promised, and thus shown to be facts.*

As our own scientists have determined by experimentation, none of the four colors most often called "primary," blue, green, yellow, and red, is perfectly opposite. Red is often thought of as the opposite of green, but when it is combined with green in equal proportions by spinning a half-red, half-green disk or overlapping red and green beams of light, yellow is produced. Were they perfect opposites, gray would be produced. Likewise, a red/blue opposition is shown, by the same kind of experiment, to produce not gray but reddish-violet. But red combined with slightly bluish-green does produce gray, so the opposite of red is slightly bluish-green (cyan). Yellow and blue are also often thought of as opposites, and despite their own data, our scientists often assert that they are. But when they are combined in equal proportions the result is greenish-white. The exact opposite of yellow is blue-purple, a slightly reddish-blue. By analogous experiments it has

* The additional data on which the following is based are taken from Bragg, Birren, Mueller and Wright, each cited in the *Bibliography*. To deepen my understanding of those materials and to experience the basic data, I also conducted experiments with colors, using the standard techniques of rapidly rotating multi-colored disks in bright "white" light and admixing "primary" pigments. Such direct experience is recommended to those who, like me, value their own sensory experience as a medium of truth.

been determined that the exact opposite of green is slightly bluish-red (magenta) and the exact opposite of blue is slightly reddish-yellow (yellowy orange). Now, if there are no exact opposites among the primary colors, it follows that each is to a degree the same as any other, including the primary color that is nearest to being its opposite. The "overlap" of red and violet, which are at opposite ends of the natural spectrum, is a direct sensory statement of this. (Note that non-oppositeness, or similarity, necessarily involves some sameness: any two things that we perceive to be *similar* have one or more *identical* features.) It follows that the primary colors are continuous rather than discrete, and relative rather than Absolute. And, again, it follows that color has no parts; rather, each color is all colors slanted in its direction, is relatively there but to an extent not there.

To continue this empirical proof: a combination of red, green, and blue in equal proportions or a combination of two thirds blue and one third yellow produces white—*all colors*. It follows that yellow is implicit in the first combination and that green and red are implicit in the second combination. This is independently demonstrated to be the case by the fact that green and red combined in equal amounts produce yellow. Thus, blue and yellow in amounts one third N and two thirds N are equivalent to blue, green, and red each in the amount one third N. One might infer from this that yellow is not a primary color, and, rather, is red+green. However, secondary colors that are half yellow (orange, yellowish-green) are *truer* when their yellow component is yellow rather than red+green. Further, in the mixing of pigments and reflection of light, as distinguished from filtering light through pigmented lenses or overlapping reflected light, as in the preceding experiments, yellow is indispensable and is combined with magenta and cyan to produce all other colors, including red, blue, and green. We have, then, two alternative sets of "primary" colors, that is, colors which in combination produce white and all other colors, and a third set (yellow and blue) which produces white but not all other colors. It follows that each of those "primary" colors is also a secondary color derived from two or three others, that each primary color does not exist per se but rather is a function of other colors of which the same is true. In just the same way, Pythagoras' "rational" sides and "irrational" diagonal of a square are also, objectively viewed, "irrational" sides and "rational" diagonal of a square. Whether a color is "primary" or "secondary" depends on the Absolute-Fragmental *human perspective* that is taken. From the perspective in which red, green, and blue are "primary," yellow, magenta, and cyan are "secondary"; from the perspective in which cyan, yellow, and magenta are "primary," red, green, and blue are "secondary." And, needless to say, a virtually infinite number of finely graded alternative sets of "primary" colors could be scientifically isolated, for example, aquamarine, reddish-orange, and bluish-violet.

So again, *empiricism*, the scientific establishing of *facts*, forces us to confront the reality that each "primary" color, conceived Absolute-Fragmentally as pure and discrete, is each other color to an extent, is both itself and not itself, only relatively exists.

Instead of confronting all the preceding, our color-scientists, as dictated by the Binary Con, have drawn two false Absolute-Fragmental distinctions. The first is a distinction between "perceived" and "conceived" colors. Since there are two theories of color-perception-cognition that "work," our color-scientists have attempted to "save" both through this distinction, positing that the eye has three directly-perceiving kinds of color-receptors, for red, blue, and green, which then pass the perceptions on to two dual "cognitive encoders," one for red/green, the other for blue/yellow. Experimentally testing the perception-based theory, the cognition-based theory, and the synthesis of the two against each other for about twelve years now, they have failed to establish that any of them are correct. The situation is directly analogous to the Binary Con which produced it: the perception-theory is the analog of Materialism, having to do with brute neural responses; the cognition theory is the analog of Idealism, having to do with ideal (or not immediately perceptual) responses; the "synthesis" of the two is the analog of Aristotle's Realist "synthesis." Such Fragmentalizations ignore the *already established fact* that human sense-cognition perfectly reflects the reality of colors and therefore cannot be reduced to an Absolute-Fragmental mechanism.

The second distinction our scientists have drawn is between "additive" and "subtractive" color-mixing, and on its basis the facts *only* about "additive" color-mixing, as with beams of light, have been taken into account in the preceding "synthetic" theory. That is, for a while, the cyan-yellow-magenta set of "primaries" suggested by pigment-mixing is simply out of their picture, left to our painters to fool with in producing a new art form each year to keep the Beautiful People of New York visually titillated. The "additive/subtractive" distinction is quite arbitrary, as the better books on color parenthetically suggest.

The converging hued beams of light "add," but the beams have already been "subtracted" by dyed lenses from white light. The pigments "subtract," by absorbing parts of the spectrum, but the reflected residuum is already the "sum" of light from two or more pigment-sources. The distinction is no sharper than the one between "dyes" and "pigments."

Just as Democritos, Pythagoras, Plato, and Aristotle, insisted on abstracting Absolutes, primary entities, from all data—Being, Non-Being, Form, and Matter—our color-scientists insist on abstracting Absolute, primary colors from all color data (but can only *appear* to do so, for these Absolutes do not exist), by *Fragmenting* their data into parts (perceptual, cognitive, additive, subtractive) that are taken into account only separately, thus violating the scientific criteria of logical consistency and universalistic empiricism. In sum, they are looking for the basic Absolute-Fragments of color, whereas they should first have determined if there are or could be absolute differences among colors, such that such basic Absolute-Fragments might exist.

And in fact, fact established by themselves, there are not: Each color consists of *all* the wavelengths of light. That is, red pigment reflects, and red light contains, not only the wavelengths that (non-Absolutely) characterize red, but also those that (non-Absolutely) characterize green, blue and yellow; and so on for each color. Our color scientists know very well, then, that each color is also every other color. It is a very short step (never taken because forbidden by our paradigm) from that fact to realizing that every color is also its so-called opposite and that opposite ends of the electromagnetic spectrum, basically energy and matter, are also each other. Of course, Einstein got at this, but he did not do so—and this is my point—via empirical observation, but via mathematical logic motivated by intuition. It is already clear, though, that the mentioned correlation between the Chinese and Einsteinian understanding of the matter energy relation is due to Chinese *science*: empirical, sense-and-common-sense-based, science. And the result, which I'll come to soon, is that the Chinese understanding of that relation is more accurate, richer, and infinitely more useful than Einstein's or any of his successors'.

The depth of unadmitted Western knowledge about color must be thoroughly appreciated before proceeding further. Western scientists will not "permit" color data to be adduced to understand electromagnetic energy-matter because color is in part "a subjective impression."* Let us take into account only their "Objective" data. What we call "color" is the humanly perceived aspect of the entire electromagnetic spectrum. "Objectively," it is nothing other than light-energy vibrating at certain frequencies and corresponding wavelengths, which have been timed and measured, by the use of instruments mediating between the observer and the phenomenon in question—which instruments are *perceived* and whose connection to the phenomena in question is theoretical—are *ideas* in the minds of the "Objective "observers." Disregarding the fact that the data in question are not in fact "Objective," they nevertheless reveal that our physicists constantly kid themselves—and everybody else—into regarding their representations of them as faithful representations of their data. These representations are Absolute-Fragmental mathematical expressions: number of waves per second and lengths of waves. They lead to the understanding that, well, colors may indeed all be each other, but each of them may be decomposed into (Aristotelean) different-lengthed-and-frequencied parts which Nature "synthesizes" into wholes. In the end, our Absolute-Fragmental worldview is upheld.

That is not the case. For example, they say that the "pure red" (the non-blue, non-green, non-yellow) "component" in red is energy-waves between, say, 550 and 650 nanometers in length, and of the corresponding frequencies. Applying the *t'ung*-principle, we must then ask (a) Is the difference between two minimally different waves within that "pure-red" range Absolute, quantifiable? and (b) If there is a pure red, is the difference between the shortest-fastest red wave and the longest-slowest yellow wave, and is the difference between the longest-slowest red wave and the shortest-fastest infra-red (micro-, heat-) wave Absolute, quantifiable? Both answers are "No"; this mathematization is based not on sense-data or logic, but on loyalty to a paradigm. Logically, the difference between any two wavelengths or frequencies *on a continuum* is an infinitely small fraction, which means either zero or the approaching-to-zero or both: not a quantifiable concept. (It may be expressed with mathematical notations, but that is only a language trick; it cannot be counted and is therefore basically non-mathematical, in the sense non-quantifiable.)

Empirically, the latest experiments with colors-as-electromagnetic waves confirm what has just been reasoned. The "components" of col-

* They remain disciples of Socrates: See Appendix: *Socrates Encircled.*

ors, specifically the "emission-lines" of color produced by the process misleadingly called "*quantum* jumping,"* have been found to be unmeasurable, non-quantifiable, thus confirming both that the electromagnetic spectrum *is* a continuum and that colors do not have "components" in the quantifiable, Absolute-Fragmental sense. As Hansch, Schawlow, and Series explain in their article in *Scientific American* (March, 1979), our instruments of measurement for this research are now finer than the components themselves. "Interferometers fail to fully separate the several components of the Balmer Lines [the color-sectors emitted by excited hydrogen atoms] not because the instruments are inadequate but because the components themselves are not sharp. Each component is distributed over a range of wavelengths that is often greater than the spacing of the components." In other words, two different wavelengths are also two same wavelengths; their difference is non-Absolute-Fragmental and therefore not measurable or quantifiable. This finding, then, should put an end to the long search for basic Absolute-Fragments of color and wave-energy, through what have been called the emission-lines and their components, the "fine structures," and their components, the "hyper-fine structures." However, it has not: the authors call for more powerful instruments to get further toward the bottom of a matter which, they already know, has no "bottom line." Ample empirical grounds for trying to generate an alternative, Polar-Complete, physical sub-paradigm are in the preceding finding alone. But, more to my present point, it is now obvious that the senses and common sense are and should always have been used as a "valid medium of truth" all along the time-spectrum of Western scientific history, as they were by the Chinese savants. Working with "mere appearances" and "ignoring the real substratum," the Chinese have understood all along—through the use of *science*, pure, human-centric, science—the true nature of the appearances *and* the substratum, which are really, Polar-Completely, an appearance-substratum.

To return to the level of "appearances": the negative effect on humans of our scientists' stubborn adherence to the Absolute-Fragmental, quantitative, color-sector of their paradigm is worth pointing out, for, as I said, intellectual validity and moral validity are inseparable. Our color-scientists refuse to mix our colors, the colors of our paints, clothes, cars, houses, TV screens, advertisements, and so on—the col-

* See *The Hinges of the Cosmos,* below.

ors of our manufactured environment—in accord with the preceding facts. Rather, they restrict their basic red, blue, and green or basic cyan, yellow, and magenta to variants that have the narrowest ranges of dominant wavelengths and that therefore are as *Absolute-Fragment-like* as possible. Not only is this approach consistent with the Absolute-Fragmental paradigm, but it is consistent with business-interests. It makes for inexpensive color duplication. The result, as W.D. Wright, a leader in this field, has admitted in his text on the subject, is that no formula for mixing a color works perfectly. Each combination must be "masked" by adding a supplementary amount of one or more primary colors, and, despite this masking, to put it in his own words, "the direction of error is towards desaturation and must be recognized as one of the factors contributing to the general degradation of colors." Masking and degradation: these are the ultimate, the final and most sophisticated, achievements of our color-science, thanks, basically, to our scientific paradigm. As the increasingly *unreal* and *dishonest* quality of our color environment declares throughout our lighted hours, each color is each other color; to deny this is to suppress truth—and beauty along with it.

It is a *fact*, then, that colors in the Absolute-Fragmental sense do not exist. Rather, they relatively exist, are relatively present in some relations to all others and relatively absent in other relations to all others. It is a *fact* that each color, in the Absolute-Fragmental sense, is both itself and not itself, including its opposite. It is a *fact* that the whole of color has no parts, and, rather, has relatively-present aspects, to which we give specific names, such as "red."

We may now scientifically characterize colors, as Polar and Complete. (The Chinese technical terms are *Chi* "Polarity, Pole" and *Cheng* "Complete, Whole.") That is, colors are non-homogeneous, non-discrete, present-yet-absent, different-yet-identical, and such that the whole of color must be taken into account if any specific color is taken into account. We often use the term "polar," especially in scientific discourse, but never do we intend it in its *empirical* sense, as above, because our worldview and paradigm neither have a place for this meaning nor contain concepts consistent with it. (By the same token, the Chinese term, *Chi*, one of the central scientific concepts, has never been accurately translated.) Indeed, Polarity in the empirical sense, the property of being itself yet also being opposite to itself, of being different from or even opposite to yet also identical to, strikes the Western mind as mystical, irrational. Such a concept fundamentally violates, uproots, our "scientific" paradigm, for our paradigm is essentially, Absolute-Fragmental, exactly opposite to such a concept. *Yet that concept is strictly empirical, factual.*

Now let us use the preceding example to affirm the crucially important connection between the non-empirical character of Western science and its underlying worldview. Parmenides, Pythagoras, Plato, and Aristotle were quite conscious of the non-Absolute-Fragmental, Polar-Complete nature of all sensory data. Finding it hard to handle intellectually—to "compost," as Douglas would say—from the start all of them agreed not to, and to rely instead of the mind, on "logic" as they (mis)conceived it, to get at the truth. As Plato put it, "Of the world of sense it is true that opposites intermingle, but this signifies only that sense-experience is evidently not a valid medium of truth." But "that sense-experience is not a valid medium of truth" does not logically follow from the fact that sense-experience is non-Absolute-Fragmental. This bit of "logic" is simply an irrational, arbitrary rejection of sense-experience and of empiricism. The motivation for it is the alienated, "this-world"-disdaining attitude that is a central feature of the Western worldview. Likewise, Aristotle, the supposed founder of *empirical* science, observed that the classes of natural things overlap, and called this "the principle of continuity." He then proceeded to "infer" that what we sense is a confusion, or "synthesis," of "higher," Absolute-Fragmental forms. The first step was to deny that what can be sensed indicates the basic nature of the universe, or is true. The second step was to erect an Absolute-Fragmental worldview, headed by "Being," for which there was (and is) no empirical basis. Aristotle's distinctive move was to then turn around and call for Natural Science based on empiricism, all the while dogmatically and contradictorily understanding that the sense-data of empiricism are deceptive, and therefore must be interpreted in "true," Absolute-Fragmental terms!

To this day, on one hand the mysticism and spurious empiricism of Aristotle is reflected by "science," and on the other hand, Aristotle, despite the preceding fundamental contradiction, is identified as the founder of our supposedly empirical science. To take the one example we have explored in some depth, to this day Western scientists describe and study colors in terms that are strictly patterned after Aristotle's mystical worldview, blithely ignoring the sensed nature of colors themselves and claiming to be doing empirical science. The concept of "primary color," of independently existing, perfectly homogeneous color, is the analog of Aristotle's primary concept, Being. The concept of "derived colors," syntheses of the "primary" ones on which they depend for their existence, is the analog of Aristotle's "derived forms." As we have seen, in empirical *fact* the so-called primary colors are no more pure or discrete, no more Absolute-Fragmental, than are the so-called derived ones, and each type of color depends on the other for its existence. More precisely, "primary" and "derived" colors are simply different aspects of the same thing—color. If all "pure red" were subtracted from the spectrum, there would remain no orange or purple, and if all orange and purple were subtracted from the spectrum, there would remain no "pure red." To deny this, to speak of "primary" and "derived" colors defined "rationally" by numerically expressed wavelength-ranges, as our scientists do, is not to exercise empirical science, but to exercise Aristotelean metaphysics, mysticism—just as the classical Greek sources of "science" dictated.

Not only do our scientists think in a mystical, 2500-year-old mode, but they do so despite extensive evidence that that mode is false. If it were true that the "world of sense" is a deceptive, confused, derivative of more fundamental, "pure" properties, the confusion of what is sensed could be untangled by the Western scientific mind into a set of universal Absolute-Fragments. We have already seen that colors, at the very least, are immune to such analysis. It follows minimally that the preceding "scientific-philosophical" proposition about "the world" of sense is excepted by colors, and maximally that it is totally false. Yet our scientists continue to deal with colors, and all other Natural phenomena, as though they *are* confusions, intermixtures, of more fundamental, pure-discrete, Absolute-Fragments (pure red, discrete wavelength-ranges . . .). In other words, purely through religious (if unconscious) belief in the validity of a mystical worldview *and despite the fact that the worldview is disconfirmed by the empirical evidence,* our scientists persist in distorting their sensed data and refusing to derive facts from them.

Now let us consider what is involved in respecting the criteria of truth and doing genuine empirical science. Let us open a long-hidden door.

To confront forthrightly the fact of color's Polarity causes an earthquake of the mind. Of course, this earthquake is strongest in the Western, philosophical-scientifically sophisticated mind, because it is trained to think only in Absolute-Fragmental terms, and the heart behind it is strictly conditioned to reject with instant reflex any alternative terms, especially if they are foreign ones. The natural reaction of a Western philosopher or scientist, therefore, is to run. Here, then, the Western philosophic-scientific reader who wishes to understand anything about Chinese science, or to understand his own science objectively, is given a choice. Looked at from the perspective we are used to, Polarity is illogical and frightening; it destroys the little Absolute walls between the Fragmental meanings of our words and blends those meanings into a single, moving, dragonlike meaning whose back is light and whose belly is *chaos*. Out of irrational fear of it, one might, and often has, labelled the dragon-meaning "unscientific" or "mystical," but to do so is to cease to be a scientist and become a mystic, to cease to be a man of truth and become a liar, for the dragon-meaning is empirical fact.

Logically consistent statements about color, or anything else, need not exclude paradoxes, as the Western mind, which likes the categories its theory strings together "self-consistent," would suppose. Logical consistency refers to a relation between basic understandings and other understandings which derive from them, and then between other understandings, and always between sensory data and theories. *If the sensory data are paradoxical, a logically consistent theory must also be.* For example, from the facts that yellow is partly green and red, green partly blue, blue partly violet, and violet partly red, it is logically consistent to deduce that yellow is also partly blue and violet. The correlation between this deduction and all relevant empirical data is also a logical consistency. Likewise, if one finds that phenomena other than colors are paradoxical in the way that colors are and generally theorizes that *all* sensible phenomena are Polar (I will show that they are), one's theory is logically consistent internally: each part of it is consistent with each other part. Logic—*Logos*—(true) Word exists only where language is scientific, is an aspect of culture that is based on truth. We are familiar only with sub-scientific (Absolute-Fragmental) language, but sub-scientific language can be *used* to arrive at logic (*Logos*) if it is kept in harmony with empirical fact or if logic is taken as far as it can go. Here, with the example of color-classification, I now *transform* (*hua*) the Absolute-Fragmental type of classificatory thought into scientific classificatory thought, by exercising genuine Complete, fully extended, logic.

The number of primary colors that people distinguish varies from one society to another. For example, the Dani people of New Guinea, according to the anthropologist Heider, distinguish only two (black and red). In many societies, as the anthropologists Brent Berlin and Paul Kay have pointed out, blue and green are not distinguished, and in their place is what we might call "grue" or "bleen." This cultural variation is made possible by the Polar-Completeness of colors. However, within that flux there is a certain regularity, for among the different peoples as a collectivity there is a tendency to classify colors in certain ways, although it may not be realized at the level of language. To make that long story short, all our empirical findings about color-perception and classification imply that there is something natural about designating "the" primary colors as black, red, blue, green, yellow, and white, as is done in both English and Chinese; and that the most basic (the relatively-basic) members of this series are black and red, or black and red-which-includes-white.

There are several ways to arrange "the" primary colors in a series, because Nature is not unilinear. For present purposes it is best to arrange the hued colors according to relative similarity of hue (as opposed to luminosity or intrinsic warmth) and to deal with black and white afterward. This gives:

REDYELLOWGREENBLUE

Now the process of Absolute-Fragmental subclassification begins. Doing it the simplest way, we distinguish each primary range into three parts: the two parts that are relatively similar to its neighbors and the central, "pure," part that is relatively not. We get:

R E DY E L L O W

bluishred *red* yellowishred reddishyellow *yellow* greenishyellow . . .

G R E E NB L U E

yellowishgreen *green* bluishgreen greenishblue *blue* reddishblue

Logically, if color is a continuum we could continue in this way eternally-infinitely, and it follows that the "ultimate" amount of each "pure" primary color (underlined above) within its own range is "infinitely small," which is to say that each is a zero-and-approaching-zero amount, and that, therefore, ultimately there are no specific colors at all. (In esoteric Sufism,* it may be noted to its credit, this is called "the transition from color to the colorless.")

* See the Appendix.

The "amounts" of "pure" primary colors, then, range between 100 to 0 percent within their own ranges. Where the "levels" of each range (primary with one color, secondary with three, . . .) are actually one in space-time, for they are products of our human perspectives, it follows that each "pure" primary color half-exists and half not-exists within its own range. This brings us directly to the question of Completeness. If pure colors only half-exist, how can they be distinguished with certainty the one from the other, especially when they are neighboring ones? Logically, it is because each actually more exists than not-exists. (Red *is* more red than it is yellow or blue, and so on.) There is something false about our initial premise, that the primary ranges do not overlap. On that premise, we subclassified each primary color by "synthesizing" "parts" of its outer areas only with its neighbors (red was Fragmented into purer red, bluish red and yellowish red; and so on). It must be that, actually, each color is distinguished not only by comparison with its neighbors but also by comparison with all its non-neighbors, which, unlike its neighbors, are more different from than same as it. Each color is identified with certainty only in relation to the *whole* spectrum.

Logic dictates, then, that the basic premise of Absolute-Fragmental classification, that there are pure and discrete forms, here, primary colors, at a "highest level," is false. Hence all the terms of such classification are false.

The logical inference, just above, that "pure" colors must more exist than not-exist, in relation to the *whole* of colors, is empirically confirmed by the fact that whereas only about 500 colors can be visually distinguished when minimal ("neighbor") contrasts are made, *millions* can be distinguished when maximal as well as minimal ones are made. Specific colors are more "there" than not when *all* colors are seen and compared.*

As to the subsensory range, that is, what is indicated by indirect empiricism, in which instruments replace the eyes as direct receivers of the stimuli that produce colors, the same pattern occurs. Actually, human eyes far surpass the accuracy of such instruments. The finest distinctions between wavelengths made by instruments is about .004 nanometers. As follows from the fact that millions of colors can be discriminated by human eyes within the range of about 370 nanome-

* See Wright, *op. cit.*

ters, people make discriminations tens of times as fine. This is so because they can take the *Complete* range and all dimensions of colors into account simultaneously whereas an instrument cannot, and because our scientists direct their instruments to the narrowest possible ranges of wavelengths, in persistent, even stubborn, hope that they will find the basic Absolute-Fragments of color.

To sum up, if we pursue the logic of Absolute-Fragmental classification to its end, we are forced to alter the assumptions that it is based on: that there are discrete color-differences and pure colors, and that there are levels of color. As explained in disclosing the Binary Con, and as exemplified above, these are the assumptions on which Materialism, Idealism, and Realism, *and all Western science,* are based. The Absolute-Fragmental primaries red, yellow, green, and blue are nothing other than the ultimate, Greek philosophical, Absolute Fragments in a specific form. But when, unlike Pythagoras, Plato, Aristotle, and company, we were logically consistent, *t'ung,* in thought and in empirical action, and pursued our classificatory logic to its end, we were forced to change our initial, Absolute-Fragmental assumptions. We were forced by exercising *science* to evolve Polar-Complete concepts in their place. In doing so, we moved from the primitive sub-scientific mode of thought to the scientific one.

Actually the word is already officially out among our own scientists that this basically Pythagorean-Aristotelean, mental-ist, aspect of our paradigm is objectionably mystical, that our "logic" does not match the data:

> To the followers of Pythagoras the world and its phenomena were all illusion. Centuries later the Egyptian mystic Plotinus taught the same doctrine, that the external world is a mere phantom, and the mystical schools of Christianity took it up in turn. In every age the mystically inclined have delighted in dreaming that everything is a dream, the mere visible reflection of an invisible reality. In truth the delusion lies in the mind of the mystic, not in the things seen. The alleged untrustworthiness of our senses we flatly deny. We frequently misinterpret the messages they bring, it is true, but that is no fault of the senses. The interpretation of sense impressions is something to be learned; we never learn it fully; we are liable to blunder through all our days, but that gives us no right to call our senses liars. . . . We not only wrong our honest senses but also lose our grip upon

> this most substantial world when we let mistaken metaphysics persuade us to doubt the testimony they bear. (*Scientific American*, July, 1875)

Unfortunately, our scientists took the preceding truth no more seriously a century ago than Aristotle did after he declared and then contradicted it; no more seriously than the successors of Francis Bacon, refounder of empiricism, took his reaffirmation of it; nor more seriously than contemporary physicists take their electromagnetic findings. There is a Hindu notion of a wheel of life-in-repeated-error on which humans revolve, reincarnating generation after generation, unless they recognize their illusions. Because of the inHuman effects of Western anti-empiricism, the social backfiring of its "side"-effects, this time the wheel is being braked.

The primitive, sub-scientific mode of thought, whose basic units are categorical, non-paradoxical, Absolute-Fragmental word-meanings, is natural. However, that does not make our science, which is an exaggeration of it, natural. That mode of thought is not the only natural one. It is natural to transform that mode of thought into scientific thought, because that transformation occurs when human knowers interact with Nature, which interaction is natural. It is unnatural to refuse to do this. Consequently, as Korzybski asserted, our Knowledge-Specialists must learn to shift from the "infantile" to the "adult" mode of thought if they are to travel as far as they wish to and deserve to. What leads to this is *direct* contemplation of Nature and the (difficult) mental exercise of transforming Absolute-Fragmental thoughts into Polar-Complete ones.

. . . AND IN BLACK AND WHITE

Now let us get at truth, truth about the "world" of sense and about the scientificness of our and Chinese science, in black and white. We think of these colors as perfect opposites, and because we do, it should be understood that they are not.

To jump to the general level which I am using colors to get to, no two things in the universe are perfect opposites. I speak here not of Polarity-*vs.*-Absoluteness but of either kind of difference. As the semanticist John Lyons has pointed out in his *Semantics: I,* what we think of as opposites are classified as having a common feature; to use the notation introduced above, the opposites X and Y-or-non-X have Z in common. For example, energy and matter have "basic entity of the universe" in common.* Being and non-Being have "logical universal ultimates" in common.** So, as Lyons goes on to point out, two things which are *not* kinds of the same "Z" may be more opposite than two things which are. For example (not his), non-Being may be less opposite to Being than is a Nobel prizewinner. Some might classify a Noble prizewinner as a kind of Being, and some might not; but, in either case, is something which shares with Being the property "logical universal ultimate" (non-Being) less like Being than something which does not (Noble prizewinner)? In short, as Chuang-Tzu observed: "For two things to be opposites they must already be the same."

So, there can be X and one Y with one Z in common and the same X and another Y with another Z in common, so that both Ys are non-Xs, X's "perfect opposites." So it is with black and white. They have in common "not-specifically-hued color" (their "Z") and are opposite in that one is minimally, the other maximally, luminous (dark-dim-*vs.*-light-bright). But *red* and white, in turn, have in common "most widespread color." Red is the most definite color; hence it is more in every other color than any other specifically hued color is in every other color. (It is "more-itself" than any other color; hence it has less of every other color and, reciprocally by the Polar-Completeness of colors, every other color has more of it, which is why they are less definite than it is.) And white, of course, being all hued colors, is more in every other color than any specifically hued color except red. In turn, whereas red is the most definite color, white, being all hues in equal proportion, is the least definite color. It follows that red is just as "perfectly opposite" to white as black is. But, in turn, red and *black* have no "Z" in common.*** And they are opposite in *several* ways—a

* Each is a kind of basic entity of the universe.

** Each is a kind of logical universal ultimate.

*** Black is not as widespread as red (and white), because, although it has all hues, its luminosity is much lower than other colors' on the average, than white's is higher than other colors' on the average.

matter of great importance to science—which will be explained in the next chapter. Here suffice it to say that, all told, red is the most *positive,* black the most *negative,* color. It follows that black and red are more different, more "perfectly opposite," than are white and red or black and white.* (Accordingly, when a "primitive" people distinguishes only two colors they are black and red or red-white; when only three, black, red, and white.)

We know, then, that the black/white opposition in a greater color-context is not perfect to begin with. Rather, like all opposites, they are *relative* opposites: opposites relative to their difference, in luminosity, and their common feature, "not-specifically-hued color."

Now let us investigate this matter of luminosity, that is, the black/white opposition. Is it Absolute or Polar? This will reveal some fascinatingly mystical features of modern physics and powerfully demonstrate how truly scientific the Polar-Complete paradigm is. What are the scientific facts about black and white?

First, the degree to which white is white or to which black is black depends entirely upon its relation to another color. Black gets blacker as the color next to it shifts from black through the increasingly luminous hued colors, or through "gray," to white, and white gets whiter as the color next to it shifts from white through the decreasingly luminous hued colors, or through "gray," to black. Likewise, the blackest black or whitest white is gray when seen by itself, without a contrasting color background. According to human eyes, then, neither black nor white has independent existence; they are a relative function of all other colors, including each other. Further, each contains its supposed exact opposite, for the same color is black in one context and gray in another (is both black and whitish black), or white in one context and gray in another (is both white and blackish white).

Second, it is impossible to produce an Absolute black or Absolute white surface color. This is because no material totally absorbs light and no material totally reflects it. It is fascinating to observe how Nature asserts its non-Absoluteness, its Polarity, as our scientists attempt to manipulate it into producing Absolute black and white surface colors. One method of producing near-Absolute black is to stack razor blades and beam light at the edges so that light is trapped between surfaces at very acute angles. This minimizes but does not eliminate

* See Berlin and Kay, cited in the *Bibliography.*

reflected light, because although the blade surfaces direct the light inward zigzagging toward the points of "V's" they form, some light directly hits the "point" and directly escapes, and, because the zigzagging light between two blades does not have an infinite distance to zigzag inward, some of it (the part not converted to heat) must reverse direction and zigzag outward. The distance between two blade edges and their intersection must be *infinite* for light to be totally trapped. In addition, if the light were trapped forever within the blades a modest input of it would heat them to incandescence: the blades would emit light. The other method is to beam light through a hole into a hollow sphere whose outer surface is white and inner surface is black. Since the inner surface is spherical, the light is reflected at all angles toward and off the inner surface—but some escapes, of course, through the hole (see Figure Four). The hole would have to be *both open and closed* for the light to be totally trapped inside. Further, the smaller the hole is the whiter the black becomes, because color contrasts weaken when the areas on which they occur are very small. Further, the smaller the hole the more parallel to the outer surface of the sphere is the light that escapes it, so that the white of the escaping light increasingly blurs the boundary within which the

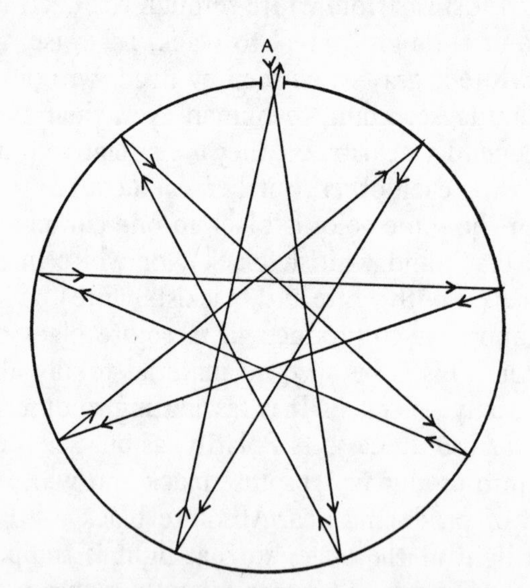

FIGURE FOUR: BLACK-BODY CONTAINER
Light-wave A is trapped while it is reflected by the inner surface of the hollow sphere, but inevitably it emerges through the pinhole through which it entered.

near-Absolute black occurs. Color contrasts are reduced when there is no "line" between the contrasting colors.

In other words, as one approaches Absolute black the conditions for producing it become weaker; Nature will not permit Absolutes: it is Polar-Complete. The two preceding light-swallowing models are called "black bodies." Whereas the empirical message they send is nothing other than the fact that black is in a Polar relation to white, our physicists have fantasized *perfect* black bodies (they call them "ideal black bodies") *and base all their theory about heat and light on the hypothetical behavior of such non-existent, impossible objects, speaking of them as though they were real.* Only through such mystical gymnastics can they preserve their Absolute-Fragmental paradigm; and their religious devotion to that paradigm is the reason for such mysticism. Calling the amount of light that inexorably escapes a *real* black body "negligible," they deny the most basic fact about Nature, its Polarity, and accordingly misunderstand Nature in the most basic way. "*A little error in the beginning, a great error in the end.*" (See Figure Five) ("There is no difference between a little error and a big error.") No error is little; no empirical datum is "negligible."

Third, black can be produced by the same colors that can produce white. Red, blue, and green in equal proportion yield white. Green, blue, and red in respectively smaller relative amounts yield black. This implies that black is nothing other than an inflection of white, or put otherwise, that black and white are variations on a same theme: opposites-yet-identicals.* This is just what a purely logical approach would imply. If black is white's (Polar or Absolute) opposite, it is non-white, and therefore non-*whatever-makes-white.* It is therefore non-blue plus non-green plus non-yellow plus non-red. But non-blue is reddish-yellow, non-green is bluish-red, non-yellow is reddish-blue, and non-red is bluish-green. Therefore black and white are identical. As follows from the fact that that conclusion is ridiculous when (Fragmentally) taken alone, there is a second logical conclusion that accompanies it: if black is non-white, it is the non-*sum of* whatever-makes-white. Hence it is blue, green, yellow, and red as separate, discrete things. But as we have understood, such Absolute-Fragmental colors do not exist. Therefore black is the total absence of all those colors. Chinese-style, without fear of paradox, we add the two logical conclusions together,

* Of course, empirically they *are* identical, in wave characteristics: neither has specific hue; and they *are* opposite, in luminosity.

FIGURE FIVE: "A LITTLE ERROR IN THE BEGINNING . . ."

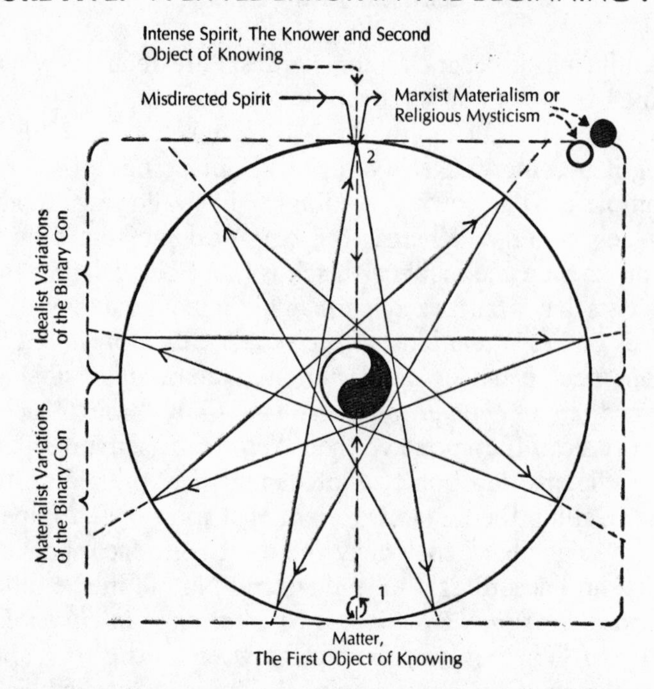

Key.

The vertical, broken-lined, two-way arrow represents the empirical approach to Nature and its Polar-Complete results. As a spirit-"ray" entering the upper hemisphere of ideas and the lower hemisphere of matter, first (1), it is directly reflected by, correctly knows, matter; second (2), it returns to its source to know spirit itself; third, Polar-Complete understanding (center), basically of Spirit-Matter, results.

The multiply reflected, zig-zagging arrow represents the non-empirical, Absolute-Fragmentally misdirected, approach to Nature. Entering the sphere at a slight angle to matter—"a little error," the misdirected spirit-"ray" impinges on matter to produce a theory of Materialist bias which, being in-Complete, "reflects" into an Idealist "opposite" to itself, which, being in-Complete, "reflects" into another Materialist "opposite" to itself; and so on. Until, after 2500 years, the truth-searching "ray" exits from the scientific sphere where it entered, Absolute-Fragmentally polarized—"a great error in the end." One pole is Marxist Materialism, by which the Binary zig-zag itself is elevated to the level of "Natural Law," to destroy the spirit which, misdirected, produced it. The other pole is religious mysticism, the respite of Free World "hard" scientists who realize they have come full circle.

and get: black is *both* identical to *and* exactly opposite to white, which is to say, the Polar complement of white. (*Complete* logic *does* reflect "mere appearances.")

Fourth, Absolute white cannot be produced even as a color that is not a reflected one, a color "in space." Absolute white would have no aspect of itself absorbed by a medium, and this requires that the light be transmitted directly to the eye through a perfect vacuum; but perfect vacuums are impossible. Further, the original light would have to be perfectly balanced in hues. As indicated by the facts that no Natural source of such light has been found and no such light has been synthesized, Nature refuses to provide conditions for Absolute white.

Fifth and last, it is impossible to produce Absolute black *even by eliminating light as a source*. The following manifold explanation will beam us into the lethal inner sanctum of Western science. At first, it seems that if one were to construct a light-excluding chamber, inside, the blackness would be total. Now, where black and all other colors are sensations, the only way to find out if that is so is to locate an observer inside the chamber. The observer, if entering it from a brightly lit room, will at first perceive what he may call total black, that is, perceive no light at all. But the black becomes whiter and whiter until it levels off at middle gray (because without white there can be no black). Further, soon the observer perceives light within the initially perceived black. Our physicists variously call it "light chaos," "light dust," "self-light," "intrinsic light," "idioretinal light," "intrinsic gray," "intrinsic brightness," or, as is preferred because it shrugs off that light more strongly than the others, the way "negligible" shrugs off the light that escapes a "black" body, "retinal *noise*." In other words, they suppose that the light exists only in the eye of the beholder, and is not natural light, but "noise"—something "subjective" that doesn't fit into their paradigm. But even if, as our scientists tentatively believe, the "intrinsic light" is entirely due to "firings" of the retinal nerves, the fact remains that Absolute black is not perceived: there is light, human-produced light, in the room. Again, Nature refuses to permit Absolute black: when conditions exterior to the observer begin to approach it, Nature produces light.

If Absolute black is defined as a perception, as a function of a relation between a perceiver and an object of perception, then, even in a light-excluding chamber there can be no Absolute black. But what if we define black "objectively," without regard for an observer, as the total absence of light? If the observer is removed from our chamber, is the chamber then lightless?

We can answer this question in terms of our own modern physical data and theories, yet, again, we will reach a conclusion that supports the Polar-Complete paradigm and undermines our own. Our dark chamber is bombarded by wave-particles which are transforms of wave-particles released by the sun, the chief source of our light, and by "cosmic rays," whose source is unknown. Such wave-particles excite the atoms in the walls of the chamber, which in turn excite the atoms in the atmosphere of the chamber; and, depending on the material of which the walls are made and the force of the wave-particles from the sun, some wave-particles directly excite atoms in the atmosphere of the chamber. Excited atoms' electrons return to their non-excited, "ground," states and release electromagnetic radiation, photons, one variety of which is light (visible light). Could we prevent this "*objective* noise"? Modern physics cannot answer this question with certainty, but it indicates that we probably could not, for it appears that each material selectively excludes kinds of wave-particles and lets others through, and that some particles, neutrinos at least, can penetrate any material. Further, the full range of wave-particles and their capacities to penetrate matter are not known. But let us assume that a thick-walled chamber of several kinds of dense, non-radioactive material would absorb all, and all types of, bombarding wave-particles. Even under such conditions, according to modern physical findings there will still be some electromagnetic radiation in the chamber, because it is unnatural for all the atoms in a given space—here, the atoms in the chamber's atmosphere—to be at their "ground" state. That a group of atoms will all be at ground state, in "perfect equilibrium," and therefore not emit photons, is only a statistical probability, not an Absolute. Specifically, not only does it appear that in an aggregate of atoms there is always at least one which will emit light, but, as Zukav observes in *The Dancing Wu Li Masters,* there are cases where *without any energetic input and in a vacuum,* three wave-particles simply emerge and then transform into light. Light is *intrinsic* to any set of conditions, even a "vacuum." The chamber contains its own light.

Let us then make the conditions of the chamber even more extreme. We can vacuum-pump the atmosphere out of the chamber to reduce the number of atoms that might emit photons, but, of course, since Nature does not permit an Absolute vacuum, some will remain.

We can also lower the temperature in the chamber as far as possible —never to Absolute zero—to further deprive the chamber of energy, but there will remain some energy. The more we make the interior of the chamber different from what is outside it, the more strongly will Nature exert pressure toward a balance between its interior and its exterior. There is a sub-Absolute limit to the suppression of electromagnetic activity in our chamber.

What is more important than the final conclusion about the "lightlessness" of our chamber is that as we have attempted to create conditions for an Absolute (Absolute black), we have created the conditions for *death*. It is such lethal and extraordinary conditions that produce data most suitable to our physics, and our physics, in turn, is of such a nature as to favor creating such lethal and extraordinary conditions. Our physics—all our sciences—are based on and favor extreme, extraordinary, bizarre conditions under which human beings would—and do—die, or under which they would—and do—become ill. Our "scientific" paradigm is based on the extreme, extraordinary, and deathly, because it seeks to establish the Absolute-Fragmental. Accordingly, it is as different as could be from universalistic or Human science. Our one-sided and side-affecting tendency is to restrict our empirical science to extraordinary-extreme conditions and then to generalize on the basis of such *deathly* findings to *vital* ordinary-moderate conditions. Our *general* "natural laws" and our applications of these laws to our social and greater Natural environment derive from *exceptional* conditions. The result is that our science is both intellectually and morally false. Let us now draw a final conclusion from this specific example of an attempt to coerce Nature into producing an Absolute—here, blackness.

Taking all the preceding data into account, it becomes fairly obvious that the occurrence of electromagnetic radiation in our deathly Western-scientific chamber cannot be Absolutely precluded. There *may* be none within a (short) period of time, but to restrict the conditions for Absolute black to a period of time is to beg the question by imposing an Absolute-Fragmental concept (a period of time) on a natural space-time continuum.* In turn, if one wishes to accept such a

* According to present Western physical theory, there is yet another source of intrinsic light: proton decay. In probabilistic terms, if one were to observe 1.5 tons of hydrogen for one year, the chance of observing one proton decay would be 50 percent (*Omni*, November 1980). To test the hypothesis, underground laboratories identical in principle to the one I have imagined here are

condition, one must also accept that Absolute black occurs only as a secondary, special case of non-Absolute black: that Polar black is *primary* and Absolute black is a *derivative* of it. This empirically stands our Absolute-Fragmental paradigm on its head.

We must now ask what kind of electromagnetic radiation there is in our dark, airless, frozen chamber. Is it (visible) light? The chamber itself minimizes energy, so the intrinsic or externally caused radiations of its own atoms are of low frequency, are relatively subtle. According to modern physical definition, the radiation is not (visible) light, for visible light radiation is medial, not low, on the electromagnetic energy-spectrum. However, just as the wave-frequency distinctions between colors within that range are non-Absolute, Polar, the distinctions within the total electromagnetic range are non-Absolute, Polar. So just as yellow is also red, violet, and blue, low-frequency radiation (or high-frequency radiation) is also (visible) light. (Indirectly recognizing this Polar reality are the terms "photon," which focally means "quantum of light-radiation" but broadly means "quantum of radiation," and "light," which focally means "visible light" but broadly means "electromagnetic radiation.") There being millions of *almost*-dead atoms in the chamber, it is conceivable that at least one photon of visible-light-within-non-visible-light is present. In addition, there is the spontaneous emergence, even in a "vacuum," of wavicles converting to light as suggested by Zukav. The physicists Josephson, Sarfatti, and Herbert have all independently speculated that human eyes might detect such events. Since more than one photon of light is involved, one should conclude that they definitely can: the incredibly sensitive human eye under dark-adapted conditions perceives light that is a mere single photon in strength—one trillionth of a candle-flame's worth.

If a naked-eyed human could be in the dark-cold-airless chamber, then, light would eventually be seen. Of course, one could place a human in that chamber only if his/her eyes were covered by a thick, resilient, vacuum-resisting translucent surface connected to the rest of a sealed suit that retains air and heat—so there can be no direct empir-

---

being built, by Americans, Europeans, and Japanese, in gold mines and under mountains. They are gigantic concavities relatively shielded by earth crust from cosmic rays, filled with water and lined with light sensors, designed to restrict the spontaneous emission of light to the variety resulting from proton decay. The predicted effect is the eerie blue-violet of nuclear-reactor cores—light which, as will be explained, is extremely *yin.*

ical confirmation. If the chamber were one in which a human with unobstructed eyes could survive, however—a chamber with air and temperate heat—there would be much more wave-particle activity, probably several photons of visible light, and, therefore, easily and frequently perceived light. How much, then, of the "retinal noise" produced by nerves in the eye is really received light when a human observer is placed in a "lightless" room? Is it not that the "intrinsic light" our physicists regard as "noise" is a Polar phenomenon: human light and object light, relatively-subject and relatively-object light, at once? Contrary to the most basic assumption of our 2500-year-old paradigm, our senses do *not* deceive us.

We have just seen an example of how it is only on the basis of extreme and therefore lethal conditions that Western science can question the Chinese scientific proposition that Form is not Absolute but Polar, not Fragmental but Complete, and that even on the basis of such conditions it can question that proposition only weakly and uncertainly, for even such conditions tend to disconfirm more than confirm the Absolute-Fragmental paradigm. Broadly put, if this example of color can be taken to exemplify a general difference between the two modes of knowledge, Western science is pseudo-empirically based on exceptional (Fragmental) data. It is indeed, like Yaqui sorcery, a half-science. Chinese science is genuine science, empirically based on all kinds of (Complete), and chiefly ordinary, non-extreme, data. Near-exceptions to the rule of Polarity that Western science narrowly selects and whose non-Absolute aspects it labels "negligible," most strongly prove that rule.

The side-effects of Polar-Complete data in the lethal labs of the Western scientists are only a special case of the resistance of energy, ultimately, human life, to the effects of their paradigm. Recognizing that their mental Narcissism had minimal survival-potential in natural society, the architects of our scientific paradigm widened their scope to all of human culture, to unnaturally alter it so that their own, parasitic, survival would be possible. Some of the cultural alterations they made were and are preoccupation with "God," a super-, therefore, unnatural concept; colonialist fervor, which is unnatural in that it is the most non-reciprocal, one-sided, relation between societies; a separation of families from society, which de-humanizes the latter and makes it susceptible to unfeeling utopian experimentation and the "Academic" association of male sex with physical superiority and Divinity, a sacredized male Narcissism, which is unnatural in that it forthrightly denies the natural, Polar relation between male and female and the natural valuative equality of the sexes. These and other Absolute and/or Fragmental, one-sided, cultural patterns established by our pseudo-scientific predecessors plainly symptomize intellectual Narcissism, now proudly called "objectivity" or "detachment," by which the primitive Absolute-Fragmental mode of thought is imposed on Nature, and by which Nature's own, Polar-Complete feedback is called "a negligible side-effect." Instead of perceiving Nature and intellectually reflecting it, they pseudo-empirically explored it "through" a mirror. Our scientists have been constrained by this Narcissist intellectual tradition to this day. They have magnified that mental mirror with instruments designed (always unsuccessfully) to reduce natural data to an Absolute-Fragmental state ("pure" hues, "ultimately small" particles, "absolute vacuums"). The architects of the Binary Con have condemned their successors to "mask" and "degrade" the objects of investigation, including human beings, rather than to yield to the facts, unite with Nature, and develop a scientific paradigm which, like the Chinese one, works with, instead of against, Nature to assure the survival and maximal material-social-ecological comfort of human beings.

Why is it hard for us to recognize and admit the Polarity of black and white? Why do we naturally persist in conceiving them as Absolute opposites? To answer these questions takes us, in one leap, from physical to *social* science. My *shih-fu* answered these questions for me as follows. These colors are associated with *personal* Absolute limits. Black is the darkness of night as compared to day, of Winter shadows as compared to Summer sun, and of unconsciousness. White is the luminosity of day as compared to night, of Summer sun as compared to Winter shadows, and of consciousness—en*light*enment. Hence, black is the color of death and white* is the color of life, our own ultimate limits as individuals. We cease to know when we die, and never stop knowing when we live (except, of course, for the little semi-deaths of dreamless sleep). Hence life itself is associated in the scientifically naïve mind with Absolutes of which black and white are representative, symbolic. "White" life is bound by "black" death. But those associations are subjective and egoistic. People living before one's birth accumulated the knowledge that was passed on to become an aspect of one's consciousness, which knowledge is shared by many others during one's life and then passed on to children. And, as is more obvious,

* And red, which is also black's opposite, as explained.

one's living form and initial life-energy are one with one's parents. From such a mature perspective it is obvious that even life and death are not Absolute limits of one's self. Again, it is a question of *perspective*—Fragmental or Complete. Just as we see "the" primary colors as Absolutes (only) at the most primitive level of discrimination, we see black and white as Absolutes (only) from the purely egoistic-individual one.

Now let us take several leaps, from "Earth" to "Heaven."

CHAPTER

4

The Dragon Stretches from Earth to Heaven

You hear the piping of men but you do not hear the piping of earth. Or if you hear the piping of earth, you do not hear the piping of heaven.

—CHUANG-TZU

[T]*hrough this prism of theirs they obtain from the white-ray only what are called "negative colored rays," and in order to understand any other cosmic phenomena connected with the transitory changes of this white-ray, they must obligatorily have its what are called "positive colored rays." Your contemporary favorites, however, imagine that the colored rays . . . are just those same "positive rays" which the great scientists obtained; and according to their naïveté they think that the, as they call it, "spectrum" which they obtain from the white-rays gives just the order of the arisings of the rays in which they issue from their sources. . . .* [O]*ne can only utter the expression often used by them themselves, "To hell with them."*

—BEELZEBUB, to his grandson
(G. Gurdjieff's *All and Everything*)

FIRE IN THE ICE

Now that I have concretely defined the Absolute-Fragmental and Polar-Complete paradigms I switch to less cumbersome terms for them. Our paradigm is basically a *mental* construct which was fixated upon by the Narcissist-elitist personality of its architects and then imposed upon Nature and Humans, including the majority of scientists, who neither invented nor reinstated it. So I will simply call it "the Mentalist" paradigm. The term is appropriate not only because it denotes the basic intellectual nature of the paradigm but also because it connotes the attitude and personality behind it: "Mentalist" calls forth mysticism, sleight of hand, charlatanism, confidence-gaming. In turn, I am simply going to call the Polar-Complete paradigm "the Tao" (pronounced *Dao!*), because there is no English (or Latin or Greek or Semitic) term that conveys the proper denotations or connotations, and because its originators, the Chinese savants, have the right to name it.

Now let us return to the Chinese, scientific, proposition that *all* Form, like color, is Polar-Complete—a proposition encoded by the term *se* ("color/Form"). In this way we will open wide the door to genuine (originally, Chinese) science. Insofar as is practically possible, I now show that all Form, or the universal pattern, *is* Polar-Complete. Then I will begin in earnest the promised trip into an alternative scientific reality (really, reality).

Let us think of all the phenomena in the universe as a set of levels and go through these levels one by one, from the "lowest," where "low" is defined as having more to do with sense than with thought, to the "highest," where "high" is defined as having more to do with thought than with sense. As we ascend through these levels it will become obvious that the levels, themselves, are Polar-Complete aspects of a manifold continuum.

This chapter will focus on the "lowest" level, but empirically derives some facts about energy and matter which begin to tie together physics, chemistry, biology, psychology, and social science. This will serve to introduce the reader to an alternative mode of thought, based on Polar-Complete concepts, and to illustrate the real meaning of "unified science."

To start: light and colors are but a special case of radiation (*feng-ch'i*), which, in turn, is the Active aspect of Energy-Matter. Since Einstein, we have begun to understand (as did Heraclitus and Lao-Tzu) that Energy-Matter itself is a Polar continuum. (See, for example, Capra's *The Tao of Physics* and Zukav's *The Dancing Wu Li Masters.*) That is, Energy and Matter are two, Polar aspects of a same thing, are opposite yet also identical. So the Energetic-Material aspect of the whole universe, as best we know it, is Polar-Complete. Trust of the senses and Nature made it possible for Chinese savants to understand it 5000 years ago. This understanding arose in the minds of physicists only in the twentieth century. Modern physics is giving us a picture of the inanimate aspect of the universe as a Polar-Complete continuum of Space-Time and Energy-Matter. Let us now see how, at the "lowest" of our levels, that picture is naturally produced through pure empiricism, simple observation—how a truly scientific paradigm can, and did, produce this picture immediately.

The natural properties that are sensed with relative immediacy (our "lowest" level) may be listed in many different ways, for two reasons. One is that our words for them greatly overlap in meaning. The other is that different peoples with different languages "cut reality up" in different ways. However, there are certain natural properties that tend to be named by all peoples, and which also serve to define many other natural properties of more complex natures. These properties may be thought of as universal primitives (basic units) of sense meaning.* The properties are: *colors, light-dark, bright-dim, hot-cold, wet-dry, straight-curved, dense-rare, loud-soft, sweet-bitter* and other taste/odors, *hard-soft, rough-smooth,* and *rigid-flexible.*

Each of these basic sensory dimensions is Polar-Complete. Each, like colors, is an aspect of a whole that has no identity or existence of its own without relation to its opposite, and that blends into its opposite.

* It is not my purpose to provide an exhaustive list or to document its validity as a list of universal primitives with cross-cultural evidence from anthropology, but to provide a list that will work fairly well cross-culturally and that covers enough of the directly sensed aspect of the universe to make the present case for Polarity powerful enough to support the central claim of this book.

Each of these dimensions behaves like a color spectrum, is Polar-Complete in the way that the color spectrum is. For example, extreme heat numbs, like cold, and extreme cold, as college fraternity initiation rituals prove when ice cubes are applied to the flesh of blind-folded initiates, produces the sensation of burning heat. (This is like the overlap of violet and red.) To reject such data as "subjective," as our scientists do, is of course to reject the observer, and thus to reject empiricism.* What makes a physicist's visual perception of phenomena in a cloud chamber more "objective" than his perception of heat during his initiation into college fraternity? There is no rational answer to that question. Either one depends on the human senses or one does not. If one does not, one is not an empirical scientist.

Of the preceding listed pairs of opposites, the only one whose Polar-Completeness is not fairly obvious is *straight-curved.* This is because we think of them, due to our Pythagorean tradition, as a perfectly straight or a perfectly curved line (the latter of which is a circle), whereas these are forms which neither occur naturally nor can be physically produced to perfection by humans, but can only be imagined. And it is for that reason that they and other mathematical constructs have been central to Western science for 2500 years, to the exclusion of obviously Polar-Complete sense-data such as hot-cold, wet-dry, and so on. Let us nevertheless deal with these imagined forms, noting that we have jumped from the "low," sensory level to the "high," abstract-intellectual level of human thought, so that Polarity is being recognized at the latter "level." The simplest proof, of which there are many variants, uses the perfectly curved, circular, line and the perfectly straight line. The circle is an "infinite series of points equidistant from one point." It can therefore be imaginatively drawn as an infinite-sided polygon. "Infinite" means "of greatest number and constantly increasing in number." Hence, mathematically, the circle is

* One may be fooled into *thinking* that the object is not ice but something on fire; neither is this perception nor does it refer to the phenomenon in question: heat. Perception is not fooled: the sudden cooling shares properties with sudden burning, and there *is* heat, because the interaction of ice and flesh which registers heat is as valid or more valid an experimental condition than is the interaction between a flame and a thermometer which registers heat. There is human heat and thermometer-heat, but this distinction is not between heat and something else. It has to do with whether heat is defined as something directly sensed, or as something sensed through the mediation of an inanimate object, or both.

an inconceivably great and increasing number of polygons with inconceivably great and increasing numbers of sides, superimposed. And they are all made of straight lines, because never do the lines become points (there is no end to smallness) (Figure Six). In other words, a circle is made out of straight lines: the concept of "infinity" does not escape this; indeed, it obliges this conclusion. In turn, the straight line is defined as "an infinite number of points along the shortest distance between two points." But because points are the smallest and constantly yet smaller circles, this gives a shape like a taut string of adjacent beads; it is all curves. Hence the straight line is based on curves, just as the perfect curve is based on straight lines. The *logic* of straight and curved—and perfectly straight and curved are nothing other than logical matter—is Polar-Complete. (Heaven is round and Earth is square, and there is Heaven in Earth and Earth in Heaven.)

Note that the straight lines in the circular one are internal to the shape, whereas the curved lines in the straight one are the external shape. Likewise, where "Heaven" (all of Nature, the universe) is "curved" and "Earth" (what is immediate, a Fragment of the whole) is "straight," Earth is internal to Heaven. In Chinese terms, then, this

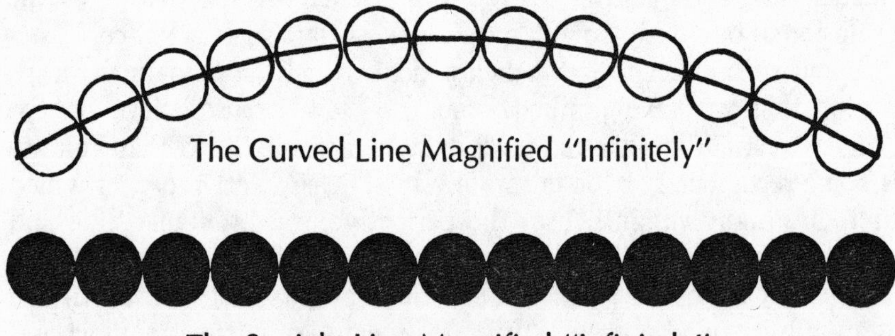

FIGURE SIX: THE CURVED AND THE STRAIGHT

Key. Lines join the "points" at which the points in the *curved* line conjoin. They are the *straight* lines that form an "infinite-sided" polygon.

Infinitely magnified, the points in a *straight* line form a "string" of *circles,* or spheres.

Straight is predicated on curved, and curved is predicated on straight. By definition and as images, straight and curved necessarily entail, and therefore in one respect are, each other.

book is making the point that Western science has developed great knowledge of "Earth" but provides no knowledge of "Heaven." It is always at an angle to Nature, much as the straight lines make angles within the curved one. "Curved" in this Chinese sense, of course, means "opposite yet identical," as in the case of straight and curved lines—Polar-Complete.

I leave it to the reader to sense and think about all the other dimensions I have listed. It will be obvious, if it isn't already, that each is Polar-Complete in the sense that colors are.

We can now understand as a scientific fact of great if not universal generality a central Chinese scientific concept, or basic unit of *Logos,** called the Quadripartite Image (*Szu-Hsiang*), first mentioned in the *I Ching* commentaries and Lao-Tzu's *Tao-Te Ching* (*Canon of the Way-and-Virtue*). This image, in turn, greatly helps to make rational sense of the Polar-Complete reality we have begun to confront. The Quadripartite Image, or, for short, Quadrimage, is simply an image of Polar-Completeness, to which a convenient visual form may be given (Figure Seven, p. 176). The two sides of the vertically divided field are any two Polar opposites, such as black and white, cold and hot, straight and curved. Each side is then divided into two aspects: major (*t'ai*) and minor (*shao*). The major aspect is what a given pole chiefly is; for example, black. The minor aspect is what that given pole *secondarily* is: its major aspect's opposite; for example, white. Further, the minor aspect of a pole is the opposite of that pole which emerges as that pole attains its extreme. For example, extreme cold is sensed as burning and extreme heat is sensed as numbing; the natural opposite of fire, heavy metal, (radioactively) burns; air and water naturally include each other; extreme fever causes chills; extreme wave activity approaches perfect quietude, that is, energy "becomes" matter, as one goes from red light to violet rays to atomic particles on the continuum of the electromagnetic spectrum; ampicillin, used to treat infections, has the side-effect of infections; and so on. This is called *fu*:reversion. We are already quite familiar with the Quadrimage, then; it is just that we are giving it a visual form.

The epistemic correlation (the theory-sense fit) of the Quadrimage with sensory reality—the *factuality* of the Quadrimage—may be made clearer by drawing it with one level, in linear form, to represent a con-

* Logic in the true and, in the West, originally Heraclitean, sense (see the Appendix).

FIGURE SEVEN: THE QUADRIMAGE

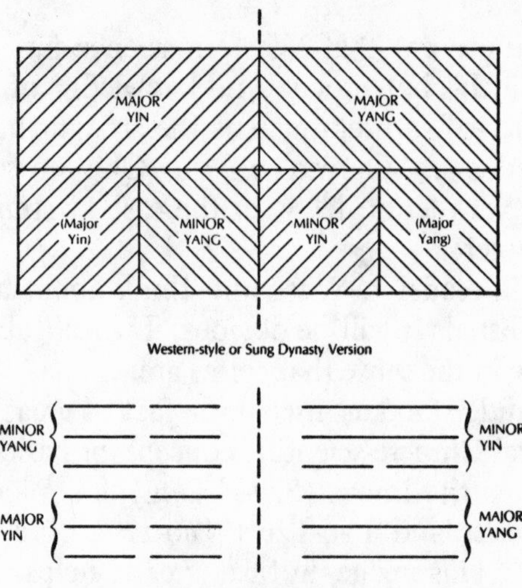

I Ching, Hexagrammatic Version*

*The lower left *tri*gram, Major Yin, represents Earth, Matter, "Mother." The lower right *tri*gram, Major Yang, represents Sky, Movement, "Father." The upper left *tri*gram, minor yang, represents lightning, material movement, "first son"; the upper right *tri*gram, minor yin, represents wind-as-a-gathering-force, moving matter, "first daughter." The left, Yin, *hex*agram represents Matter-beginning-to-be-energized, relatively-material Energy-Matter; the right, Yang, *hex*agram represents Energy-beginning-to-be materialized, relatively-energetic Energy-Matter.

tinuum as we usually conceive of *and perceive* one—that is like, or is, a color-spectrum (see Figure Eight, p. 177). The minor aspects of two poles, then, are the aspects in and through which the two poles meet as one, in and through which the two poles are ultimately identical.

It is these minor "middle grounds" that the Western, Absolute-Fragmental paradigm mystically, anti-empirically, excludes. Our science is based on a Binary Image which corresponds to two major aspects—unnatural, non-empirical, "halves"—of any two Polar opposites: an Absolutized Fragment of the empirical Quadrimage (Figure Ten, p. 243). This was stated by Parmenides as the rational but non-empirical proposition: "What a thing is, it is not also not," and called by Aristotle "the Law of the Excluded Middle" and "the Law of Non-Contradiction" (of Non-Paradox). The excluded middle grounds are then called a "higher-ordered unity." For example, there is "+ charge," "− charge," and their so-called unity "charge," as though "charge" can mean anything that is not also + and −, as though it were something in but other than + charge and − charge. This amputation of the Polar-Complete facts is nothing other than an arbitrary proposition with appeal for those who have amputated themselves from sensory reality. I choose that surgical metaphor for good reason. The next section leads to the correlation of the Quadrimage and its hexagrammatic alternative, *T'ai* (Figure Nine), with harmony and life, and that of the anti-Quadrimage and its hexagrammatic alternative, *P'i* (Figure Ten), with disharmony and death.

Not only is each "low," immediately perceived, dimension Polar-Complete, basically a Quadrimage relation. The many "low" dimensions themselves make up one Polar-Complete continuum. Each *dimension* is like a color on a greater spectrum, because the distinctions between one dimension and another are not Absolute. For example,

FIGURE EIGHT: THE QUADRIMAGE AS ONE LEVEL (THE PEACE-HEXAGRAM, *T'AI*)

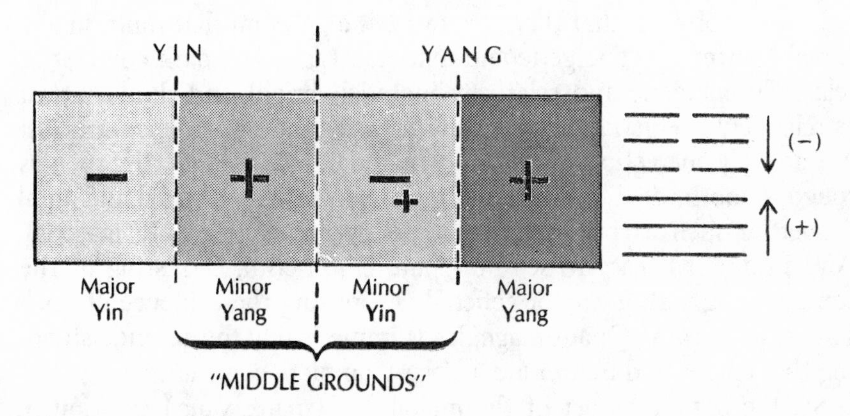

KEY. The hexagram corresponds to the Quadrimage. It is *t'ai*, the image of harmony, fundamental unity, living.

FIGURE EIGHT A: THE WESTERN ANTI-QUADRIMAGE (THE CATASTROPHE-HEXAGRAM, *P'I*)

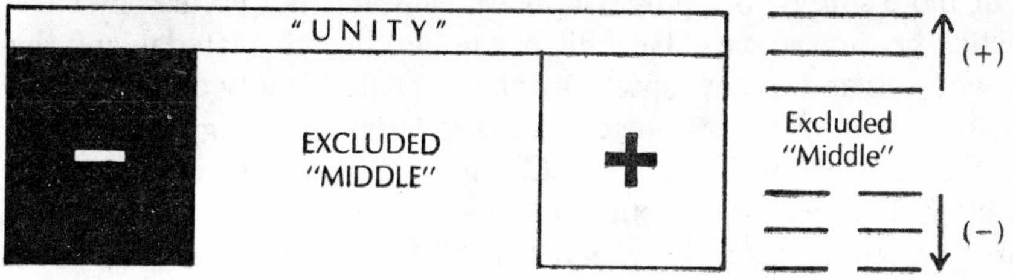

KEY. ⁻+ stands for "the minus (yin) that is in the plus (yang)"; +₋ stands for "the plus (yang) that is in the minus (yin)." The hexagram exactly corresponds to the anti-Quadrimage; it is *P'i*, the catastrophic image of fundamental disharmony, separation, dying.

there is a light in bright and a bright in light, a dark in dim and a dim in dark, a light (as opposed to dark) in rare (diffuse) and a rare in light, a dark in dense and a dense in dark, a hot in rare (dispersing, radiating) and a cold in dense. There is a rough-or-smooth and a rigid-or-flexible in dense-or-rare and in tastes and odors; there are tastes in odors and odors in tastes; there is a straight in rough and in rigid, and a curved in smooth and in flexible. There is a dense in loud and a rare in soft. Our terms for opposite, immediately perceived properties actually refer to ultimately indistinguishable dimensions on a multidimensional Polar-Complete spectrum. It is for this reason that the definitions of "levels" and their "natural properties"—the ways that this manifold spectrum is cut up into as-if-Absolute-Fragmental pieces, vary from one society's culture to another's.

It is also obvious that there are two *basic* poles on that multidimensional continuum: energetic and material. Light and dark contrast as relatively energetic and relatively material; bright and dim contrast as relatively energetic and relatively material; rare and dense contrast as relatively energetic and relatively material; tactile properties such as rough/smooth, rigid/flexible, and tastes contrast with visual and aural properties such as bright/dim and loud/soft as relatively material. And so on. It is easy to see how pure empiricism, a trusting of the senses, along with a way of generally expressing the evidence of one's senses, such as the Quadrimage, leads immediately to the understanding that energy and matter are a Polar-Complete continuum.

So, the basic member of the major-plus (major yang) position in the Quadrimage is energy, the basic member of the major-minus (major yin) position is matter, the basic member of the minor-minus (minor yin) position is highly material energy, and the basic member of the minor-plus (minor yang) position is highly energetic matter. Put more simply, the plus pole is basically but not exclusively energetic, the minus pole is basically but not exclusively material, and the minor plus and minus aspects are the ultimate identity of energy and matter. The relatively concrete Chinese terms for energy and matter in these senses are *ch'i* and *shui*. *Ch'i* also stands for the unity, energy-matter. The relatively abstract Chinese terms for energy and matter are *yang* and *yin*. *Yin-yang*, and *t'ai-ch'i*:the extreme pole, also stand for the unity, energy-matter. All those pairs of terms have yet fuller meanings, but it is useful to identify them at first in the concrete terms of one's immediate perceptions. (Where else to start to *know*?)

A RED SHIFT

Now I will indicate how the Chinese physical sub-paradigm that derives from the preceding *empirical* approach to sense-objects is more scientific than ours. At the same time, I will introduce the reader to the Chinese scientific mode of thought as applied to a natural-phenomenal manifold—something much closer to what actually goes on in the Chinese savants' mind than the simple and highly abstract explanations I have used so far. In short, here is where the departure from our paradigm and mode of thought, the Trip into another reality-galaxy that I promised, starts. I begin by following our physics into one of its impasses. Then I map the territory beyond that impasse in Chinese terms, for which there are no Western correlates. At the end of this Trip lies a predictive scientific theory which to the Western scientific mind is inconceivably powerful, that is, inconceivably inclusive, of phenomena of several varieties, each of which occupies the attention of one or another isolated Fragment of our science. The reader will travel most naturally if he or she attempts to use the Chinese way of thought to understand the Chinese system of knowing. It is like switching from an automobile to a spaceship, when it becomes obvious that the automobile will get to the launch pad but the destination is another galaxy—in this case a galaxy which was really *here* all along.*

Recognizing that various sensed properties are somehow interrelated, our physicists posit correlations between colors and "levels of energy," correlations central to all of modern physics. These correlations are made within the narrow confines of extreme energy-levels in inanimate phenomena such as heated metals and burning gases—basically, fires. Now, although our physicists speak of *color*:energy correlations, their "color" refers to the wave-properties of phenomena that have colors. Hence, actually they never take *colors* into account at all.

* Lao-Tzu said that one could learn everything from one's backyard, never leaving it. Here is one way of demonstrating that.

This bit of sleight of hand is well worth understanding, because it is at the core of our physical science, and because its genuinely scientific alternative exists in this gap between Western theory and empirical reality. Our scientists understand that the qualitative difference between colors and other electromagnetic phenomena (microwave heat, x-rays, and so on) is "purely subjective"—a result of the fact that we can see colors but not other electromagnetic phenomena. On that basis they "objectively" reduce all electromagnetic phenomena to wave-characteristics, expressed in definite, mathematical form. But, as we have seen, those waves are in reality Polar-Complete, not subject to counting and measurement; and what they actually are, in turn, is immediately accessible to the senses as color. The reasonable assumption to make, then, is that *non*-visible electromagnetic phenomena are also Polar-Complete. They are, and our scientists already know it. Nevertheless our physicists work with their mathematical wave-characteristics to formulate "natural-lawful" color:energy correlations, and the result, naturally, is several dead ends and theoretical self-contradictions.

The correlation is as follows. Where there is red there is long wavelength and low wave frequency; where there is violet there is short wavelength and high wave frequency. Long-slow waves (red) correlate with low "energy-levels" and short-fast waves (violet) correlate with high "energy-levels." This correlation not only ignores color itself but also is inComplete. Since white cannot be characterized by a wave-range, its association with a certain "energy-level"—the one between green-hot and yellow-hot—cannot be accounted for by this correlation. Indeed, the underlying assumption that discrete energy-levels produce specific wave-ranges is already undermined by the fact that white has no specific wave-range. To be sure, this hole in the system can be patched. By assigning it the *average* wavelength of its total range, a mathematical fiction, white can be located on the wavelength continuum between yellow and green, where it occurs on the natural color-spectrum. This is consistent with the understanding that white-hot things are intermediate in temperature (energy-level) to yellow-hot and green-hot things (gaseous torch-flames, for example). The fact that white is an exception to the assumption of a correlation between *specific* wave-properties and discrete energy-levels, however, is obscured by this artificial, mathematical assignment to it of specific wave-properties.

Indeed, this "treatment" of white is only a special case of an equally mystical treatment of all the other color-energy correspondences. The basic assumption is that a certain wavelength-frequency range correlates with a certain "energy-level," or temperature. On this assumption, white, then, should have all temperatures within the visible spectrum, not a restricted range of temperatures, because it is a "synthesis" of all the hued colors. Of course, it does: what the thermometer records as a restricted range is the *net* effect of all temperatures of hued colors, in equal proportion. This is consistent with the basic assumption, but further consideration shows it to be fundamentally false. As documented above and as our senses tell us, every other color is also every other hued color. And every other color-temperature is also every other color-temperature. What distinguishes one color: heat variable from another, then, is *not* a wave-range but *the relative proportions* of every color in the color in question. Of course, this more sophisticated understanding is implicit in the Western correlations. But as we have seen, these proportions are not a simple geometrical or quantifiable matter, for there are colors within colors, whole spectra within whole spectra, in an infinite series. Quantitative expression of dominant or complex wave-ranges and of average or "netted" temperatures serves to predict with great quantitative accuracy, but it fundamentally differs from, ignores, and obscures the reality of the matter. Both the temperatures, be they unitary or composite, and the wave-characteristics, be they of the dominant waves or, in a more sophisticated approach, of the proportions of wave-ranges in a color, are illusions. Not only the inclusion of white in the Western color:energy correlations, but also the remainder of the correlations are "a patch-job."

Black is also outside the mathematical wave-characteristics of that correlation. The (ideal) black of our physicists has no wave-properties at all. Its wavelength, therefore, is zero, and where energy increases as wavelength decreases, this (wrongly) predicts that black-hot things have the *highest* energy-levels. In turn, black's wave frequency is zero, and where energy decreases as wave frequency decreases, this predicts that black-hot things have the *lowest* (zero) energy-levels. The latter is (ideally) true, that is, of pre-incandescent matter, but the implication, from black's zero wave frequency, that black-hot things also have maximally high energy, remains. Further, *real* black, since it is never Absolute, has the wave-properties of white. The aspect of black that does have wave-properties is the aspect of it that is white, luminous with balanced hues. Black differs from white not in wavelength or wave frequency but intensity, a dimension our physicists deal with independently of this correlation. That is, they recognize two dimensions, wave-character and luminous intensity, but black is arbitrarily excluded from the former one because they cannot handle it. In this way they totally obscure the fact that it has a wave-character, which, because black-hot is far from white-hot, perfectly violates their basic assumption: that dominant or averaged wave-range is a universal index of heat, or "energy-level."

There are further, simply empirical, problems with those correlations. For example, to humans the red-hot coals of a wood fire are much hotter than the violet flames over those coals, as anyone who wishes to test this fact through his or her sense of touch will confirm. Likewise, to the eye, aesthetically, red and yellow are hot colors, green and blue are cold colors, and violet, because of the red in its blue, is ambivalent.* Since our science denies human sensing, it simply eliminates such human data as "subjective and negligible." Because of the difficulty in taking the temperatures of different colored things which are adjacent or intermixed, such as the red coal and the violet flame, it bases its correlations on the surface temperatures of a black body, (for example, a wood stove) which heats up to one color at a time, ignoring the complexly colored phenomena inside it. And it ignores the facts that the thermometer, a senseless thing, must give much less information about heat than does human touch and that the thermometer, in turn, is perceived by human vision. Simply put, there are two "heats," one defined by instruments, the other defined by human-sensing. For no "reason" other than anti-empirical and anti-human denial of human sensing and the anti-scientific requirement that the data be a selected, small, and simplified fragment of the whole which is under investigation, our physicists "know" heat through human *vision mediated* by expanding metal instead of the *direct* human *touch* according to which heat was naturally defined, in the first place. The conclusion is that green-blue-violet-hot is much hotter than yellow-red-hot. The color-heat complexity of the fire internal to the black body is dispensed with by the mathematical-logical deduction that regardless of that complexity the internal temperature of a black body is uniform. This deduction, in turn, is based on an assumption our physicists

* This is not to be confused with another distinction physics makes between heat and light, whereby heat is distinguished from light as invisible—infra-red. In that context, red heat *is* the coolest.

know to be false: that the black body is an Absolutely black, closed system (a system closed off from its environs). In these ways, the contradiction between human perception of color:heat correlations and selected thermometric data is simply ignored.

Another "logical" implication about black-body energy is that it should be emitted all at once, in what was called "an ultraviolet catastrophe." This conclusion was based on the assumption that light is a wave-continuum. Planck therefore concluded that light is *not* a continuum, but discontinuously emitted packets of light, or quanta. But the Western conception of a continuum of course was not and is not as the sensory data dictate. All a continuum is to a Western scientist is something without gaps in it, whereas a full-fledged continuum is Polar-Complete: infinitely subdivisible and not truly quantifiable. Had the originally assumed continuum been such a *true* one, the implication of the "ultraviolet catastrophe" would have been avoided. Being a whole of infinite parts (*and* not being a system closed off from its environs), an energy-continuum in a heating black body does not as it were pile up and explode, as was concluded. Rather, to put in a primitive but empirically sound way, because it transcends space-time with its infinite structure, because it is metaphysical in this respect, it can, and does, continuously radiate gradually increasing energy when heated.

In sum, our physics' color:energy correlations are not factual; instead, mathematical constructs are substituted for empirical data. Nor do they account for all the relevant data. Extrapolations from them, therefore, are not sound. So one is scientifically free to consider an alternative approach to and interpretation of the data and to make new extrapolations from the results.

Now we develop an understanding of color:energy relations through alternative, Chinese-scientific (Human) means; we switch from the automobile to the spaceship. Each color, except black and white, co-occurs chiefly *but not exclusively* with a definite but non-Absolute range of wave-properties, so that in fact each color is each other color. To put it another way, each color (save black and white) is white (all hues) inflected in a relatively specific direction. Therefore, each color is a *kind of manifold multicolored relation,* a complex *quality,* and a quantitative, linear coordinate sytem such as the preceding Western one cannot reflect, indeed, directly contradicts, the reality of color-distinctions. Now, colors and energy-levels are aspects of same things, so it stands to reason that energies are also Polar-Complete. If colors are defined as manifold relations, then so are energy-levels, and each color correlates with a kind of manifold relation that defines an energy-level. Let me make that subtle matter clearer. To define blue-hot, for example, as faster-shorter waves of energy than red-hot is a misleading abstract statement. Actually, as our physicists know, blue-hot should be defined, if defined in wave-terms, as having more fast-short waves than does red-hot, but including the slow-long waves which red-hot has more of than does blue-hot. So in truth, to say that there is increase in frequency and decrease in length as heat increases, is to give only a Fragment of the actual pattern. Rather, it is that the *relative proportion* of faster-shorter waves increases: the longer waves do not cease to occur. The phenomenon is not accurately expressed by a single-dimensional and linear scale; it is accurately expressed as a continuous shift of proportions in a manifold.

Now, a crucial implication of this fact is that red-hot, the lowest-temperatured hot, is an *essential* property of all hots, including the hottest, and, conversely, violet-hot, the highest-temperatured hot, is an *essential* property of all hots, including the coldest, red-hot. This is a "middle amputated" by the Western color:energy correlation, or "law." All the relative degrees of energy are an interdependent, manifold complex which, as explained with reference to colors alone, cannot be broken down into ultimate wave-components, but rather are an infinity of manifold complexes within manifold complexes. It follows that the true nature of the *energy* scale must *also* be expressed as a manifold relation between or among relative-colors (colors scientifically defined as Polar-Complete phenomena) and whatever they represent in terms of energy and matter. This, the Chinese savants have done, as follows.

According to the Chinese system, the basic colors are red and black, that is, the basic color-dimension is red-to-black, and red correlates with energy and black correlates with matter. So, each energy-level that co-occurs with a color is a *kind* of manifold relation between energy (*yang, ch'i*) and matter (*yin, shui*). This conclusion is arrived at by dealing not with selected, idealized, phenomena associated with colors (wave-properties), but with colors themselves and energetic conditions themselves, by letting Nature communicate instead of by declaring it deceptive and erecting an abstract system that ill-reflects it.

The natural spectrum directly communicates that all colors save black are relative phenomena on a continuum whose extremes are red and violet, with yellow, then white, at the red end, and blue, then green, at the violet end. Given the fact that colors are Polar-Complete, this means that yellow can be defined as minimally-violet red, white as medially-violet red, blue as minimally-red violet, and green as medially-red violet.* There are further, equally valid definitions, such as, yellow is maximally-white red, and so on. As long as the *whole* spectrum and the relative positions of the colors are taken into account, all such definitions are valid. But by selecting definitions in terms of the *extremes*, red and violet, we capture the Polarity of the spectrum as a whole.

Now, to complete the system, or image, we must locate black in relation to all the other colors, and by equally natural means. Either it is at the red end or it is at the violet end, or it is at both ends, because Nature does not locate it within the bounds of the red-violet extremes. (It may be said to be some colors' relative lack of luminosity, but this relates it only to white, not all other colors, as we need for a Complete system.) Black is more different from red than is any other color, and black and red are more different than any other two colors. Red has the most distinctive hue, aesthetically is very hot, and is fairly bright; black is without hue, aesthetically is very cold, and is maximally dim. This is the dimension of "positiveness" mentioned above, the most powerful dimension that can be abstracted from color-relations, analyzed into three aspects. Red is the most positive, black the most negative, color. (Hence the necessity of red to advertisement; Coca-Cola is positive, "adds life," by its red association.) Black and red are super-opposites.

Violet, in turn, is more like black than is any other color: violet and black share a deep-recedingness, a mysteriousness. Accordingly, Chinese use the term *hsuan* (mysterious) to denote both of them.

Plainly, black belongs next to violet at the opposite extreme of the spectrum from red. As the non-super ("z-sharing") opposition of

* This at first strikes the Western mind as circular. We fail to question the fact that when we define colors, not in terms of colors but in quantitative terms of instrumental reactions to them, we have made a radical translational jump for which there is no scientific motivation—only the overwhelming desire to mathematize. We do not *ab*stract, generalizing from exhaustive empirical data; we unnaturally *ex*tract from Fragmental data. The Chinese way, in contrast, is from the natural (all accessible sense data) to the abstract.

white to both black and red predicts, white is just intermediate to the two; and so it is, between green and yellow on the prismatic spectrum. All of this is to state in inevitably inadequate words what is obvious to any people whose perceptions are not adulterated by pseudo-empirical theories—people such as the Dani of New Guinea, whose primary colors are black and red, or the Chinese savants, who were and are able to produce sophisticated science without alienating themselves from Nature and their direct perceptions of it.

Now, to capture the Complete Polar dimension of colors, we define all the colors in the same way we defined all of the colors save black. We substitute black for violet and recognize that the basic color-dimension is red-*black*. So, yellow is maximally red red-black; white is minimally red red-black; green is minimally black red-black; and blue-violet is maximally black red-black. In Quadrimagic terms, yellow and white are *basically* red (yang) colors that are *inflected* in the black direction, and violet-blue and green are *basically* black (yin) colors that are *inflected* in the red direction.

Now we can derive the color:energy correlations of Chinese physical science, and compare the results with the Western ones. The crux of the matter, as in the previous section, is to think aesthetically, paradoxically, and at several levels simultaneously, to faithfully reflect the sense-data that Nature provides. Color is plainly an energetic phenomenon and red is the most positive color, the color that represents all other colors, color maximally present. Hence, red is the color, or color-direction, of *energy*, that is, of energy as distinguished from matter —it being understood that energy and matter are Polar, ultimately indistinguishable aspects of the same thing. Black, then, is the color of *matter*—of what is still, dense, opaque, receptive—the *real* Black Body. It follows that as energy increases, red increases, as humans perceive. But, *paradoxically*, as energy (as manifested by thermometrically measured heat, for example) increases, the proportion of red *de*creases through yellow, white, green, and blue *until violet* where it *in*creases until the fuel is consumed to leave black (or dead white) ash, the quintessence of matter. This paradox can only imply that as energy increases, red *does* increase, but, through the energy's *changed relation* to matter, it becomes less apparent until an extreme, violet, is reached. It "mixes" with black to become other colors. Reciprocally, as energy and red increase, matter becomes more colorful-energetic, so that basically-matter-colors increase in proportion to the basically-energy-color, red. So, as energy increases, it is an energized, internally *redder* and *redder,* black that increases. The *internally* reddest, most energized black is violet.*

The shift from red to violet, then, is symptomatic of an increase in the interaction, mixing, *intimacy,* of energy and matter, and the terminal shift from violet to black is symptomatic of the loss of matter-energy interaction. Red is the color of energy maximally separate from, just beginning to get hold of, or gently dominating, matter. And, strictly speaking, violet is the color, not of high energy, as the Western correlation states, but of matter deeply invaded by energy, and of matter and energy about to become only matter: it is "black light," the material "shadow" of energy.

Let me elaborate on this foreign way of looking at energy-matter, adducing some Western data which have not been taken into account by Westerners in their own Mentalist and reductive correlations. Implicitly I have distinguished between energy and heat—heat as that which is registered as high temperature. Energy is the active cause of heat and matter is the passive cause of heat. Heat, in turn, is an energy-matter interaction which has color and wave-characteristics. Westerners have looked only at those characteristics and thus have equated "high energy or heat" with fast-short waves and "low energy or heat" with long-slow waves—whereas in fact these are Fragments of the actual, richly manifold phenomenon in question. By combining the Western electromagnetic wave-data and direct human aesthetic perceptions recognized and reflected in the Quadrimage by Chinese savants, the pattern of the underlying reality of energy:matter interaction can be made clearer.

We must at first think of matter and energy as Absolute-Fragmental parts of a manifold phenomenon, but with the understanding that neither *is* a part in the Absolute-Fragmental sense. Conceived of as perfectly homogeneous absolute opposites, energy is movement (*chou*) and matter is a space-time container of movement (*yu*)—pretty much as Einstein said. (*Yu-chou*** means 'universe.') The essence of

* There are two kinds of black, then: energized black, which we call violet, and the almost-absolute absence of red, which we call black. In the cycle of increasing energy and burning, maximally energized black (violet) becomes red-absent black as energizeable matter is exhausted, and ash, matter without energy, black-without-red in color, ends the cycle. The use of the term *hsuan* for both kinds of black, colorful and color-absent, energized and just-material, encodes this underlying identity of what we call "black" and "violet."

** From the 2000-year-old *Huai-Nan-Tzu* (again see *The Sea Turtle and the Frog*).

yang and *ch'i* is formless-substanceless movement, or energy, and the essence of *yin* and *shui* is motionless containing, or matter. Now, *a lot of waves in the same place* is the equivalent of matter, as mass. As the electromagnetic spectrum ranges toward higher-faster wave frequencies, it passes from violet color to ultraviolet "rays" to soft then hard x-rays to atomic particles. What our physicists call energy-waves are continuous with what they call matter. Energy, the opposite of matter-as-compact-energy, then, is at the red end of the general wave-spectrum: infrared-colorless-heat—sound—radio and beyond (the sector least explored by our physicists.*) And it is defined by opposition to matter, hence, as *few waves in many different places,* or very long and slow waves extending very far.

In this rigorously logical-empirical understanding, then, color and all the other electromagnetic phenomena are a mixture of matter (that which contains, impedes, and compacts energy) and energy (that which moves matter). And the violet-black end of the spectrum is relatively material and the red-hyper-red end of the spectrum is relatively energetic. More precisely, the material pole is relatively material in that matter, containing, *is dominant* and the energetic pole is relatively energetic in that energy, movement (extending-far) *is dominant.* It is not that "there is more energy" at one end and "more matter" at the other, but that the energy-matter *relations* differ at the two poles in dominance of one or the other over the other. In Chinese terms, in the red direction, energy—*yang, ch'i*—is 'the host' (*chu*), and matter—*yin, shui*—is "the guest" (*k'o*). And in the violet direction matter—*yin, shui*—"is the host" (*chu*), and energy—*yang, ch'i*—"is the guest" (*k'o*)—for, as modern physics now pretty much recognizes, in all or a particular fragment of space-time there can be no more energy than there is matter.

It should now be obvious that what our physicists call "great heat" or "high temperature," and identify with short-fast waves and greater mass, or matter, is a matter-dominant interaction between energy and matter which does not equate to "higher energy." Rather, it equates to "higher *containment* of energy": energy occurring more in same time-place. Energy *itself,* in fact, is the opposite of that: it is long-slow. The movement, the actual energy, in the super-hot violet fire, then, is basically *red.*

Our physicists confuse the *conditions* for high temperature

* We will explore it here.

("heat") with energy *itself*. They confuse formless-substanceless movement, the essence of energy, with the extreme effect of matter upon it: movement-in-a-same place. Reciprocally, they confuse motionless containing, the essence of matter, with its effect, density of movement, or mass. The heat in the very high temperature of violet fire, then, is movement, is energy, is red; the energetic aspect of high temperature is the red energy in the black matter.

As that effect reaches its natural extreme limit, the interaction between energy and matter weakens: the matter's capacity to contain the energy is exhausted and the energy begins to escape it, to go on and again begin the cycle elsewhere. That condition is nothing other than a verbal expression of violet, what violet virtually tells us through our senses, of its own accord. Red is the color of energy entering into a relatively obvious interaction with black matter; violet is the color of the energy-matter interaction at its peak and, consequently, its turning point, its decline. One *sees* in violet the red reemerging, escaping. Between the two poles the other colors are symptoms of energy-matter interactions intermediate to the red and violet ones. From green to black, *yin, shui,* matter is dominant. From white to red, *yang, ch'i,* energy, is dominant.

Now pause for a moment and observe or imagine the natural color-spectrum, as through a diamond prism or as in the heating of a metal to red-, yellow-, white-, green-blue-, and violet-hot. The *fit* of the preceding Chinese theory will be aesthetically obvious, felt in the heart.

Put simply, movement (energy) and space-time (matter) meet, and as they blend more and more deeply, the matter forms, impedes, the energy and the energy enlivens the matter. A result is faster and faster, shorter and shorter waves and, in a measuring instrument extended into the energy-matter, higher and higher temperature (the red-energy enters into the instrument's matter and accordingly expands it).

Further Western data, which lie outside the simplistic Western color:energy theory, are accounted for by the manifold Chinese analysis. First, the reappearance of red in violet is accounted for: it has been implicit in green and blue as it maximally entered into black-matter, and as the relation declined by reaching its extreme (*fu*), it reemerged. Second, as an object is heated from 2854 degrees Kelvin to 7500 degrees Kelvin, the amount of so-called cooler color (red-yellow) does not decrease in direct proportion to the increased amounts of so-called hot color (green-blue-violet), as would be expected if increased energy (in a given place) directly correlated with higher wave frequencies. Red-yellow decreases in amount by roughly half of what green-blue-violet increases by. What really goes on, then, is that the *amount* of red-energy is *constant,* but the *relation* between red-energy and black-matter changes, so that black-matter increasingly dominates. The increasing amounts of black-end colors are a symptom of that. That is, red gets blacker (energy becomes more material) to give yellow and white, and black gets redder (matter becomes more energetic), to give green, blue, violet—and, as the interaction peaks, red-in-violet, red escaping black. Third, likewise, as the temperature increases there is more and more white, which means more and more of *every* specifically hued color. This is a symptom of increased energy-matter involvement, increased blurring of their (already non-Absolute) distinctions, increased mixing of red and its hue-opposite, cyan (blue-green), and, as a product of that red-black mixing, of yellow and its hue-opposite, violet.

The contradiction between the human perception that red is hotter than blue-violet, on one hand, and the physical finding that blue-violet-hot is hotter, more energetic, than red-hot, on the other hand, has been resolved, then, in Tao terms. That is, human beings perceive energy-matter relations whereas physicists, through their thermometers applied to real black bodies which are then transformed Mentalistically into "ideal black bodies," perceive only the material aspect of energy-matter phenomena (the number of waves per time-unit, the length a thermometer expands). Accordingly, human perceivers know that energy, and its specific form of heat, is basically red, whereas our physicists do not.

One question remains: why is it that humans *feel* that a red-hot thing is hotter than a violet-hot thing? One possible answer is that they do not only feel, they also think, so that their perception is influenced by their understanding that energy, heat, is basically red. A better answer, which incorporates the preceding one, is that human beings, unlike thermometers-and-their-physicist-extensions, operate in the natural Polar-Complete space-time continuum. Specifically, there is no such thing as "now," or "present time." Rather, there is past time, an abstraction from completed events, and future time, an abstraction from what is expected to happen: "I have planted half of my garden; I will plant the rest of my garden." "Now," as in "I am planting the rest of my garden" really refers to nothing other than the past-future middle grounds, where past and future are indistinguishable because they are continuous with each other: there is no gap, no "instant," between them, hence they *cannot* be Absolutely distinguished. Past and future being, in fact, Polar-Complete in this way, with some future in the past and some past in the future, it follows that humans can, according to their individual abilities, perceive—or perceive-deduce—some of the future, future events. Now, if in all relevant time a red-hot thing burns longer than does a violet-hot thing, it follows that in space-*time* the red-hot thing may be hotter, produce more heat, than the violet-hot thing. Humans, then, might be perceiving this. They know that in space-*time* the coal is hotter than the flame, has-will burns more than the flame. Of course, the red-hot coal does burn much longer than does the violet-hot flame, but let us use some of modern physics' rigorously standardized data to see that my inference is perfectly confirmed.

By analogy to observed nuclear reactions our physicists infer that the overall, surface-and-internal, temperature of a violet-hot star "is" (in non-existent "present time") about 10 times that of a red-hot star. But the burning cycle of the elements characteristic of a red star (which are relatively light) is 1000 times as long as that of the elements characteristic of a violet star (which are relatively heavy).* If the masses of the burning elements in the stars are held equal, then, in same space-*time* the red star is 100 times as hot as the violet star. In turn, the masses of violet stars are computed to be about 50 times those of red stars, so all told the red star is about twice as hot as the violet star. Mass, as I said, is a lot of energy in a same place, energy dominated by matter—the violet condition, a fast burn. Energy, as I said, is movement itself, uncontained by, dominating, matter—the red condition, a slow burn. The slow burn, being an energy-dominant phenomenon, has more heat than the fast one.

One may contest this by observing that, where what we are really getting at is the true nature of the energy:matter relationship, the violet burn yields more energy than has been taken into account, for much of it is not in the form of heat but momentum: rays, particles, shot out at high velocities. It yields much more of this energy than does the red burn, (in space-*time*) probably about twice as much, so that the totals of red and violet energy are equal. That makes my basic point precisely: there is no difference in *amounts* of energy and matter at the "red" and "violet" ends: "without yang, no yin; without yin, no

* See Mueller *et al.*, cited in the *Bibliography*.

yang"; matter and energy are equivalent in "amount." Rather, there are differences in the energy-matter *relations,* basically qualitative and manifold phenomena which the Chinese system recognizes and uses with great success, but which the Western system partly ignores and thoroughly distorts in the ways shown.

In general: energy, yang, *ch'i,* per se is movement, matter per se is containment; energy is in the red direction, matter, yin, *shui,* is in the black direction. As energy increases *per time-unit* from red to violet, matter goes from a subordinate ("guest") relation to a dominant ("host") relation: but energy remains *basically* red (really, hyper-red) throughout.

Confucius called what I have just done "rectifying the words." Even in China it has been repeatedly necessary, when dynasties and mental culture decay, to rediscover the real meanings of central words, to redefine them by reference to the realities that they originaly referred to and must be understood to refer to. Here I have "rectified the words" matter and energy, their associations with colors, and their relations to each other. I have shown that they are Polar-Complete, and identified the basic poles of the physical aspect of the universe as red and energy (yang) and black and matter (yin). In doing so, I have recapitulated in both traditional Chinese and modern terms the most basic information in the *I Ching,* which, in its most recent, 3000-year-old version at least, describes yang (*Ch'ien*) as active, moving, dry (which implies hot and basically means non-material, for water stands for matter), and red; and it describes yin (*K'un*) as passive, quiescent, and black.

As Confucius made clear, the most important thing about rectified, or real, words is not that they make knowing possible but that they make Human reality possible—knowing is only a means to that vital-moral end. That is true of this specific, energetic-material, consideration. Red energy, that is, red energy and black matter in gentle, optimal and therefore long-term interaction, is a central condition for life. In contrast, violet energy, that is, red energy and black matter in violent, extreme, and therefore short-term interaction, is a central condition for death, even to the "black-lighted" streets of *Clockwork Orange* adolescents. That red-hot lasts longer than violet-hot, as in wood fires, nuclear fires and stars, that there is more heat-energy in red space-time than in violet space-time, is one aspect of this general pattern, of which more will be described in the following section. The "choice" of a "cool" red sun, not a "hot" violet one, for our lively planet is consistent with that pattern. The Ultraviolet Catastrophe is not a cute theoretical error of pre-Planckian physicists, not a mere furnace explosion, but a scientific paradigm which confuses energy and matter by trying to Absolutely distinguish them and by ignoring human perception, and which is experimentally based on energy in space but not also time—energy for us now not our children later, violet without its red complement and active source, matter and energy separating, the Catastrophe Hexagram *P'i*. The real Ultraviolet Catastrophe is the creation of unnatural radioactivity on earth, a sorcery-like lowering of the sun into the earth which puts yin and yang into conflict and thus separates them, with the side-effect of mutant babies and horrible illnesses, such as cancer, whose central characteristic is physiological disorganization. Our nuclear physics is a form of *huan*: a hasty and violent human interaction with Nature with extraordinary, bizarre, and injurious results.

The implication of this section, then, is not that the Chinese way of looking at energy-matter would be useful to modern physicists but that we must make a Red Shift whereby physics would be altered to become one with the *life*-sciences. To that matter I now turn.

UNIFIED SCIENCE

With regard to our Trip, what we have done so far is roughly map the galactic territory and cruise in our space-time ship through a local zone where Western scientific reality and Chinese scientific reality touch and may be understood to a great extent the one in terms of the other—basically, in terms of the Quadrimage and the anti-Quadrimage. Now it is time to switch to intergalactic warp drive. What makes this leg of our Trip a "warp" is that, from the Western scientific perspective, it collapses space-time. In other words, here we will get to the level of genuinely unified science which, by definition, uses a single theory to predict, or make sense of, several sectors of the universe considered separate and fundamentally different by Western scientists. It is like being in several place-times at once, or traveling in several different directions at once. For such an "adult form of travel," a more sophisticated sub-paradigm than the Quadrimage is used. In learning it, we will be recapitulating for the first time how Chinese scientists actually think. The Quadrimage is actually a "kindergarten course," a shuttle to the real space-time-ship, simply a way of introducing Polar-Complete thought so as to align the mind with empirical reality. (It is because of its oversimple nature that it may be compared with the Western scientific paradigm.)

The Quadrimage is further specified to become a Sextimage,* and the result is much greater intelligibility and prediction, of all sorts of natural phenomena. To explain the relation between the two: whereas the Quadrimage distinguishes two kinds of yin and yang (major and minor), the Sextimage distinguishes three: major, minor, and extreme-reverting. The last, unlike minor yin and yang, are within yin and yang of the *same* value as themselves; where minor yin is "the yin inside the yang" and minor yang is "the yang inside the yin," they are "the yin inside the yin" and the "yang inside the yang." That is, they are yin at its extreme interaction with yang and therefore almost indistinguishable from yang, and yang at its extreme interaction with yin and therefore almost indistinguishable from yin—hence, "extreme-reverting." The Chinese terms are *chueh-yin*:exhausted yin and *yang-ming*:yang florescing. For short, I will call them "reverting yin and yang," but the Chinese qualifiers precisely indicate the yin and yang states referred to. Yin, matter, is exhausted as such when it retains only minimal energy-containing capacity. We have already seen examples of this state: radioactive substances, violet-hot metals. And the color assigned to reverting yin in general is, indeed, violet. In turn, yang, energy, floresces, declares itself, just before it loses its dominance to yin. In heating metal the brightest stage is yellow-white, which occurs just before the middle of the heat-color-spectrum, as we saw. And the color of reverting yang in general is yellow-white.

Because there are six, not four, sectors in the Sextimage, major and minor yin and yang are defined more precisely than they are for the Quadrimage. In the Quadrimage, major yin is what we can call "matter" in the broad sense which includes substance. In the Sextimage it is distinguished from substance as matter in the narrow sense: the space-time container, receiver, of energy. In the Sextimage *minor* yin is substance, for substance is understood to be a derivative of major yin (matter) and major yang (energy): materialized energy. The com-

* Called *Liu-Ke Yin-Yang*:The Six-Yin-Yangs, in Chinese.

mon-sense example of Sextimagic matter is earth, including its atmosphere, for it is the earth which receives and contains cosmic energy and to which all dynamic phenomena, be they plants at the end of their life cycle or orbiting objects,* return. The common-sense examples of minor yin are cloud-gathering-wind and water, which, as a fluid, changing substance, openly exhibits its basically energetic (yang) nature. Just as Sextimagic major yin is narrowly material, major yang is narrowly energetic. It is called *T'ien-wai-chih yang*, literally "the yang that is outside Nature," which is to poetically express its almost formless-substanceless nature and its general presence in the cosmos. It is almost pure movement, the Polar complement of the space-time container of matter. Its common-sense example is sunlight. Minor yang, in turn, is energized matter and its common-sense examples are wood and, more generally, life-forms; and, where electricity is understood to be an aspect of the life force, lightning is associated with it.

The colors stipulated as characteristic of the six yin-yangs are assigned both on the basis of the distinctive colors of the common-sense phenomena which exemplify them and by differentiating the four Quadrimagic colors into six, as follows. Major yin is dark-dim yellow-white. As the space-time container of energy or movement, distinguished from energy or movement, it has no color of its own. Rather, it takes on the colors of what it receives: all other yins and yangs, chiefly the yang which most closely approaches it, reverting yang, which is light-bright yellow-white. The central, most neutral and mixed, colors are indeed yellow and white. Dark-dim yellow, that is, brown, results from a great number of mixtures of different colors; and that is the color of the earth's soil. Dark-dim white, or pewter, is yin white, all hued colors darkened and dimmed, and that is the color of the earth's atmosphere during its most yin phase, winter. Minor yin, materialized energy, substance, is distinguished by density and

* When, to our physicists' surprise, the orbit of Skylab decayed, I was in Taiwan where my *shih-fu* observed that their surprise was due to not having recognized this fundamental principle. In calculating only the balance of centripetal and centrifugal forces which should yield perpetual orbits, they ignore the fact that the orbiting object is originally part of the earth, which is balanced against the rest of the *cosmos*. The orbiting object has a place on earth according to that balance. Hence it returns or it is attached to another, earth-like, object with an established orbit, such as the moon, and the overall cosmic balance is slightly disturbed.

opaqueness; hence it is black. Black is the color of water as thunder-clouds, the image of minor yin. Major yang, almost-pure energy, of course is red, more exactly hyper-red (infrared and beyond). Minor yang, energized matter, would be on the yin half of the spectrum close to the yang half, therefore green, and green is the chief color of wood and, where all non-plant life-forms are ecologically and structurally based on plants (hemoglobin is much like chlorophyll), of life-forms in general.

Now I align the Sextimage with the Quadrimage in terms of colors and yin-yang sectors to show how the Sextimage's assignments of colors to yin-yang sectors follow not only from observation of its common-sense correlates, above, but also from differentiating four colors into six. (See Figure Eleven.) Actually, the Sextimage has several orders, each for a specific variety of phenomenon: animate, inanimate, both animate and inanimate, healthy animate, unhealthy animate, and so on. Here I give the most inclusive and therefore most general order, the cosmic, or animate-and-inanimate one, which refers to the Four Seasons in the inclusive sense of climate, land-and-sky-scape, and the life cycle. Quadrimagic yin and yang correlate with the Four Seasons as follows: major yang: Summer (red); major yin: Winter (black); minor yin: Fall (white); minor yang: Spring (green). This is shown in the square internal to the circular Sextimage (Figure Nine). Extreme-reverting yin and yang occur between the major and minor sectors and thus overlap with one third each of the major and one third each of the minor yin or yang sectors.* So a "subtraction" of their colors from those of Quadrimagic major and minor yin and yang is implied, as follows. If the red of Quadrimagic major yang has "one third" of its yellow-white "subtracted" (by reverting yang), it becomes hyper-red, as on the Sextimagic figure. If the white of Quadrimagic minor yin has "one third" of its yellow and "one third" of its white "subtracted" (by reverting yin), the subtracted yellow becomes bluer, blacker, and the substracted white makes it dimmer, blacker, hence, black-which-is-just-on-the-black-side-of-black-white, or, simply put, black—as on the Sextimagic figure. In turn, if Quadrimagic major yin's black has one third of its violet substracted (by reverting yin), it becomes dark yellow-white, as on the Sextimagic figure. That is, where black has all hues in equal proportion and minimal luminosity,

* The prismatic or "heat-" spectrum is inanimate, not also animate, and therefore has different order, as explained below.

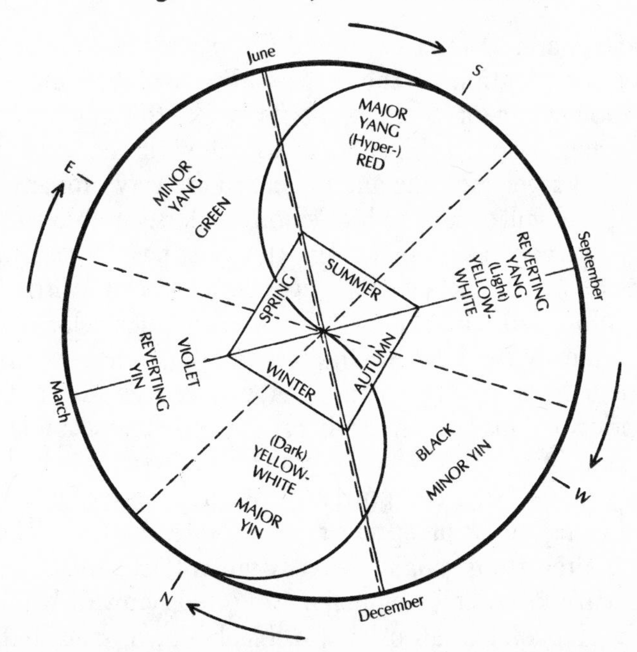

FIGURE NINE: THE SEXTIMAGE OF ANIMATE-INANIMATE PHENOMENA, CORRELATED WITH THE QUADRIMAGE OF THE FOUR SEASONS

if its red and its blue are diminished, it acquires their hue-opposites, which are the area between blue-green and reddish yellow: yellow-white; and it remains dark; hence, dark yellow-white. And if Quadrimagic minor yang's green has one third of its violet subtracted (by reverting yang), it becomes a lighter-brighter green.

To move from simple Polar-Complete color-logic back to empirical reality, observe that the Seasonal color-coordinates of each of the six yin-yangs are empirically correct. Of course, each Season has all colors; it is a question of determining the distinctive color of each. On the basis of the *non*-Seasonal data (common-sense examples of general yin or yang concepts) and the color-"subtractions" above, according to the Sextimage the chief color of Summer is red and the secondary color is light-bright yellow-white. Where the distinctive feature of Summer is the strong sun, this is correct. The chief color of Fall is black and the secondary one is light-bright yellow-white. The black is of the water evaporated into the sky in Fall, which returns as snow to darken the sky in Winter, and the yellow-white is the color of the sunlight which does the evaporating, and of the dried leaves and stalks. The chief color of Winter is dark-dim yellow-white, that is, the brown of Fall leaves and Winter earth and the pewter of Winter sky. This color is a combination of the water-darkness of Fall and the light-bright yellow-white of Summer at its peak—the color of reverting yang, darkened, yinned. This is consistent with the posited general principle that reverting yang and major yin are very similar, consistent with the temporal order of the Seasons' colors, and consistent with the general principle that major yin, the most passive sector, takes on the forms of what affects it. The secondary color of Winter is violet, as in the Northern Lights and blue ice. The chief color of Spring is green, specifically the light-bright green, *ch'ing*, which is stipulated in the Chinese texts; *ch'ing* means "living green, as in plant-sprouts." The propriety of this color to Spring is obvious. The secondary color is violet, found in the subtly violet brown (dark yellow-white) of wet, fertile, energy-releasing composted plant-earth-matter, and in the (reverting yin) lightning of Spring storms. Note that, like the relation between the reverting yang and major yin colors, violet, the color of reverting yin, is the color of major yang, red, yinned, blackened, or, conversely, the color of all other yin, black-brown, reddened, yanged.

To generalize, Summer, and all Summer-like phenomena, such as our sun and human arterial blood, that is, optimally and long-termed heating, or energy-dominant, phenomena, are chiefly red or hyper-red and secondarily light-bright yellow-white in color. Fall, and all Fall-like phenomena, such as cigarette ash and human feces, that is, burnt-out phenomena, are chiefly black and secondarily light yellow-white. Winter, and all Winter-like phenomena, such as metals in general and human bones (hard, cold or cool, dense, slow-changing phenomena), are chiefly yellow-white and secondarily violet (like heavy metals, human bones fluoresce violet in ultraviolet light). And Spring, and all Spring-like phenomena, such as new plant life and wood, are chiefly green and secondarily violet (wood is partly mineral—metal—violet; Spring's characteristic thunderstorms have violet lightning).

Of course, it appears, superficially at least, that there are exceptions to the preceding Sextimagic color-pattern and room to argue for other color-assignments to Season correlates, and so on. This by no means implies any weakness in the sub-paradigm. First, it is a normative frame: what usually and most generally is the case, not what is Absolutely the case. And in this it is no less rigorous than our most rigorously predictive science, physics, which is also normative (probabilistically)—but, unlike Chinese physical science, encompasses only certain varieties of inanimate phenomena. Second, the predictive range of this sub-paradigm has not yet been demonstrated. Third, and as is understood but may easily be forgotten, this is a basically empirical, therefore Polar-Complete way of understanding things. There are no "lines" between any two adjacent Sextimage sectors: there is minor yin in reverting yang, therefore black in light yellow-white; there is major yin in minor yin, therefore dark yellow-white in black; there is reverting yin in major yin, therefore violet in dark yellow-white; and so on, so that in reverting yang there is minor yin, major yin, reverting yin, minor yang and major yang; and in its light yellow-white there is black, dark yellow-white, violet, green, and red; and so on for each sector. Each color assigned is the Polar-Complete *focus** of its yin or yang sector.

All told, each color assigned to one of the six yin-yangs is assigned with aesthetic-logical-empirical exactitude. Of course, such a naturalistic constructing of a paradigm strikes a Western scientist as naïve, primitive, and Medieval. But this is a mere impression, and one without any grounds save that the closest the West has come to Chinese science since Heraclitus was indeed during the so-called Dark Ages. Then, it so happens, Western science, because of the weakness of the Mentalist paradigm and the consequent closeness to Nature during that period, was at its peak, and analogy among all things was posited. It must be remembered that the aesthetic-logical-empirical mode illustrated here is the only way to deal with Nature as a whole system, and that Western science does not do so at all. Rather, it erects blind little near-vacuums within Nature called "laboratories," in which an extremely limited set of data are manipulated toward Absolute-Fragmentality. A patronizing attitude toward the natural method of knowing, whose scope is enormous, Complete, derives forthrightly from the disdain for Nature and the human senses which is prescribed for any Western scientist, and from the feudal loyalty to the Mentalist paradigm, which must not soil itself with phenomena such as soil, sunlight, Spring, and thunderstorms. Rather than study soil, it deals with Ph-balance; rather than study sunlight, it counts numbers of waves which cannot, actually, be counted and are not sunlight itself; rather than study Spring, it computes the angle of sunlght during that period; rather than study thunderstorms, it determines the electrochemi-

* *Ching-sui:* essential "gravity-center."

cal conditions under which water droplets amass in a laboratory. In this way it preserves its Pythagorean mathematical expression of "everything"—at the cost of ignoring almost everything.

The sub-paradigm I have just outlined actually more recalls our most modern, innovative, and promising proto-paradigm, the ecological (von Bertalanffy's Systems-Theoretic) one than it recalls Medieval Wholism. The Sextimage is consummately ecological. Each sector is part of each other, as the use of color-coordinates emphasizes; the implication and intent is that no sector has independent existence and that each sector is a necessary condition for each other. If it weren't for the heat of Summer, there could be no Spring at the other "end" of the cycle; and the heat of Summer, transferred to the energy of Fall locked as frozen compost in Winter and released in Spring, is present in Spring; and so on. Just as there is red in green, there is Summer in Spring; and so on. Likewise, as already documented, no subatomic "particle" has an existence of its own, but rather is a part of every other "particle," and reciprocally. Similarly, no organ in a living body functions independently of any other, and as follows, each organ function is, in part, every other organ function.

If the Sextimage serves to predict such diverse phenomena, its scientific power is so much greater than anything Western science has produced that to take objection to the non-Mentalist and therefore "naïve" nature of its construction would be petty to the point of irrelevancy. The proof of a paradigm's "scienticity" is its superior ability to predict, or make intelligible. Let us then engage the warp-drive of our space-timeship.

There are several general orders of the six yin-yangs, both in time and in terms of immediacy or mediacy of relationships among them.* Here, for simplicity's sake, I will bring our ship into warp-drive first by showing how the six yin-yangs themselves, without regard to their relative order, take the knower through cosmic, microcosmic, inanimate, plant, mammal, and human space-time with uninterrupted ease, revealing that they all have the same basic pattern. Then I will explain

* I will not completely describe and explain them in this book because it would require too long an exposition and because these orders, which must ultimately be understood and, as in Chinese medicine, used in diagnosis and prescription simultaneously, are an extremely difficult matter. The medical text which focuses on orders is called the *Nan Ching:The Classic of the Difficult.* In his cited book, Porkert translates some of the material on orders from other texts, to which the English-speaking reader might refer.

the basic difference between animate and inanimate, one of the central mysteries for Western scientists, in terms of different yin-yang orders, uniting inanimate science with life science in a Red Shift.

We begin at the lowest level of the inanimate-animate scale, with metals, most distinctly inanimate things. The most stable, long-lived, metal is gold; it is the natural inanimate standard. The most unstable, short-lived, natural earthly metals are radium and uranium, radium being derived from residues of uranium ore. Hence gold is maximally animate , or life-like, is close to the animate/inanimate "border," and radium-uranium is maximally inanimate-, or death-like, is extremely inanimate: gold conserves energy-matter, radium-uranium loses it. Sextimagically, it follows that gold should have an animate-like, Spring-Summery, color-configuration: red for the major yang energy it conserves, green for its minor yang Spring-rebirth-like stability, yellow-white for its reverting yang, slightly energy-dominant yin-yang balance (energy holds things together, is the "unbroken line" of the *I Ching*; hence, yang-dominant). Gold, of course, is a red-yellow-white metal, and it absorbs, contains, green light. Indeed among metals it has an exceptional capacity to do so. It is, as it were, maximally alive, Spring-like, internally. (Accordingly, all civilizations use it to symbolize life, immortality, and so on.) In contrast, Sextimagically, radium-uranium should have an inanimate, Fall-Wintery configuration: dark-dim yellow-white for the major yin matter which is its substratum, black for its minor yin density, violet for its reverting yin energy-releasing and substantial decomposition, which is its most distinctive characteristic. Uranium radiates, loses, violet energy-matter, and, where its residual substratum would be opposite to what it loses, it would be internally, therefore dark-dimly, yellow-white—gray-brown-plus-black, leaden. Radium, oppositely to gold, radiates green specifically a *blue* green, a reverting yin, quintessentially uranium-violet, green: an image of "life-"loss resulting from extreme yin structure. In short, the Sextimage roughly predicts the colors and yin-yangnesses of gold and radium-uranium and, more to the point, it makes their colors and yin-yangnesses intelligible within a universal, unified-scientific, frame.

Now we shift from the particular inanimate to the cosmic-inanimate level to penetrate the fascinating matter of "creation." The Sextimage is based on the understanding that hyper-energy is in the red direction, is movement per se almost without form, and therefore has waves so long and slow as to be almost straight, opposite to matter's, which are so short and fast as to be perceptible only via the highest-amplifying modern instruments. (But they are deducible, I remind the reader, from the sense-based Polar-Complete paradigm.) This hyper-energy is supposed to be primal and basic, perceptible only in the shorter-faster forms that are produced by its interaction with matter (sound, heat, color). The sun is understood to be a concentration of it and matter, is distinguished from it as one of its effects. The sun is major yang but it is not also "the yang that is outside Nature." Hyper-energy is understood to be present in the universe as a whole at all times during all the cosmos' phases. We, also, understand the sun to be our immediate but, cosmically, a secondary source of energy. So let us trace that energy "back" as far as we can to see if the implications of modern physics' findings are consistent with the Sextimage's picture of hyper-energy. This will take us beyond the "Big Bang."

The total energy of a star includes its luminosity and the "potential" energy in matter that escapes it, which is a function of the mass and velocity of that matter (energy = mass × speed of light squared). All these forms of energy are transforms of atomic energy, that is, are the force that binds electrons to nuclei, and the force that binds protons and neutrons (nucleons) to form nuclei and repels incoming protons and neutrons. The chief and by far most powerful force among these is the nuclear binding-repelling force, which is what is released in the process of the nuclear fusion of star-fires. Roughly speaking, in nuclear fusion nuclei are forced to grow larger, to acquire more nucleons. The nuclear force between nucleons and their environment is violated and escapes, as nucleons are forced to unite with excess nucleons. The general condition for nuclear radiation, then, is *amassing, condensing.* The element that amasses, condenses, in nuclear fusion is deuterium, or abnormally massive, super-yin, hydrogen. Nuclear fusion is the effect of forced amassing of what is already abnormally massive. The energy that *causes* this, then, is a force that brings *like* things together. It is therefore unlike magnetism or electricity (which are "plus-minus" phenomena) and like gravity. That force, or energy, is also, at least in one stage of this amassing process, *heat,* for great heat is the initial condition for fusion. Atomic fission, the splitting of nuclei, produces such heat. It, in turn, is again the effect of the amassing of abnormally massive, nucleon-emitting, radioactive elements. So, all told, nuclear fusion is an amassing of like particles of abnormally massive elements which is a transform of heat which is a transform of an amassing of like particles of abnormally massive elements which is an effect of a force that brings together, amasses, condenses, like things. That force is energy in its causally original, purest, form. And that force is like gravity. It is reasonable to suppose, therefore, that it is characterized by long-slow waves, is in the red direction, for gravity is the weakest (in the short run) of the Four Forces, just as red-hot is the "coolest" (in the short run) of the hots. The effect of that force then goes in the violet-black direction, becoming heat then light then x-rays then nuclear radiation then "cosmic" rays as it increasingly interpenetrates with matter.

In sum, modern physics leads to the hyper-red energy of the Sextimage as the primal cosmic force. But how did the Chinese figure this out?

A phenomenon observed by the ancient Chinese which has the same general pattern as nuclear fusion is the lightning storm, one that is particularly interesting in that lightning is naturally supposed to have had something to do with the origination of life. The lightning storm is the natural phenomenal example of the *I Ching* trigram, or configurational pattern of yin and yang lines, *Chen,* which represents the first effect of the meeting of cosmic yin (the trigram *K'un*) and cosmic yang (the trigram *Ch'ien*): the meeting of matter and energy, of "earth" and "heaven," of the cold-wet-receptive and the hot-dry-active. (The color associated with *K'un* is black; of *Ch'ien,* red; and of *Chen, hsuan* [violet] and yellow—the *extreme* yin and yang colors.) The energetic phenomenon of lightning is the effect of an *amassing* in the sky of water from the earth and of a separation of like, heavy things (earth and water); and it accompanies a return of those like things to each other (rain onto earth). (Heavy hydrogen is derived from sea*water,* and nuclear fallout, like rain, returns to earth.) There are two kinds of force or energy that account for this amassing of water in the sky. One is the vaporizing heat from the sun. The other is what attracts water droplets toward each other—a gravity-like force that amasses like with like.* The same forms of energy are recognized as ultimate causes on the basis of the empirical data of a storm as are recognized on the basis of the (theoric-)empirical data of an atomic explosion. In both cases the energy is "red," of the long-slow variety.

Likewise, the very origin of the universe, that is, of the universe of galaxies and solar systems that we know, would be the result of gravity or a gravity-like force, a "red" energy. For those disinclined to believe the fantastic proposition of our physicists, that the universe began as

* Wind, of course, is also an effect of the sun, with water.

an exploding cosmic egg of dense matter that emerged instantaneously out of "Nothing," the rational Chinese hypothesis may be held, that the universe is eternal, always was and always will "be there," and that the present, expanding phase of the universe is the effect of a prior condensation—substantizing, therefore energizing, of matter—major yin becomes minor yin. All the known forms of energy, then, except, perhaps, gravity, would be materially conditioned transforms of an energy that amassed all the matter in the universe into a cosmic egg (or eggs) of critical mass. That prior form of energy would be or be like gravity, would be in the red direction. This confirms the hypothesis, above, that the nuclear binding-repelling force that is released by fusion in star fires, since it is a transform of a prior, amassing energy, is basically red; that is, as it interpenetrates with matter to acquire form, it becomes red.

Actually, according to Chinese science this original and purest energy is not gravity, for gravity exists only when there are cosmic concentrations—substance, cosmic bodies—and here we speak of a cosmic phase prior to the existence of such "eggs." The energy or force we are trying to get at *creates* substance, and all forms of energy other than itself, out of the undifferentiated energy-matter that results from terminal entropy. This gravity-like, pre-gravitational force or energy, energy in its purest and original form, then, is the Chinese "hyper-energy," that is, the energy that causes the cosmic Tao-pattern of Nature as a Polar interaction of energy and matter, Nature as Polar differences of all kinds. What makes it gravity-like, "red," is that it brings like and like together. This is associated with warmth (long-slow waves) as distinguished from heat: warmth promotes chemical combining and organic functioning; high heat cooks, burns, de-structs. Amassing matter, on one hand it makes gravity and the strong nuclear force possible by creating mass, and on the other it Polarizes energy and matter, creates maximal tension between them, sets the conditions for the other, plus-minus, forms of energy, that we (now) call electromagnetism and the weak nuclear force. As Lao-Tzu laconically put it: "The one gives birth to the two." As follows the preceding Polar-Complete analysis, the color of space, which hyper-energy and primal matter fill, is said to be dark red (*chiang* or *chih* – the *I Ching*)—red plus black.

I suspect that contemporary physics is sitting right on top of the preceding Chinese understandings. Our cosmologists-physicists have quite possibly discovered both hyper-energy (*T'ien-wai-chih yang*) and its primal material substratum (major yin). The preceding deduced characteristics of hyper-energy are those of the "black-body radiation" which has been discovered through radiophotography in space and taken to be an echo of and evidence of a "Big Bang." Its waves are in the Absolute-Fragmental terms of our physics "three meters in length" and its temperature is three degrees above "absolute zero." Its color is precisely dim infrared. According to the Tao, one would posit that these waves are movement plus matter: that there is a material substratum, that they are of major yang *and major yin.* In the 1970's our physicists finally abandoned the notion that there is absolute space, substituting for it a "neutrino-sea" of tiny particles which, until spring of 1980, were thought to be "massless" and now appear to have extremely low mass. A minimal interaction or blend of almost energyless and therefore apparently motionless movement per se and almost massless and therefore apparently immaterial matter is precisely the image of hyper-energy and major yin, the unbroken line and the broken line, at the entropic extreme and/or at the most basic, cosmos-causing state of the universe.

Assuming that the universe had a beginning, so that an Absolutely metaphysical, supernatural, cause is implied, our cosmologists suppose that the "black-body radiation" is energy at or approaching the entropic extreme, like the energy in Aristotle's run-down cosmos, which I have likened to a watch. But there is, of course, no reason to suppose that the universe had a beginning, that it is not eternal. And there is reason to suppose that it is eternal and therefore not linear but cyclic. First and foremost, it dispenses with the supernatural mysticism on which all Western science, with its cause-effect, is based, and provides, instead of a schizophrenic paradigm in which Natural "laws" dead-end at supernatural causes, a consistently scientific paradigm. Second, among all known phenomena there is not one which is not cyclic, not one which has a unique beginning or permanent end. Inanimate processes have temporary beginnings and ends: metal heats then cools, for example. But they repeat. Eventually the energy lost in the cooling of the metal, or equivalent energy, will reheat it. Such temporary beginnings and ends, then, are subordinate to the circular-cyclic processes of which they are aspects, punctuation marks. Animate phenomena, in turn, do not even have temporary beginnings or ends, for they are self-reproducing. If all known phenomena are cyclic, the logical thing to posit is that the whole cosmos also is, for they are aspects of it and, assumedly, effects of it, extensions of it.

Accordingly, Chinese science posits that the universe is eternal and cyclic. The observed hyper-energy-primal-matter mix, or the "black-body radiation" plus "neutrino-sea" that I suppose symptomizes it, is therefore *both* the state of maximal entropy (or close to it) *and* the original, pre-cosmic (pre-structured) nature of the universe (or close to it). The phase of maximal entropy/cosmos-origination is called the *T'ai-Chi*: Great Extreme, Great Pole (the Pole itself, *almost* without the Polar opposites contained within it).

It is worth pointing out that "black-body radiation" could not have resulted from the explosion of a Cosmic Egg, in any case, and its discovery does not necessarily imply that the cosmos had a beginning. The Western rationale boils down to this. The radiation emanates in all directions and propagates uniformly, Therefore, it cannot have solar or galactic sources; it must have been emanated by an explosion long ago. But, as should be obvious, that conclusion does not follow from that finding. To cause uniform movement in all directions an object, which must be spherical—here, the Cosmic Egg (a metaphor, like our entire paradigm, borrowed from ancient West Asia and Greece)—cannot simply explode, that is, expand, move outward. It must *simultaneously im*plode, contract, move inward, to enturbulate energy-matter is all directions.* Such Polar-Complete movement, unlike the simple explosion our cosmologists unrealistically posit, is not necessarily a beginning-movement, a cause on a time line. It can be posited that it occurred once to establish the universe or that it constantly occurs, is a "given," a basic eternal Natural property, or both. (We will see that this pattern is true of atoms; it constantly occurs.) What other than a Polar-Complete movement could underlie a Polar-Complete array of natural phenomena?

In turn, the Chinese understanding that simultaneously expanding-contracting hyper-energy is the spatial-temporal axis of the cosmos, the source of all its movements, accounts for an important datum which our cosmologists cannot account for. The center of our galaxy and all others known is very bright infrared, and there is as much of it

* One might posit explosion, concentration, and, instead of simultaneous implosion, reflections of exploded energy from the concentrations, but the result would not be perfectly uniform, because the concentrations incorporate and trap some of the prereflected energy, because they are not distributed with perfect uniformity in space, and because such concentrations—stars for example—fluctuate in radiation.

as there is visible light in the whole galaxy. Likewise, most extra-galactic objects in the universe radiate more infrared light than any other form of electromagnetic energy. In other words, hyper-energy seems to exist in the same amount as visible light, which suggests a complementary relation between the two, and it tends to occupy the central position in the cosmic structure. In short, it seems to be the hearth of the cosmos, the sun behind the suns, of which the relatively material and, in space but not time, relatively energetic suns are energy-containing effects. This is just as Chinese scientific theory has it, and almost exactly opposite to the way Western physical cosmological theory, with its God-caused Ultraviolet Catastrophic Big Bang origin, has it. This is not to reject the possibility of a Big Bang; it is to say the "egg" or "eggs," in turn, must have been an effect of the *subtlest* energy—of which black-body radiation may be a symptom.

The fit of the Sextimage's major concepts with the central contemporary cosmological data is plainly superior to that of Western cosmological theory, if only for the fact that it eliminates the mysticism essential to the Western paradigm. Yet, it is but one of many such superior fits: a minor capacity of a genuine unified scientific theory.

To continue our Trip in-all-directions-simultaneously, I steer our Tao-space-time-ship toward the biological level. According to the Sextimage life is basically red, major yang, and reaches its prolongation-culmination, its rebirth, its Spring, in green, minor yang, a red-black mixture, with violet, reverting yin, to a secondary extent. Living things, like gold, should be chiefly red and green, and being animate, they might differ from gold in the positions red and green have in them. Gold, as we saw, is externally red and internally green. Of course, plants, the basic life-forms, are externally green and their chloroplasts absorb basically red energy from the sun. And it is, indeed, chiefly the red sector of the sun's light-spectrum that they absorb.

More precisely, plants absorb chiefly the red and violet light in sunlight, the two poles. According to our Chinese principle, the red light should serve relatively energetic plant processes, for example, reproduction, and the violet light should serve relatively material processes, for example, the multiplication of cells. This is precisely the case: the red light has the greatest effect on flowering and extension; the violet, on growth. Further, it is the red light that gives plants the energy to produce carbohydrates. The energy in plant food chiefly is red light. Further, looked at through a red lens, *living* green (*ch'ing*) leaves are red whereas dead green (*lu*) leaves are black. Living green, which the architects of Chinese culture had the sensitivity to distinguish with its own term (*ch'ing*) from dead green, is internally red, is black-plus-red, whereas dead green is internally only black: substance without life (minor yin). The protein, phytochrome, which responds to red light to initiate bright-green photosynthesis, is specifically *blue*-green, a *black* green, as precisely conforms to the Sextimagic pattern.

Consistent with that, the fruits of plants, the ultimate repositories and expressions of their vital force, should be red or at the red end of the spectrum, in contrast to the green bodies of their plants. And they are. Would Western biology dare even to pose the question: "Why are plants green and their fruits red or relatively red?" Such questions may be posed and answered in the West only in the fantastic etiological tales we read to our children. (And I for one have wished to have answers to them since I was a child. Today, the open-mindedness and unbridled self-assurance of children, tempered by the skepticism and probing-in-depth of adults, is what Western scientists must cultivate in themselves, without embarrassment, if such questions ever are to be answered.)

Sexual intercourse is a Polar-Complete opposition-neutralizing phenomenon. Plant flowers, as sexual organs, should be, according to the Sextimage, of extreme yin and yang color: red or light-bright yellow-white for male, dark-dim yellow-white or violet for female; or, more generally, in each species the female flower should be more at the yin end of the spectrum than the male flower (and reciprocally), and both sexes should be reddish (highly energetic). The most focal exemplar of a plant would be a tree, because of all plants it is chiefly trees that emit the substance most immediately necessary to mammalian life—oxygen. It would be an archetypal tree, an historically early one, and ecologically the most widespread, therefore the conifer. Here is a list of the flower colors of conifers that grow in my own area, each color followed by the number of its instances. Similar results are obtained from a list of all the conifers.

MALE (YANG)	FEMALE (YIN)
yellow (5)	purple (5)
red/yellow (2)	red/yellow/*green* (1)

MALE (YANG)	FEMALE (YIN)
red (2)	green (1)
purple (1)	scarlet (1)
	orange (1)
	violet-pink (1)

In 9 out of 10 species, the male flower is at the red end of the spectrum; in 8 out of 10 species, the female flower is at the violet end of the spectrum; and in 9 out of 10 species, the female flower is further toward the violet end of the spectrum than is its male counterpart. What's more, the most common male and female colors are yellow and purple, the *ming/chueh, extreme* yang and yin colors, the colors of energy and matter in maximal reaction to each other, one specific manifestation of which, of course, is sexual reproduction, the function of the flowers. No aspect of Western biological theory could make such a prediction. Much less could any general, physical-biological, Western theory. None exists.

Let us head toward the inner planet of mammalian life. The basic blood structure, hemoglobin, is allied to that of plant chlorophyll, and its color is chiefly red and secondarily green—precisely as would be expected in higher-than-plant life-forms, which process relatively refined, purer, therefore redder, energy—and which, in their highest, human, form are able to aesthetically-logically tune in on, deduce, and empirically find hyper-energy, to engage in genuine science. According to the Sextimage, violet is matter yielding or exhausted of energy, black light, waste. The color of blood in the veins, waste blood, the by-product of vital energy-matter interaction in mammals, is violet. In contrast with the Tao-ship's all-directional travel, Western biology can explain the color correlations, but it cannot do so on the basis of any scientifically *general* theory or even on the basis of an aspect of biology not directly pertaining to the blood system.

To complete the predictive analogy between plant and mammalian life: mammalian sexual organs and/or signs are predicted, like those of plants, to be relatively yin and yang in color and reddish (as are red, yellow, and violet). Mammalian genitals and their surrounding area are red; so are the lips, whose erotic function is to signify kinship-like relationship, through the kiss. Where maleness is yang, the markings of male mammals are almost unexceptionally brighter and redder, more yang, than those of female animals.

Human beings, because their nature is to understand and use the Tao and because, being the most animate among beings, they most complexly integrate yin-yang oppositions, are not simple. But they are equally Sextimagic. At their animal level, and as psychologists working for the advertisers who subliminally manipulate humans have concluded, men are relatively attracted to blue (yin, female) objects, and women are relatively attracted to red (yang, male) objects, just as the Sextimage, applied in the simple manner used just above, would predict. However, men and women most attract each other by wearing clothes of the colors *opposite* to their own sex. Attracted to a red, yang-male-colored, dress, a woman puts it on and unconsciously attracts men by, as it were, being red-hot with input male sex. Attracted to a dark blue yin-female-colored suit, a man puts it on and unconsciously attracts women by, as it were, having his compulsive male energy contained and controlled by yin, communicating self-control, reliability, and authority. These colors of the clothes tell the opposite sex: "I like you, I am like you, I can unite with you."

To digress into the social science of this matter: the female-red and male-blue association being established and exaggerated, in modern society the same trick is played with opposite tools to get people down to an animal level and sexually confuse them at the same time. Progressive girls wear blue jeans and send off two signals: the brute yin one and "I am like you in that I dress like you." "Swinging" men wear red disco-shirts and drive red cars and send off two signals: the brute yang one and "I am like you in that I dress like you." Boys dress like girls dressing for boys, and girls dress like boys dressing for girls. Such sexy-antisexy imagery is a specific form of the "dead neutralization disguised as equality, which results from extreme opposition," a general feature of our scientific paradigm and its social effects, as explained above. Our society separated the sexes by overemphasizing sexual differences. Clothes have been strictly marketed for sex connotation—an aspect of the hyper- and therefore unhealthy sexual syndrome as a whole. One symptom of it is an overabundance of prostitutes in red dresses and of dark-blue-suited politicians whose real objective is to own the best of those prostitutes. There followed an extreme opposite reaction, toward "unisex" and "trans-sex," of which the present state of confused clothes symbolism is an aspect.

To return from social to biological science: Kirlian photography reveals precisely the vital energy-patterns predicted by the Sextimage and the concept and mapping of *ch'i*, or configurative, vital, energy. (*Ch'i* is discussed in the following chapters; here I deal with it minimally, concentrating on colors and yin-yang patterns.) Kirlian photography was invented by the Soviet Armenian Semyon Davidovitch Kirlian. The auras or coronas it reveals, in conjunction with the Chinese-medical mapping of vital energy, are now one of the central objects of Soviet and enlightened West Coast American medical research. It records color-shape images produced by an interplay of electricity set up by the researcher and an unknown form of energy which surrounds the human body the way flames fringe wood coals or the sun. Excellent examples of such photographs may be found in *Electrophotography*, a book by Earle Lane. According to Chinese medical theory, there is an energetic exchange between the human body and its environment through the human skin, which involves both intake and elimination, paralleling the gross physiological system to which our medical scientists and biologists exclusively devote their attention. According to the Sextimage, then, if this energy were materialized enough to acquire visible color, it would be red for the intaken, relatively energetic, energy and violet for the expelled, relatively material, energy. This is precisely the Kirlian finding. Lane's photographs of human coronas are chiefly of people in extraordinary emotional or drugged states, but include coronas of people in normal states and the corona of a person who has ingested *red* ginseng, which according to Chinese medical theory intensifies vital energy by strengthening its blood substratum. The corona of the red-ginseng-eater is red and violet and, of all the examples provided, the colors are the brightest, and the Polar opposition of the colors most pronounced. People in normal states, in turn, have the same color-pattern, but less brightness and balance. In contrast, when subjects are burning their *ch'i* (energy) unnaturally fast or directing it unnaturally by ingesting cocaine, nitrous oxide, or marijuana, or when they are energetically involuted (as an effect of introspective meditation or Western psychotherapy), the colors are asymmetrically distributed and shift toward black. More specifically, images arise either of rapid loss of chaotized energy to the external environment (nitrous oxide), or of a locking-inside of energy and subnormal resistance to the external environment (cocaine, introspection). The flesh itself registers black, or dull (black) red. In general, these results confirm that human vital energy (originally, *t'ien-wai-chih yang*) is basically red, that in its relatively material form it is violet, and that in a healthy state it balances red and violet in a Polar manner.

The success of the preceding Chinese prediction has a specific empirical basis which is worth pointing out. At this point the credulity of the most open-minded Western reader will have been strained, not by any violation of the criteria of truth, but by the diversity of the phenomena which have just been made intelligible by a single scientific sub-paradigm. The Chinese empirical basis is also worth pointing out because it further supports my most basic scientific argument—that genuine science is based on *sense*-data, and that sense-data lead to a paradigm which operates successfully with regard to imperceptible, indirectly sensible data, such as precede. As spatial characteristics, yin and yang are defined respectively as inner and outer. Substance is yin, its color is yang, and so on. The human body has been mapped accordingly: generally speaking, the surface of the body, which is concave when a human contracts torso and limbs, like a foetus, is yin; and the surface which under that condition is convex is yang. (Front of neck to groin is yin; head to buttocks is yang; outer surface of arms and anterior surface of legs are yang; remaining limb surface is yin.) Even without tanning, the yang body surface is "darker," that is, red-yellower, and the yin body surface is "lighter," that is, blue-whiter. What's more, according to general Tao theory, the yang aspect of any body sends energy outward and the yin aspect of any body brings energy in (and the basis for this generalization is also obviously sensory). It follows from these basically common-sensical-sensory generalizations that the body emits extreme yin energy-matter and intakes major yang energy-matter, precisely what the Kirlian photographs show.*

Independently supporting those conclusions, of course, is the belief (substantiated below) that there is an invisible energetic aspect of the body which is continuous with the visible one. Given visible red arterial and violet vein blood, one deduces invisible incoming red yang vitalizing energy and invisible out-going yin waste energy. And there is every reason to make such an analogy, because where logic dictates that the cosmos as a whole is based on invisible hyper-red energy, and humans are obviously extremely energetic entities, it is to be expected that they very actively and extensively use such energy.

* All the preceding is a gross simplification of the Chinese theory, and many of its details have been confirmed by more detailed Kirlian data, some of which are given below.

THE CONDITIONS FOR LIFE

We have almost completed our Red Shift from inanimate to living phenomena. We have seen that energy is in the red direction, and that primal cosmic energy, hyper-energy, emerges as human vital energy via a super-ecological chain. The chain is traceable from the first or continuous-and-basic Polar-Complete, gravity-and-substance-creating expansion-contraction of the Great, major yin-yang, Pole, through nuclear sun fires, through the red energy in green plants, constituted by hydrocarbons (hydrogen and oxygen are red and carbon is black, just as the Sextimage predicts), to the red oxygen, red plant-animal food-energy, and directly received red cosmic energy ingested by humans.

That red energy completes its circle when it is released at organisms' deaths, again becoming free, primal, cosmic hyper-energy. But more precisely, it is a residuum of it, which has not been transferred to individuals' offspring, that is released. As should be obvious but as is totally ignored by Western biology, the energy ingested by an individual organism cannot during its lifetime be sufficient to its living. The energy ingested powers the organs which do the ingesting and processing. Hence, originally, there has to be energy for ingesting and processing; the organism must be "primed" prior to its first ingesting and processing. That intrinsic vital energy comes from the parents.

In the foetus, the first loci of reproductively transferred vital energy are the kidneys, whose function it is to circulate minor yin—materialized energy, water, and minerals—and the heart, whose function is to circulate, in the water provided via the kidneys, major yang—what we call "oxygen" and refined food energy. As Western biologists know, the first organs to form in the foetus are, in fact, the kidneys and heart. And it happens, not by arbitrary design or accident but as a result of universalistic aesthetic-logical-empiricism, that the *primal couplet* in all the Sextimagic orders is black minor yin and red major yang. These intrinsic energies and organ functions are distinguished from those which result when the other organs have formed and the infant breathes.* Likewise, as we saw, the protein, phytochrome, which reacts to *red* light to initiate the basic life-process of plants is *blue*-green, that is, a *black*-green—in perfect accord with the general Sextimagic principle that the basic vital couplet is red-plus-black.

As the original organs, the heart and kidneys have Quadrimagic values: major yang (heart) and major yin (kidneys); as members of the fully developed organ-function system they and the other organ functions have different, Sextimagic, values. But in both contexts their respective energies are major yang and major yin. The Sextimagic organ values are determined topologically (yang is external, yin is internal); functionally in terms of internal or external direction of vital energies processed; with respect to solidity (solid organs are yin, hollow organs yang); yin-yang nature of the energies processed; colors of organ tissue; and colors of symptoms of organic-functional health and illness. To go right to the end of that exceedingly complex and subtle aesthetic-logical-empirical story and on to the question of its scienticity, the ten major organs** are Sextimagically classified as follows:

major yang: bladder and small intestine

reverting yang: stomach and large intestine

minor yang: gallbladder

major yin: lungs and spleen

reverting yin: liver

minor yin: kidneys and heart

The simplest, theoretically "lowest," formulation of the correspondences between physical diseases of the organs, their symptoms, and their medicinal cures, is found in Chang Chung-Ching's 2000-year-old *Shang-Han Lun.* (The *Shang-Han Lun* is the theoretical basis of the Japanese version of Chinese medicine—Chinese medicine in its very simplest form.)

Here is one simple example out of over one hundred Sextimagically classifiable ones, of interest because the mysterious Legionnaire's Disease is a variety of it: illness of the major-yang type originates in the

* As *Hsien-T'ien-te:* "of the Original Nature" *vs. Hou-T'ien-te*: "of the Derived Nature."

** The eleventh and twelfth are discussed at a later point.

bladder and small intestine and, where the bladder directs water out of the body, is centrally felt in the external area of the body—the skin and flesh and, by virtue of their own externality and their direct coupleting with the major yang small intestine, ultimately the lungs (major yin). The symptoms are fever, headache, stiff neck, general aching, an impression that the bones are out of place, insufficient sweating for the amount of evident heat; "floating" pulse, that is, a pulse able to be felt only with a light touch of the fingertips, imperceptible when pressed; bright red (major yang, hot) tongue, often with a wet white (major yin) coating indicating an internal coldness and wetness centered in the lungs; at the extreme, dying, a black (minor yin) coating indicating terminal kidney dysfunction (the kidneys form a couplet with the bladder). The medicines prescribed for illnesses of the major-yang type Sextimagically correlate with its major-yang nature but are not themselves of major-yang nature. This is in keeping with a general rule of Chinese medicine, that the diseased organs, being diseased, cannot process the medicine and therefore should not be directly addressed. (The direct address of Western medicine is one of the reasons for its side-effecting character.)

Further examples need not be given here. It is sufficient to the present purpose, which is to demonstrate the universality and validity of the Sextimage, to point out that the Sextimagic correspondences among symptoms, organs affected, and medicinal cures constitute the most basic (and simplest) aspect of a medical theory, which is applied to heal a much broader range of diseases than does Western medicine, and does so *without side-effects*. In short, the scientific validity of the Sextimage and the Tao behind it as applied to human biology, health, illness, and healing is confirmed by the efficacy and non-side-effecting Human-ness of Chinese medicine.

To posit that although the healing techniques may be more efficacious than ours the theory behind them is false, is absurd. The techniques are exact and forthright realizations of the theory. Nor does Chinese theory contradict the *empirical* aspect of Western medicine; rather, it goes beyond it. What's more, Western science and medicine are not legitimate grounds for determining the scienticity of any alternative discipline. First, Western science has not achieved theoretical unity. Second, being devoted almost exclusively to inanimate physical conditions associated with it, not only does our biology admittedly know least about life itself, but the "scientific" medicine that is based on it declares its fundamentally unscientific character and its ignorance, even opposition to, the living process every time it produces a disease-causing side-effect, which is almost unexceptionally. Chinese medical theory is not unscientific, it is non-Mentalist, Polar-Complete, natural, and alive. To say that it is unscientific is really only to say that it is non-Western. That is to announce the possibility that it is genuinely scientific and to imply the great difficulty that Western scientists, because of their conditioning to an unnatural and restrictive mode of thought, experience when they attempt to understand Chinese medical theory. "I can't understand it; therefore it is false." To follow such "logic" further: "Yet it works better than ours does; therefore Chinese doctors have for 3000 years been using a form of *Western* medical theory more advanced than ours, without knowing it."

We have seen that the Four Seasons, heating phenomena, metallic properties, cosmic structure and process, and plant and mammalian life-forms and processes are all made intelligible by the Sextimage. In addition, it makes possible Human, non-side-effecting, medicine. By implication of that, all those phenomena are basically similar: the universe is, indeed, organized analogically. It was the Medieval European recognition of this, a result of non-Mentalist aesthetic-logical-empiricism, which made the Dark Ages our most enlightened scientific period and the Enlightenment, in contrast, the beginning of a dark one. What that science lacked, because of its Christian Absolute-Fragmentalness, was the concept of Polar-Completeness, is indispensable to achieving theoretically unified science with Human effects.

Now that the structural-analogical Sextimagic unity of animate, inanimate, and animate-inanimate phenomena has been indicated, I conclude by explaining in Chinese terms what the distinctive feature of animate phenomena is. Distinctive features are differences, which can only be truly recognized after samenesses have been.

Because Western science is non-vital (specifically, because it regards physical-structural data as primary and qualitative-functional data as secondary*), it ignores one half of the nature of animateness. Also, because of that, it will be difficult for the Western-educated reader to grasp the following concepts; but, if he has absorbed all that has preceded, a pattern that makes sense will form in his mind. Western science recognizes two animate features: that animate things are

* Also see Porkert, 1975, on this difference. (He does not, however, say anything about vitality or animateness as between the two sciences.)

more highly organized than are inanimate ones, they go beyond molecules to cells and, if higher organisms, organs; and that animate things self-reproduce. Western science goes beyond Chinese science microscopically to describe the reproductive division-and-multiplication, and then, in higher organisms, sexual fusion, of cells, but it does not begin to explain the conditions which make this aspect of self-reproduction possible. Chinese science recognizes not only higher-organization-and-self-reproduction but also the reason for those conditions. It is to those conditions, then, that we must attend.

In general, what distinguishes animate from inanimate things is that they are self-perpetuating. This self-perpetuation is true not only of species' regenerations through the generation of new individuals, but also of the individuals themselves. Once established, their metabolisms continually function: all the cycles within metabolism repeat themselves. Likewise, once the cellular-and-organic structure is established it self-regenerates. Indeed, it seems obvious that self-reproduction is made possible by the self-perpetuation of metabolism and organic structure, for they constitute most of the individual which does the self-reproducing. So the datum in question here, ignored by Western science until it recently produced tentative Systems- and Cybernetic theories, is the self-perpetuating capacity, or *non-entropy*, of the living individual. That is, animate things conserve their energy and therefore—because it is energy which holds physical things together—their physical structures. In contrast, inanimate things simply lose their energy and physical structure, as the Second Law of Thermodynamics accurately states. The extreme example is radioactive substance. The borderline inanimate example is certain crystals, which "grow" and therefore, when they lose part of their physical structure by entropy, can "regenerate" it. (The animate-inanimate difference is of course Polar-Complete, not Absolute-Fragmental.)

Nothing about the physical structure of organisms, which Western science exclusively attends to, explains the energy-conserving capacity. And this stands to reason, because higher organization, or holding-together, requires more energy; it is animate *energy*, not physical structure, which is the relatively independent variable. But something about the Sextimagic, qualitative-functional, structure *does* explain the energy-conserving capacity. The functioning of the yin-yang sectors (the correlate of our organs) of a living individual is non-sequential, simultaneous. In this sense it transcends time. This is obvious without regard for the yin-yang sectors: the organs, regardless of their yin-yang values, are observed to function simultaneously. The recognition of the Sextimagic structure makes it possible to *formulate* this specifically animate condition and to relate it to the universal Sextimagic pattern, thus making biology an aspect of unified science. First let me illustrate the non-simultaneity of inanimate processes and the simultaneity of animate ones.

Inanimate phenomena have relatively definite yin-yang stages. For example, when red energy and black substance interact at the initial stage of metal heating, there ensue relatively definite, that is, temporally separate, stages of red-hot, yellow-hot, white-hot, green-hot, and blue-violet-hot. Then the energy of the heat is lost through cooling, or the energy of the molecular composition of the metal is lost through further heating—ultravioletness—atomization. Likewise, to take a more general phenomenon, metals are formed (first stage) by earthly pressure (major yin produces minor yin) and decompose (second stage) either through superheating or water-and-oxygen erosion or friction or nuclear radiation (focally, reverting yin energy-loss and therefore also substance-loss). Again, there are definite stages. Likewise, all machine cycles (really the models for Western "biology") have definite yin-yang stages (combustion is yang, exhaust is yin, and so on). Machines—unfortunately, our models for understanding life-forms—are an excellent example because they are maximally human- or animal-like yet inanimate. In contrast, organisms, with respect to their most animate aspect, the life process, or what we call metabolism, have no yin-yang stages. Every organ function operates constantly. One does not digest, breathe, circulate blood, and so on in any order. Such an order would bring death. Otherwise put, where there is temporal order of yin-yang sectors in inanimate phenomena, there is none in animate phenomena in this centrally animate respect. In this sense, animate yin-yang transcends time and is free, therefore, of *cause-effect* relations.

That is worth elaborating on. Based on the Absolute-Fragmental mode of thought, Western Systems- and Cybernetic Theory attempt to formulate the animate conditions as complex feedback systems. These systems are simply cause-effect relations dialectically multiplied. These theoreticians find that X causes Y, but also that Y causes X, which neutralizes the concept of cause and effect. Rather than recognize this, however, they attempt to formulate what goes on as a *series* of feedbacks: X-causes-Y and Y-causes-X are regarded as spatially and temporally distinct events. The simultaneity of the actual animate process is then *approximately* expressed as an overlapping in time of those events. Initially, either X causes Y or Y causes X; then there is alternation: X causes Y causes X . . . ; and overlapping: just after X begins to cause Y, Y begins to cause X. The resulting cause-effect pattern can be complicated infinitely: there is an X-caused Y inside a Y-caused X inside an X-caused Y . . . The number of variables involved may be many more and causation simple (X:Y) or multiple (X:Y-plus-Z), or direct (X:Y) or indirect (X:Y:Z), but the basic, mechanical structure of the system remains. And that structure is a Mentalist avoidance of the recognition of the actual Polar-Completeness of the animate process under investigation. Just as it follows from the susceptibility of color to infinite subclassification that colors are Polar-Complete, it follows from the susceptibility of animate "cause-effect" relations to infinite subclassification that they are Polar-Complete. In one respect, X and Y are *relatively* causal and relatively caused in relation to each other, and in the other respect, neither is causal or caused in relation to the other. This and only this conclusion is consistent with the fact that feedback within animate systems is total, so that all organic functions are simultaneous. Plainly, the Mentalist paradigm cannot handle this aspect of reality.

The respect in which X and Y, for example, heart and lungs, are neither causal nor caused is the respect in which animate processes are simultaneous. Where there is no cause or effect there is no time, for time—con-*sequence*—is necessarily implied by cause and effect. Without time there can be no entropy. The nature of the animateness in question lies here: movement which is non-temporal. Movement which is temporal is recognized as such when the spatial characteristics of the phenomenon change. We infer time from changing *spatial* characteristics: the sun is here, now there, and we call that "one hour" or "Summer," ascribing movement to the sun, or the earth. It follows that non-temporal movement is independent of space. Where space, or containment, is nothing other than major yin, matter, it follows that non-temporal movement, or animate energy, is energy that is in some way relatively free of matter. One can then infer that in animate things either matter (major yin) is relatively weak as compared to that of inanimate things or that energy (major yang) is relatively strong—penetrating, through-going (*t'ung*)—through-going like high intelligence, one of the most distinctive potentials of the most animate beings, humans. Of course the matter of animate beings is not relatively weak but relatively strong, for it contains energy (in a specially animate way, explained below) to the qualitatively superior extent that there is no entropy.* So it is not that matter is weaker in animate than in inanimate things, but that energy is stronger in animate than in inanimate things. Of course, the strength of matter or energy is not gained at the expense of its Polar complement. On the contrary, yin and yang are only as strong as each other.

Since it is animate energy which accounts, by its special *t'ung-te* (through-going, property) for the simultaneity of animate process, then, it is animate energy which is the relatively independent variable of animateness. In colorific terms, animateness, vitality, is redder than inanimateness. Consequently, where the opposite of the major yang optimal-energy-condition is violet, animateness is less violet than inanimateness.

But animate matter is also specifically animate, has a special animate quality. It follows from the deduction that animate energy is relatively free of animate matter, yet that animate matter is as strong as animate energy, that the exceptional energy-conserving property of animate matter is not due to the material property of energy-impedance, but to something else, which permits animate energy to be relatively unimpeded. This is a very subtle concept, but the Sextimage makes it explicit and intelligible. In general terms, energy could be conserved, contained, in two possible ways: impedance by matter, which substantizes and temporalizes it, or *re-cycling* by matter, which would not substantize or temporalize it. To illustrate by analogy: steam can be contained by a strong-walled container, which would make it more substantial, a fluid, by putting it under pressure; or it can be contained by being conducted through a circular pipe, which would let it course at its own speed and would not make it more substantial. Of course, much more heat is lost in the first instance. The latter "solution" is like what animate matter does in this case. "The weakness of the flesh is also its strength. Now let us see in Sextimagic terms how the human organism, for best example, does this.

As hinted at with reference to the stability of gold as life-like, therefore Spring-like, the distinctively vital sector of the Sextimage is reverting-yin-plus-minor yang—specifically the extremest "half" of reverting yin and all of minor yang, for the initial "half" of reverting yin is Wintery. This is known through common sense, then marvelously

* The vital process is self-repeating until death, before which it has extended into the individual's offspring; hence, no entropy.

confirmed empirically. As we saw, violet (reverting yin) is the condition of maximal entropy. It follows, by common sense, that in animate things the reverting yin sector is coupled with another sector so that it does not simply lose energy and, therefore, physical structure as well, but rather recycles it to replenish the energy-matter supply of the organism. Violet-reverting-yin is not a terminal sector in animate things. We have an animate-inanimate model for this: the transition from Winter to Spring, wherein violet-brown compost releases its energy into minor yang green wood—plants and all reborn life-forms. These two sectors are, of course, an aspect of the basic, major yin, matter in question.*

Just as the Sextimagic correlations state, the violet-tinted organ is the liver and the green-tinted organ is the gallbladder. Bile is yellow-green: yellow for the major yin basis of minor yang (yang in yin); green for its yangish yin. This is not to say that the liver-gallbladder is the center of vitality any more than any other organ or organ couplet, but it is the coordinate at which animate yin-yang, as compared with inanimate yin-yang, is a closed circle rather than a line with a beginning and an end, and therefore is self-reproducing. According to Chinese medicine, the liver (made energy-*conserving* reverting yin by the minor yang gallbladder) chiefly serves to organize all the other organs, to make them operate smoothly and harmoniously by distributing vital energies. It is called the "General" of the organs. The liver-gallbladder couplet includes the only yin-yang color generally associated with a living concrete referent: green, for wood, plants, life-forms.** The liver is chiefly violet and secondarily green; the gallbladder chiefly green and secondarily violet. As Marc Duke has pointed out in his *Acupuncture*, modern medical research has confirmed in

* Ilya Prirogene, the enlightened chemist who is making revolutionary connections among physical, chemical, biological, and social data in terms of "coherent energy," has stated that somehow life "eats" entropy. How it does so is what I have just explained in Chinese-paradigmatic terms including as a central concept one similar to his "coherent energy." It is qualitative-aesthetic dimensions such as the Chinese savants use, tied to a universal scientific paradigm, that seekers such as Prirogene are approaching. I regret that this book was completed before I encountered Prirogene's thoughts, but I am encouraged by the fact that our confrontations with this matter have yielded parallel results. (See Prirogene in Ferguson, cited in the *Bibliography*.)

** Of the *Wu Hsing*:Five Evolutive Conditions which, with the sun, compromise the concrete representatives of the six yin-yangs, only wood is a living one.

Western terms that the body's metabolism is, indeed, controlled by the liver, just as the ancient *Internal Classic* specifies.

> The lungs, for example, take in oxygen that eventually fuels the liver, and the heart, no more than a pump, circulates the enriched blood to the liver. But it is the liver that dictates the rate at which other organs perform their jobs and maintains the proper flow of nourishment to every part of the body. Billions of cells in the body undergo chemical changes that create energy. The energy is used to produce new material that replaces dead cells. The liver regulates this creative process and starts the chain reaction that eventually yields a living human being. (*Acupuncture*, p. 67)

Actually, as pointed out, and as Western biology recognizes, the heart and kidneys (Quadrimagic *major* yang and yin) evolve before the liver does, so that Western confirmation of the Chinese theory is potentially misleading. The liver-gallbladder does not initiate but *perpetuates* the life process.

The vital-organizing function of the liver was implicitly recognized at the "folk" level of our medicine long before the preceding findings were established. Probably the most central symptom of old age is liver spots and a dulling and paling of the healthy youthful red-yellow skin tones: altogether, a dark-dim yellow-whitening, "parchmenting," a symptom of major yin, whose correlates are the spleen and the lungs, a symptom precisely of "their *particular* colors emerging." Of the several connections between the liver-gallbladder and the skin, on one hand, and the dark-dim yellow-white coloring of age, on the other, the following is the simplest. According to the *Huang-Ti Nei-Ching*, "the liver rules the lungs" and the lungs control the skin. Sextimagically the color of the lungs-skin is dark-dim yellow-white. Hence when the vital nexus of liver-gallbladder begins to weaken from age, there is the directly connected skin-symptom of that color—specifically, a non-vital, parchment-like variety of it.*

To put that into universal scientific context, the *general* image is that of major yin, matter, insufficiently energized and therefore defi-

* As the cited text states, there is a vital and a lethal form of each color of each organ. The freckles of pregnancy, for example, are of a vital color—a red-black symptomatic of amplification of kidney-womb energy. The ultimate diagnostic in all genuine medicine is vitality itself, and the ultimate medical talent is vital-aesthetic sensitivity.

nitely symptomized, almost Absolute-Fragmentally, by its own color. If the hyper-material "neutrino-sea" within the red-black combination of terminal-regenerating cosmic energy-matter were deprived of yang and its substantizing black effect so that it had its own visible color, it would be precisely weak red (red-yellow-white) plus weak black (gray): dark-dim yellow-white. This is also the central color of the totality of decomposed metals. Both animate and inanimate processes, cosmic and microcosmic, begin with major yang action upon major yin and, with entropy, end with major yin relatively alone—a lack of red.

In sum: animateness is due to an especially through-going, non-temporal major yang energy and a complementary major yin matter which combines with major yang to form a minor-yang-reverting-yin, *muted, green-*, violet couplet which contains vital energy by recycling it. Together, animate yin and yang defeat entropy.

We have made the Red Shift. We have united the inanimate sciences with biology and recognized that redness-non-violetness* and what it symptomizes is the basic condition for living. Not for nothing is red the Chinese color.** Of course, we have also made a Red Shift at the intellectual level itself. Where red is the color of through-going, we have gone through a sufficient amount of maximally diverse data to recognize a scientific paradigm literally thousands of times more powerful, general, than ours. That paradigm may well be universal. We have also begun to think like Chinese doctors, who, I cannot overemphasize, prove the validity of all that precedes by using it to heal without limitation as to variety of disease, and without side-effects.

* De-entropized violet conditions.

** And not for nothing have the Communists, after Coca-Cola, appropriated it: it literally turns people on, for many reasons, all of which we now understand.

PART TWO:

WAN-CHIN YAO-FANG

Prescription Worth Ten-Thousand in Gold

CHAPTER

5

Reappropriating the Ultimate Force

I think I have seen the Western mistake. You are very able to distinguish things, but you are unable to put all things together. Your scientific conceptions therefore all have holes in them, and numerous incomplete principles are set forth. If you continue in this way, you will never be able to repair this.

—HSIA PO-YAN

In brief, then, . . . in this still quite young China, . . . [*it was*] *made clear that all the separate and, by their exterior, independent phenomena . . . are in the totality of their manifestations again . . . secondary independent units . . .* [*which*] *in their turn, consist of—tertiary units and so on to infinity.*

—BEELZEBUB, to his grandson (circa 1930)*

[W]*holes in an absolute sense do not exist anywhere. . . .* [*There are*] *systems of sub-wholes containing sub-wholes of a lower order, like Chinese boxes.*

—ARTHUR KOESTLER (1972)

* Gurdjieff, cited in the *Bibliography.*

The mere knowledge of a particle's existence . . . implies that the particle possesses internal structure.

—GEOFFREY CHEW, avant-garde physicist (1974)*

The consciousness of the observer must be altered, . . . reoriented toward the "unbroken wholeness" of which everything is a form.

—GARY ZUKAV, after the avant-garde physicist DAVID BOHM (1979)

SCIENCE THAT BREATHES

As most people who would be interested in this book know, Chinese science, centrally its medicine, extensively uses a concept called "*ch'i.*" I follow Dr. Porkert in translating that term as "configurational energy." Knowing that this concept is central to "acupuncture," our medical scientists stubbornly persist in trying to determine "how acupuncture *really* works," in terms of the anatomical and physical concepts of the West. This orientation is unfortunate not only because acupuncture is only a minor aspect of the sociological-psychological-physiological whole of Chinese medicine, to all of which *ch'i* is central. Unless they accept the scientific validity of the concept of *ch'i* they will never—could not possibly—fully understand how acupuncture or *anything else* "really works."

Ch'i is a concept essential to *all the sciences*' understandings of change (*pien-hua*), organization, and, most important, the full range of human powers of knowing and action. It must take the place of what we call "cause," and it makes possible *scientific* explanation.

Many of our medical scientists have witnessed successful Chinese therapies, which are based on theories entailing *ch'i*—often, of diseases they themselves cannot cure. But much more than that is desired and deserved by a Western scientist before he might accept the validity of such theory. There are several proofs of the existence of *ch'i* of which we are not aware, because the Chinese medical perspective

* Capra, *op. cit.*

has not been applied to formulate them for such an audience. So I'll do that here. As a side-effect, these proofs indicate how several of our sciences could become non-side-affecting. Here are some of my *shih-fu*'s defining statements of *ch'i.* (The Chinese medical texts are consistent with them but, being extremely laconic and intended for use only as a supplement to oral transmission, are much less useful.)

> *Ch'i* is a force. All things have *ch'i,* so it is not life. Life is a blend of *ch'i,* basically *t'ien-wai-chih yang* [the ultimate hyper-red energy deduced above] and certain substances. Ultimately, those substances, and all others, are also *ch'i,* but here we want to distinguish between the subtle, "floating" *ch'i* and the gross, "sinking" *ch'i* which we call *shui,* and which, in the human and animal forms, we call *hsieh* ["flesh-bone-blood"]. *Shui* stands for body-substance. *Ch'i* without body has no place to be-at; body without *ch'i* does not live. *Ch'i* is yang, and body is yin. *Ch'i,* being an almost-not-there thing, has only one physical property of its own: weight. . . . The functions [*yung*] of all things are determined by their *ch'i*s.

Ch'i is of two basic kinds, then: *ch'i* as energy-matter, the basic substance-and-force* of the universe, and *ch'i* as energy, or better, *force,* as distinguished from *shui* (substance). It is the latter that will concern us here; it is also the more common referent of the term. In living things, *ch'i* is what might be called "vital force," but, it is important to understand, it is not life itself. Life is *ch'i* (vital force) in combination with certain physical properties, is an energy-substance with a metaphysical as well as a physical aspect. According to Chinese medical theory, the coronas evidenced by Kirlian photography of animate things would be life itself. (It is life's relatively-physical aspect which makes life Kirlian-photographable by interaction with electricity.) *Ch'i* taken alone has only one physical property in the sense we attribute to "physical": weight, or mass. It has no shape, color, size, sound, smell, or taste. Beyond its mass, the existence of *ch'i* is recognized through its effects, for example, the *shapes* of Kirlian coronas. To put it otherwise, it is demonstrated by the perfect fit of the concept of *ch'i* into (a great number of) explanatory gaps in each

* In Quadrimagic terms, "substance" refers to both matter and what, with hyper-energy, it constitutes: substance. For simplicity's sake here the Quadrimagic perspective is taken.

of our sciences. *This applies equally to the atom, the living body, and the social body. Ch'i* organizes them all.*

Ch'i, then, is a demi-physical force whose relatively physical aspect is mass and whose basic property is energetic, spatial-temporal organization. I turn now to empirical evidence of it, beginning with physics and ending with social science—and the indispensability of this concept to Human science.

THE HINGES OF THE COSMOS

Arthur Koestler, who has the talent of tying existing scientific data and theories together to produce new and more powerful theories, wrote a book called *The Roots of Coincidence*, in which he examines the proposition that all phenomena are interconnected in a way which is not understood by Western science, a way which appears to involve a force or forces which are not directly perceived and which perhaps are metaphysical. With reference to physics, he compresses a statement by Professor Margenau of Yale:

> At the forefront of physical research, we find it necessary to invoke the existence of "virtual processes" confined to extremely short durations. For a very short time, every physical process can proceed in ways which defy the laws of nature known today, always hiding itself under the cloak of the principle of uncertainty. When any physical process first starts, it sends out "feelers" in all directions, feelers in which time may be reversed, normal rules are violated, and unexpected things may happen. These virtual processes then die out and after a certain time matters settle down again.

"Reversing time" refers to the traveling of a "virtual particle" at

* I refer the reader to Dr. Porkert's book for his list of many qualified forms of *ch'i* (for example, *jen-ch'i*:"human ch'i": "The physiological energy of the microcosm resulting from the synthesis of ch'i *caeleste* and ch'i *terrestre*" [cosmic and earthly ch'i]). Here I will deal with *ch'i* more generally, specifying only some of those forms.

greater than the speed of light, and its return, with "information" gathered from its voyage. As Zukav explains, this concept is set forth by quantum physicists in order to preserve their scientific sub-paradigm. If a particle can travel faster than light, then the constant of the speed of light, on which all relativistic physics depends, is lost, but if the faster-than-light particle—or something like it—is not posited, then all of quantum physics is lost, because it appears that all physical processes entail coordinations, which quantum physics specializes in predicting, among "actual" particles when they are at a distance from each other—what used to be called "action at a distance." It is worth knowing two of the experiments on which this new, speculative, theory is based.*

If light is beamed at a polarizing filter, only some of it will emerge at the other side: the part of it which was (or has) aligned with the molecules in the filter. If two such filters are combined with their axes at right angles, no light will emerge. If two such filters are combined at a non-right angle, some light will emerge. With series of three filters, the following occurs. Except in the case where the first and second filters and the second and third filters are at non-right angles to each other, no light passes through the filters. Otherwise put: in a series of three filters, one horizontal, one vertical, and one diagonal, if a horizontal and a vertical filter are adjacent, with the diagonal filter at one or the other end, no light emerges; but if the diagonal filter is between the horizontal and vertical ones, light emerges. When our physicists established these data, they were amazed. Assuming that light is nothing other than waves whose axes are (to use the filters as reference frames) either vertical, horizontal, or diagonal, it followed that regardless of the order in which three filters, one vertical, one horizontal, and one diagonal, were placed, the same results should ensue. The theoretical explanation for the exceptional combination, in which the diagonal filter is between the vertical and horizontal ones, was set forth by von Neumann and Birkhoff in 1936 and became the basis of quantum logic (as though there are different logics for different phenomena—hardly an assumption which promises theoretical consistency and unified science!). The basis of this "specific form of" logic is nothing other than the concept of Polar-Completeness, which they call "coherent superposition." For the case of the strange behavior of

* They are described in greater detail by Zukav, to whom I am indebted for knowledge of them. The reinterpretations, of course, are my own.

"polarized" light, it was supposed that with respect to its wave-characteristics light is not made up of horizontal and vertical and diagonal waves (waves whose axes have definite degrees), but rather is made up of waves whose axes are *at once* horizontal, vertical, and diagonal. This is just as in all my Polar-Complete demonstrations: each color is *at once* itself and all other colors; each of the six yin-yangs, and whatever phenomenon has its value, is at once itself and all the others; energy is at once energy and its opposite, matter, and so on. It follows from the empirical indication that the axes of lightwaves are all-angled that in one aspect light is metaphysical, and possibly pre-configurated. (All-angledness transcends space; the aspect of light which does may be responsible for its spatio-physical characteristics.)

Our physicists have ingeniously developed a way *around* the preceding implications to predict the behavior of light impinging on filters under various specific conditions. But their method is a Mentalist one, which fails to express the Polar-Complete "structure" of the light (which they themselves recognize) and has led to mysticism. Since their predictive solution, based on the preceding experiment, is the very basis of quantum theory, these fallacies should be understood and corrected. Although light is understood to be "coherently superpositioned" with respect to the axes of its waves, the basic variables of the predictive equations are Absolute-Fragmentally conceived waves (individual waves with definite axes) corresponding to the angles of the filters. The behavior of the light is then calculated by the probability of each of these Absolute-Fragments occurring in a given time sector. Consequently, probability ("chance") is thought of as the factor which determines the light's behavior. But probability of course is nothing other than a mental construct; it has no empirical reference (as Einstein, criticizing quantum theory, pointed out). Accordingly, our physicists have become mystical Idealists. They believe either that there is a "real probability" in the external world which acts as a force on light (Objective Idealism) or that the universe is actually nothing other than a projection of their own minds (Subjective Idealism).

What, other than *ch'i*—demi-physical energy with intention and configurational effect—is implied? This alternative is superior to the Western one, because it is not mystical: instead of the universe as an extension of a megalomaniacal mind (the physicists') or probability-as-a-natural force, one posits an extremely subtle form of energy (for which we have evidence from other phenomena) which seeks to through-go. And, unlike quantum probability, this interpretation perfectly fits into a *general* scientific theory.

As to the inability of a single-angled lightwave to pass through a filter at a 90-degree angle to itself, it need simply be stated as a "law" to account for the mystery of the three filters, once it is recognized that there must be a configurative force, *ch'i* ,which seeks to adjust a lightwave so that it will pass through a filter. It is the assumption that the lightwave does not adjust itself, *is a totally passive entity,* which makes these experimental data surprising and mysterious to the Western-physical mind. The real mystery, or intellectual challenge, is to habituate the mind to the Polar-Complete concept, "fraction of infinity," which is implied by the transformation of energy at infinite, and therefore no, angles (a non-numerical, *qualitative,* concept) into energy at one angle (a numerical concept). Mathematical thought, basically depending on the concept "probability," suffices to predict, but not to *describe, the variables implied by* the experimental results. In this way the existence of light-*ch'i* is neglected at the cost of empirical validity and of unified science—science which has *Logic,* not separated and mutually contradictory "logics."

Quantum-theory is based, then, on Absolute-Fragmental misrepresentations of its data, for the sake of preserving the mathematical nature of our paradigm, and for the "practical," Ultra-violet-Catastrophic effects that such mathematics make possible. That is the chief point I wished to make in reinterpreting the evidence of the filters-experiments, for to account for the results of yet another experiment, our physicists themselves posit something like *ch'i,* without any help from outsiders.

I refer to the Clauser-Freedman experiment, for knowledge of which I am, again, indebted to Zukav.

When two photons are beamed in opposite directions, each toward a polarizing filter, if the filters are parallel, some light, and the same amount of, passes through each; if the filters are at a non-right angle, some light, and the same amount of, passes through each. But if the filters are at right angles, some light passes through one and no light passes through the other. In other words, in the last case two photons going in opposite directions, each to impinge on one filter, behave like one photon going toward and impinging on two consecutive filters at a right angle. It is just that in the one-photon case, two filters at a right angle stop all the light, whereas in the two-photon case, there is a compromise by which they stop half the light. The fascinating implication is that the two photons somehow communicate: "OK, filters at right angles are coming up; let's say you make it through and I don't." *Together* they are presented with a 90-degree turn between horizontal and vertical; being two, they half-pass through.

The implication of this new evidence from physics has revolutionary potential of a magnitude which has yet to be recognized. Parmenides, original proponent of the Death-Formula, recognized that the universe is a "plenum," that it has no "spaces." Logically, there follow two opposite implications of which one must be chosen. One is that if there is any movement, the entire universe must move. The other is that there actually is no movement at all, which is as much as to say that living beings are actually dead. The first implication entails denying that there are Absolute-Fragments; the second entails affirming their existence, from which follows the Death-Formula on which all Western science is based. Avant-garde physics, then, is *empirically* rediscovering the first implication, the logical possibility of which has been suppressed for 2500 years. It leads to the Life-Formula. In physics, it is David Bohm who is at the leading edge of this incipient revolution. If the "Red Shift" is made, so that all of science is transformed from a Fragmented lethal into a Complete vital system, a living Human social reality may be restored to us.

There are two possible types of interpretation. One is that an aspect of the photon-pair travels faster than light to do the communicating. (This is the lawbreaking referred to in the Koestler quote.) This is undesirable to our physicists because they wish to regard the speed of light as the fastest possible. The other, proposed by the avant-garde physicist Jack Sarfatti, is that the photons are connected not by signals, but by something which transcends space and time altogether: a "correlation" between the two photons (and the filters) "at a higher level of reality." The difference between these two interpretations is not significant here, for in both cases a demi-physical configurating force, *ch'i*, is implied. It is demi-*physical* because it affects physical behavior—has a kind of "weight" or mass, as stated by the Doctor.

In terms of Bell's theorem, one of the hottest items in contemporary physics, the preceding data imply, as Zukav wrote, "the principle of local causes must be false." There is "action-at-a-distance," and the action has a specific form and cannot be attributed to the physical (aspect of the) variables in question. To quote David Bohm, an avant-garde physicist: "Thus, one is led to a new notion of *unbroken wholeness* which denies the classical analyzability of the world into separately and independently existent parts." This is nothing other than the basis of the 5000-year-old Chinese theory of causation: there are no Absolute-Fragments, no "parts"; everything is connected to everything else so that cause is also effect; everything is Polar-Completely energy-matter and it is the energetic aspect of it which connects everything. So, the energetic aspect of everything gives each "part" of it its behavioral pattern—a pattern which is related to that of everything else. As our physicists begin to think this way and explore its potential, they will begin to resolve their probabilistic problem, emphasized by Koestler, that they can predict the behaviors of numerous variables of the same class, but not the behaviors of *individuals* of the same class: the individual "choices" are made in relation to a physical-metaphysical whole greater than the numerous variables in question. Prior to that, our physicists will have to define those variables in a way they themselves know is not false. Rather than the concept of individual light waves of different, specific angles—an analytic, Absolute-Fragmental definition of unfiltered light—the concept will consist of relatively-individual light waves, which as individual phenomena are chiefly at specific angles and secondarily at all others and which, as a whole, are *one* light wave at all angles simultaneously: a Polar-Complete definition.

The *ch'i* in "gymnastic" light (the filters being the jungle gym) is not the only "feeler" which "transcends time," that is, which *truly* moves, in modern particle physics. *All* subatomic interactions are now thought to be mediated by "virtual particles," which in turn, are increasingly indistinguishable conceptually from "actual particles." The major-yin-major-yang, energetic-material, metaphysical-physical dragon in the cloud chamber is demanding attention.

The very atom, Vintage Basic Physical Unit that it is, seethes with virtual particle behaviors whose source is a demi-physical configurative force. For example, an atom's electrons continually emit and reabsorb virtual photons. That is, as Zukav puts it, "first there is an electron, then there is an electron and a photon, and then there is an electron again." What makes these photons "virtual" is that they spontaneously appear: they are not the effect of an input of energy (energy in the relatively gross, materialized, sense that our physicists, in contrast with the Chinese savants, define it). "Actual" photons are apparent effects of input energy. Sarfatti explains that, in contrast, a virtual "particle," or process, is triggered by a "jump of negentropy," that is, an instantaneous informational input which organizes "some of the infinite vacuum energy" that is space. This is as much as to say that configurative hyper-energy, *ch'i*, organizes hypermatter (major yin) to form a virtual particle. (It is essentially the same image as that of the creation of the Cosmic Egg—or Eggs.)

The amount of input energy which results in an actual photon emitting from an atom is equivalent to the energy of an emitted (actual) photon minus the energy of a virtual photon. The implication is that the non-input, intrinsic, basis of an emitted photon is a virtual photon, whose basis, in turn, is an action of "negentropy"—*ch'i* on hypermatter—major yin. Visible photons emitted from atoms are *colors*. Hence modern physics implies that the source of colors is *ch'i*. Conversely, the Polar-Completeness of colors implies that their source is *ch'i*, that is, the unity-within-difference and movement-continuum of "infinite components" which is directly sensed by a careful observer of colors. There is no mystery as to how the Chinese savants discovered *ch'i* as the source of all physical behavior without the aid of bubble chambers. They used their eyes and harmonized their theoretical thinking with the empirical evidence.

The nucleons (protons and neutrons) of the atom also emit and reabsorb virtual particles. The "strong," nucleus-binding, force *is* exchanges of virtual pions. The "charge" of particles is also now accounted for by virtual-particle-interactions. Two electrons, for example, are "minus-minus," repel each other, by exchanging virtual photons. Electromagnetic force *is* virtual-photon-exchange. This is to say that both the nuclear force and electromagnetism are behavioral patterns configurated by *ch'i*. Physicists now assume that the remaining two of the Four Forces are also due to virtual-particle-interactions.

In sum, according to avant-garde modern physics, the entire physical aspect of the universe (at the least) is governed by *ch'i* (negentropy jumps), whose immediate effect is "virtual particles."

The Polar-Completeness of the energy-matter which is the basic variable in all the preceding is also being affirmed by modern physicists. The difference between virtual (subtle, relatively energetic) and actual or real (grosser, relatively material) particles is dissolving. As Richard Feynman, formulator of the diagrams that are the basic shorthand of quantum physics, has said, ". . . what looks like a real process from one point of view may appear as a virtual process occurring over a more extended period of time." As I have explained, according to Tao, major yin (hypermatter) is a space-time container. What is relatively enduring (in a given space) is relatively material, that is, relatively dominated by matter. Obviously "virtual particle" is a way of saying "the middle ground between energy and *matter*—minor yin-yang."

To round out the picture I attack the last bastion of quantum physics, the supposedly Absolute-Fragmental, quantifiable, nature of the heights of electron-orbits over atomic nuclei, of the levels of energy of atomic electrons, and of the "jumps" made by atomic electrons—all three of which are aspects of a same thing. And I replace it with the Tao concept, foreshadowed above, of Polar-Complete atomic movement. Let us begin with the Chinese sense-based, genuinely empirical, perspective. Whatever energetic phenomenon underlies colors (what Western physicists call atomic quantum jumps) must be Polar-Complete, because colors are. Color-differences are symptoms of those movements; hence those movements must, like the color-differences, be Polar-Complete. Applied to the Western concepts, there are no definite orbit-heights, there are *relative*-heights; there are no definite levels of electron-energy, there are *relative*-levels of electron-energy; there are no definite distances "jumped," there are *relative*-distances jumped. Neither does an electron have a definite level nor does it jump a definite distance: like the color it produces when it has jumped and returned to its original level, it has chiefly attained a certain level but, secondarily, it has also attained all other electron-levels. The electron (now conceived as a wave propagated in an "orbit"-circle around a nucleus) is chiefly, not exclusively, in a certain, quantitatively expressible, place. This is consistent with the quantum-physical treatment of the electron as a particle somewhere within an orbit: its position is expressed as a probability, analogously to the axis-orientation of a light-wave within a photon of light, as discussed above.

The probabilistic expression of its position and the electron itself are often collapsed so that the electron itself is called a "probability-wave." Again, to probability is attributed existence in the external world. My solution for this mysticism, again, and as follows from all, both Western and Chinese, that has preceded, is that the real variable operative in Polar-Completely positioning the electron is *ch'i*—which, as the interconnecting variable, coordinates the at-once-here-and-there aspects of the electron, and coordinates those aspects, in turn, with a greater pattern our physicists have yet to take into account.

This is not to suggest that Chinese theory or a use of it in this direction would result in exact, non-probabilistic, prediction of individual atomic events. It will not, because it is not designed to do so. Chinese physical theory is designed to communicate understanding of the physical aspect of the universe which can be put into the service of humans at the biological, psychological, and social levels of science. It is not designed to finely dissect a microscopic aspect of the physical aspect of the universe. Such dissection is desirable only to those who, in the service of insane industries, increasingly seek to further derange the natural balances to intensify the Ultraviolet Catastrophe already upon us. That is the only concrete end such dissection and sharpened prediction can serve. If Chinese theory could serve that end, it would prove that it is not scientific. (Science, by definition, has *Human* use.) What I seek to show is that the Chinese scientific understanding of the basic nature of the physical aspect of the universe, centrally that *ch'i* governs it, is valid; and that present modern physical theory about it confirms this, yet is self-inconsistent and inaccurate in its representations of the variables involved. In other words, what leads to no good is also false, what leads to good is also true.

To return to the question of atomic structure, the electron is now understood—uncertainly in quantum physics, but with certainty in the present, empirical, view—to be in several places at once, and the energies involved, therefore, to be a *continuous flux*, just as the Chinese have it; and the colors, resulting from this flux, are known to "be as same as different," again as the Chinese have it. Nevertheless, present quantum theory retains the Mentalist concept of discrete orbit-levels, energy-levels, and jumps—discontinuity, the anti-Quadrimage. Supposedly, this is because Planck's Constant, a definite amount of energy per a definite amount of time which defines electron-jumps—in short, the basic analytic unit of atomic behavior—has been verified to be dead accurate, and because it is essential to almost all of physics' theoretical predictions. Let us suppose that Planck's Constant *is* dead accurate. If it is, it is accurate only as a predictive concept, but it cannot be accurate as a concept of atomic reality, for that reality is not discontinuous. Colors are not discontinuous (their wave-characteristics are ultimately indistinguishable). So how could the electron-orbit-levels, energy-levels, or "jumps" be? The virtual photons which electrons emit are ultimately indistinguishable from the actual ones they emit, so even if electrons could be mathematically represented in non-probabilistic, not at-once-here-and-there terms, it follows that orbit-levels and energy-levels continually fluctuate and, therefore, that electrons "jump" non-definite distances non-discontinuously. Planck's Constant "works," therefore, despite the fact that the atomic-structural discontinuity on which it is based does not exist. It is an average of approximate and imaginary discontinuous components, just like the definitely-angled axes of light waves in probabilistic equations correctly predicting the behavior of (actually Polar-Completely structured) light impinging on polarizing filters. The concept "works" (predictively), but it is false as a representation of reality.

Actually, the predictions performed with Planck's Constant are not perfectly accurate. As our physicists stubbornly and ingeniously persist in trying to get to the bottom of this matter, their theories are increasingly riddled with corrections. The basic variables in these equations are the Constant, the energy-difference between two electron levels, and the frequency of color-waves emitted by the atom. Our physicists still seek to refine the results of these computations, attempting to account for their inaccuracy by, for example, positing that the electron "perceives the nucleus as being somewhat smeared in space." Such an hypothesis already admits that the atom is not discontinuously structured; it is just that it does so in mystical, obscurantist, terms. Electrons do not "perceive," and the "smeared" nucleus refers to continual fluctuation of the distance between the electron and it. Regardless of the fact that such hypotheses implicitly admit the point I am making, there is no scientific reason to assume that only variables other than the Constant are responsible for the inaccuracy of these predictions. Any one, two, or three of them could be.

In sum, our physicists have no proof whatever of the alleged discontinuous structure of the atom, or of the "light quanta" it emits, and have ample evidence of a continuous structure, specifically, that the electron-movement of the atom is simultaneous expansion-contractions. That is precisely the Polar-Complete movement which was inferred, above, to be the basic cosmic one when it was shown that the "Big-Bang" concept (an explosion without simultaneous implosion) is an illogical and unrealistic one. This Polar-Complete movement is exactly the kind of movement which is ascribed to the T'ai-Chi (the Ultimate Pole, the Generator of the Cosmos) by Chinese savants.*

* As I prepare the text of this book for the printer, the May issue of *Science 81* reports that some of our physicist-cosmologists now question the rationality of positing a simple explosion of the Cosmic Egg. Its results would be uniform, and it therefore cannot account for galactic, solar, and planetary

The Tao picture of intra-atomic behavior removes the inconsistency of modern physical theory. The quantum theory of faster-than-light communication or space-time-transcending coordination among atomic particles, cited above, violates the relativistic theory that the speed of light cannot be exceeded. That quantum theory is motivated by the assumption that quantum jumps are discontinuous: that they involve discrete amounts of energy.* An electron is either at level 1 or level 2, and so on; it cannot move between any of them. Hence its movement is an "instantaneous," that is, faster-than-light, transition. Communication between the energy which causes the electron to jump and the electron, then, is a faster-than-light probability-computation which leaves "enough time" for faster-than-light "actualization." But, unless mysticism is desired and the empirical data are ignored, all the preceding assumptions must be rejected. As we have seen, the empirical evidence obliges us to understand that an electron is never at any definite level 1 or 2, and so on; therefore the "amounts" of energy involved in its movements are not discrete. And since the electron is always between as well as at all levels, its movements cannot be instantaneous. Rather than a probability-computation and transition to actuality, there is a governing *ch'i* which is just as "actual" as the physical aspect of the phenomenon it governs and which, since it is an aspect of the phenomenon, moves at the same velocity as does the physical aspect of the phenomenon. Nor should we expect inanimate hyper-energy to transcend time, for as we saw, nontemporal hyper-energy is a characteristic of *animate* things. Inanimate hyper-energy is relatively material as compared to animate hyper-energy. That is, it is more dominated, in the sense impeded, by matter; hence it has velocity. More exactly, *its effects* have velocity. Analytically viewed as something separate from hypermatter, hyper-energy, or *ch'i*, has no physical characteristic other than a subtle form of mass.

---

concentrations. Accordingly, almost exactly along the lines I set forth with regard to the creation of the Egg (or Eggs), they posit a second variable: a subtle patterning force of some kind operating upon extremely diffuse and fine matter—the exact correlates of *ch'i* and major yin, which I said are symptomized by black-body radiation and the neutrino-sea. I rest my case.

* Since the atom is the core object of quantum theory, I do not also deal with "faster-than-light communication" between twin photons impinging on filters (above); but essentially the same arguments as are required for that case are contained in what follows.

So, Einstein's theory that the speed of light cannot be exceeded is saved by the understanding that the atom is a Polar-Completely fluctuating continuum.

It should be noted that there is an argument to the effect that the electron can only occupy definite orbit-levels, which is adhered to despite the cited empirical evidence. It is therefore worthwhile to show that the logic behind the argument is weak and circular. Here, the electron is pictured as a wave, identified with the probabilities which predict its position as a particle within a given orbit. In Capra's words:

> In the orbits, the electron waves have to be arranged in such a way that "their ends meet," i.e., that they form patterns known as "standing waves." These patterns appear whenever waves are confined to a finite region, like the waves in a vibrating guitar-string. . . . It is well-known from these examples that standing waves can assume only a limited number of well-defined shapes. In the case of the electron waves inside an atom, this means that they can exist only in certain atomic orbits with definite diameters.
> (p. 70)

The argument is weak in that it is assumed that electron-waves behave as do those of a guitar string, which is to assume, for one thing, that there is an analog to the string: a substantial medium for the wave. Physics specifies no such analog. The wave is the *whole* phenomenon. If physics specified such, the analog to the string is hypermatter, which, since it is Polar, is indistinguishable from the energy which commutes movement to the phenomenon and with which it interacts to yield spatial-temporal (wave-) form. Since, as seen with respect to colors and heat (phenomena which directly result from electron-movements), the relation between hyper-energy and hypermatter may *continually* change, there is no reason to posit that the waves on this elemental guitar string are limited in shape. Second, the argument is blatantly circular in that it proves that electron-orbits have definite diameters by positing wave-behaviors of a variety which occur in media, such as guitar strings, which have definite lengths! (The fluctuation in the hyper-energy:hypermatter relation corresponds to continual *change* of length.) As stated earlier, all arguments to the effect that electron-orbits, and so on, are discontinuous should be traced to a belief in Planck's Constant, for which there is only approximate-predictive evidence and against which all the other, quite concrete, evidence goes.

For the record, if one were to attempt to draw a diagram of an atom to express the Polar-Completeness both of its content (hyper-energymatter) and of its movement (simultaneously inward-outward), perhaps the most natural alternative would be the T'ai-Chi Diagram (Figure Ten). Of course, it represents not atoms in particular but the cosmic substratum in general: the proto-cosmos, proto-subcosmoses (cosmic concentrations such as proto-galaxies), and microcosmoses (what we call "atoms" and "subatomic particles")—the "Hinges of the Cosmos." One can analyze the T'ai-Chi Diagram into components to produce a predictive geometry (see my own doctoral dissertation), but the results are really a Western (and Sung-dynastic*), Mentalist perversion of its intended meaning. The T'ai-Chi Diagram is a "mandala": a crude image of reality which is to be perceived as a whole and which acts on the super-verbal, higher aspect of the mind (mistakenly called "the unconscious" by Western scientists), to which Polar-Complete paradox is not paradoxical and great complexity is simple, so as to render it more powerful by harmonizing it with external reality.

The power of Tao science ultimately lies not in any verbalizeable formula, as in the West, where all science, all knowing, must be externalized as an object which can be computerized and automated so as to make human participation and responsibility and, therefore, self-worth minimal. The Western scientific paradigm exploits the scientist, by "detaching" him to its life-denying 2500-year-old purpose, whereas the Tao-paradigm transforms him into a savant whose life-energy united with all other movement is maximal. The power of Tao science lies in the scientist, the knower, who, consequently, is maximally involved and responsible and has great self-worth. Man, not Planck's Constant, is the measure of all things.

Our paradigm is shifting in the Chinese direction as I write. Some data given me by my philosopher-friend, Charles Holmes, report that Robert Penrose of Oxford, the mathematical-physical theoretician,

* As stated, during the Sung dynasty official science deteriorated into Mentalist, Western-like form. Consequently, Western Sinologists regard the Sung as China's period of Enlightenment. And Chinese savants may use such forms to communicate Chinese science to Westerners at a "kindergarten," Western, level which Westerners find pleasing.

FIGURE TEN: THE *T'AI-CHI* FIGURE: Schematic, Stylized Version and Esoteric Version Accompanying Verbal Instruction

is constructing a new image of the *whole* universe on the basis of mathematics hitherto used exclusively for the *micro*-scopic aspect of the world.* The central notation and concept of that mathematics is complex numbers, which combine "real" numbers (for example, 1, 5) and "imaginary" numbers (for example, square root of minus-1)—the old Pythagorean Formed-Formless dichotomy which was neutralized above in Chinese terms is now synthesized, in an Aristotelean manner, by avant-garde physicists. "The behavior of the real part is controlled by the imaginary part and vice-versa." If we substitute for these "parts" (which, being numbers, are Mentalist constructions) the realities they refer to, we again find Chinese physics hidden under mystical Western avant-garde physics: the electron's orbit is definite and in one place in one aspect, and indefinite and in all places in another aspect, and so on.

The talent to which Penrose's extraordinary problem-solving is credited is recognized by himself and others as *aesthetic* more than logical, as recalls the Chinese aesthetic-logical (but also *empirical*) mode of thought illustrated above.

One implication of Penrose's theoretical synthesis is that neutrinos, before believed to be massless, have slight mass, just as I have predicted, in terms of Polar hypermatter, by Chinese theory, above. (Empirical experiments have recently confirmed this theory.) Penrose is pointing at the Polar-Complete "border" of *ch'i* and *shui*, or hyper-energy and hypermatter.

The central concept in Penrose's theory is *twistors*, entities much smaller than atomic particles which are configurated *space* (compare *t'ai'chi*, above) as well as matter (compare hypermatter, *shui*, above). A twistor is a "fuzzy particle." Again, the at-once-formed-and-formless, both-here-and-there, mathematically definite-yet-indefinite, Polar-Complete pattern of all cosmic concentrations is found hidden just below the mystical surface of avant-garde physics. (Penrose says that if he is right, we are all abstract mathematical entities—mysticism. The Polar-Complete dragon breaks through that mystical surface.) And Penrose extends his vision to include not only subatomic phenomena, but the cosmos as a whole, much as I have in Chinese terms ("all cosmic concentrations, including the cosmos itself").

His twistor-difference with quantum physics' view of space and time is precisely the one I have illustrated and documented with refer-

* *Science 80*, December 1980.

ence to a 2000-year-old Chinese text. In one aspect they are distinct, but in the other they are indistinguishable: *yu-chou.*

I concluded, above, that the universe originates (and atomic and subatomic wavicles move) with a simultaneous expansion-contraction in which cosmic *ch'i,* or hyer-energy, configurates hyperenergymatter (*t'ai-chi*); that the expansion-contraction is *prior* to, *creates,* mass; and that the first product of this phenomenon is gravity. Penrose's post-Einsteinian theory is to almost exactly the same effects. The twistor has simultaneous twists on two axes at right angles to each other. And, "in the twistor picture, gravity is also caused by mass giving rise to a curvature in the universe; however, instead of just a warpage in space-time, it is now a warpage of twistor-space." *Twistor*-space *includes* matter, hence, mass; mass, therefore, is an *effect* of the warping. Penrose suspects that different types of twists in twistor-space produce not only gravity but also electricity, magnetism, and nuclear forces. (Again see my Chinese explanation of the arisings of these forces after the initial "twist.") In other words, to quote the Chinese theory: "First there is the *t'ai-chi,* then there are the two yin-yangs, then there are the Ten-Thousand Things," and "All change is due to *ch'i,*" a demi-physical configurative force.

It is worth pointing out that the chief architect of Western science, Aristotle, did a better job of formulating the conditions of physical causation than have his successors who, after 2500 years, are beginning to sound just like him. Aristotle inferred the existence of what might be called *ch'i*-distorted-into-a-shape-which-fits-into-the-Mentalist-paradigm. Recognizing that the evolutive sequence (changing forms) of physical objects cannot be attributed to their physical aspect, he logically deduced that a metaphysical will-like factor (formal cause plus God or, in animate objects, a soul) which is, as it were, ahead of the physical object in time must determine the evolutive sequences. Today, instead of a will-like, or as I have put it, "intention-like," factor, our physicists impute to inanimate phenomena an *intellect*-like factor: probability.* In Aristotle's time, the Narcissistic savant identified with God's will. In modern time, the Narcissistic savant identifies with God's mind, and is a bit more Pythagorean than was Aristotle. What I have proposed does away with the fundamental split, the Binary Con, on which both Western versions are based. The factor in

* The original formulation, Einstein's, was "a guiding ghost-wave," which is closer in intended meaning to *ch'i.*

question is not and does not derive from the supernatural; it is not metaphysical. It is demi-physical and an aspect of Nature. After all, super-Nature is nothing other than a garbage can for data which "refuse" to fit into an unnatural worldview.

That old, Binary Conning, pattern is made explicit by the present alternative to the inference that an intellect-like factor, probability, governs subatomic and atomic behavior: the Many Worlds Hypothesis, which is that no "choices" are made as to which atom will emit a photon, which "part" of a photon will make it through a filter, and so on; rather, all the possibilities are actualized, but "in different worlds." Of course, these two theories are basically the same. Where probability is the operative, "choosing," factor, a metaphysical-supernatural level of reality is implied. It is best expressed by Sarfatti because he follows the implication through, as "a higher reality." This is as much as to say that there is more than one world. In both cases, Nature is split into a knowable half and an unknowable half. The latter is a non-empirical area "created" for data that do not fit the paradigm—a self-delusion as old as Babylon. The paradigm, it cannot be overemphasized, is the actual problem, because it is anti-empirical. As soon as empiricism is exercised, the Polar-Completeness of all phenomena is recognized, all data can be taken into account, and there is but *one*, whole, world—which is what a world *is*, by definition.* The concept of *ch'i*, an aspect of *Nature*, then replaces the concept of supernatural cause, or its inferior alternative—the concept of "many worlds." The latter is inferior because it proposes that, rather than there being an unexplained patterning, there is chaos which appears to be patterned from the perspective of one of the "worlds" within it. In the first (mystical-Idealist) case, there is intelligence but it is attributed to God instead of man; in the second (mystical-Materialist) case, intelligence of any kind is rendered "unnecessary." In both cases, man's potential is diminished.

Let me make the actual nature of causation clearer, with reference to the increasingly weak distinction between "virtual" and "actual" particles, to get to the *Human* point of all of this. Virtual particles are distinguished by the fact that they occur without energy-input. But the concept of input (perceptible cause) is a fallacy in the first place,

* The world is all. If there are "many worlds," then, they are *sub*-worlds of *the* world. One cannot logically disguise one's lack of understanding as "another" world.

which is due to the Fragmental nature of our paradigm. If, as our avant-garde scientists now realize, all phenomena make up an "unbroken whole," there can be no input, no cause. If one link in a chain moves, *simultaneously* all links in the chain move. What our scientists, in contradiction to their uncertain understanding of unbroken wholeness, insist on calling "input" is simply movement, change, at an aspect of the whole upon which they have hypnotically fixated their attention. So, as Sarfatti suggests, the only difference between virtual and actual processes is that the latter are relatively enduring, which, as explained, is to say that they are relatively matter-dominant, obvious, gross—like the empirical sensibilities which our scientists seldom make use of. Cause, then, is found not in a change in the relation between billiard ball A and billiard ball B. It is found in a change in the pattern of the entire universe and all specific manifestations thereof. In this light, cause need not be distinguished from its effect: it is movement, patterning-change—configurative energy. It is an aspect of phenomena, *in* the phenomena.

From that enlightened perspective, it is obvious that the nature of true science is not a set of "laws" which relate causes and effects, but *understanding*. That is, the notion of "laws" derives from the notion of an Absolute difference between phenomena and their causes, which is due to a Fragmental, microscopic view—a psychical illness. As soon as it is understood that "cause" is simply the dynamic aspect of phenomena, the scientific objective can no longer be the formulation of "laws"; it becomes, simply, understanding. And for yet another reason, the objective becomes *the scientist himself* who, instead of being dwarfed by a god of one sort or another lying logically behind the notion of cause, realizes that as a human he is a being of a higher order than any god a mystic might dream up. There being no cause, a god cannot be said to create. So, humans and gods are equal on that score. But a human, unlike a god, can acquire understanding and exercise responsible action. Understandings such as *that*, in turn, amplify *Human ch'i.** *Genuine* science is also medicine, a pro-survival supertonic which amplifies human image and self-governing force.

Polar-Completeness and *ch'i* also exist at the higher-inanimate level of chemistry and so-called biochemistry. Physics overlaps with chemistry in identifying kinds of substances. "Elemental" ones are defined by atomic structure, "complex" ones are defined by molecular structure

* The existence of which is demonstrated in the next chapter.

—specific combinations of kinds of atoms. Elements, the province of physics, are identified by the colors ("atomic spectrums") they emit when heated. Since every color includes all colors, it follows that the elements which emit them are not Absolute-Fragmentally, but rather are Polar-Completely, different. Hydrogen's spectrum is redder than uranium's, but uranium's includes red and hydrogen's includes violet. It follows that these extremely different elements are not different in a discrete way. To say that uranium has a definite, high, *number* of electrons and nucleons and hydrogen has a definite, low, *number* of electrons and nucleons must be false, for the result, in such a case, would be color-spectrums of which numerical statements of difference could be made. They cannot. Imaginary numerical differences can be specified and used predictively, but this is not to represent the structural *reality* of the matter. And accordingly, the predictions, although very fine, are always inaccurate even with reference to the imaginary Absolute-Fragmental frame that is used. The truth of the matter is that uranium is a relatively dense concentration of energy-matter which can be imaginarily and falsely portrayed for predictive, Ultraviolet Catastrophic purposes, in numerical terms, and that hydrogen is a relatively diffuse concentration of energy-matter which can be imaginarily and falsely portrayed, for Ultraviolet Catastrophic purposes, in numerical terms. In reality, the difference between them is qualitative-relative.

Since complex substances are molecular combinations of different elements, it follows that they, too, form a Polar-Complete continuum. The kinds of substances are structurally just like the colors which physics and chemistry take to identify them. The senses *are* "a valid medium of truth" and those who really use them are not "naïve."

Just as for intra-atomic structure, it follows that there is a unitary, "connecting" aspect to molecular structures which is metaphysical or demi-physical in that it transcends spatial boundaries, both "is" and "is not." If it also patterns (the physical aspect of) molecular structures, again it is *ch'i* that is indicated.

Not to speak of new empirical evidence to this effect, logic dictates that *ch'i* does pattern molecular structures. As von Bertalanffy, founder of Systems-Theory, pointed out, any organization which is not predictable from the (physical) parts, or properties, of the object in question must be basically of a non-physical, or formal, as distinguished from material, nature. He called such properties "emergent properties." Add to that impeccably logical inference the general fact that the more organization there is, the more energy there is, and an organizational, or configurative, demi-physical energy is implied. Of course, the structure of a molecule is not predictable from the properties of its components, atoms. But there is more than logic to this. It has been found that certain crystals, at least, have an electron structure, which bears no apparent relation to the atomic or molecular structure of the crystals. In other words, a (physically) un-caused "ghost-structure" surrounding and overlapped with the gross atomic-molecular structure has been found. And it fluctuates continually just as one would expect. The analogy to "virtual photons" emitted by electrons is obvious. This electron-structural flux would be the energy-matter phase intermediate to molecular-structural *ch'i* and the gross physical aspect of the molecular structure, a direct symptom of the demi-physical "cause"—relatively energetic aspect—of chemical structure.

Now let us approach biology by seeing how *ch'i* is implicit in present knowledge of the so-called biochemical level. "So-called" because, as will be made clear, there is nothing *living* about the chemicals of living organisms as they are portrayed by our biochemists. Here are two particularly interesting examples among many. The blue-green, major-minor-yin protein phytochrome, which initiates the process of photosynthesis, not only is activated by light, but changes its molecular form as light of different colors impinges upon it. Since there is nothing known about its basic molecular structure which would predict this response, it appears that the independent variable is the light: the light-energy configurates it. Even if it is found that physical properties of phytochrome are such that they respond to light of different colors in certain ways which affect its overall structure, it is reasonable to expect, in light of all that has been established so far and in light of the general tendency for energy, not matter, to organize, that energy will prove to be the *chief* configurative variable. Of course, the configurative variable must be traced back to *ch'i* in any case. If the atomic-molecular structure of phytochrome is sufficient to these formal fluctuations, nevertheless it, in turn, as just established, is formed by atomic-molecular *ch'i*. Likewise, the colors of the light. This is what the Chinese doctor means when he says that *ch'i* may be regarded either as the energetic and/or exterior (yang) aspect of a phenomenon or as *both* that *and* the material and/or *interior* (yin) aspect of a phenomenon (*ch'i* as just-*ch'i*, or *ch'i* as just-*ch'i*-plus-*shui*).

There are similar findings about the amino acids, at this time called "the building blocks of life." When ammonia is radiated electromagnetically, in imitation of the hypothesized originating conditions of life-forms on this planet, it transforms into amino acids. This is a terribly complex transformation. The result is far more intricate than the molecular structure of ammonia. Again, by the same reasoning as precedes, there is a strong implication that a demi-physical variable in the electomagnetic radiation is chiefly responsible for the configuration of the result; and it is undeniable that even if that is not the case, light-*ch'i* and ammonia-*ch'i* are, for they must be posited as soon as the ammonia and the light are introduced.

THE QUANTUM JUMP OF THE SPIRIT

Contained in the previous section, thanks to the efforts both of our quantum physicists and of the Chinese savants, is the potential for a quantum jump *in science itself*, and it is time, just before entering into the realm of biological *ch'i*, to make that jump. As an energy-gathering exercise toward that end, let us reexamine the rationale which led to the conclusion that atomic electron-pulsations (quantum jumps and energy emissions) must be understood to be caused by a demi-physical configurative force.

Although they know better, our physicists state that when an atom emits a photon it is the result of an "external input" of photon-energy. Inconsistently with this, and more intelligently, they also maintain that in a group of many atoms the number of such emissions, and which atoms do the emitting, is the result of "probability"—not photon-input but a metaphysical force. With respect to photon-output as the result of an external input of photon-energy, the circularity lies in the fact that the photon-energy input to one atom is nothing other than the photon-energy output by another (or by one or more cosmic concentrations other than but previously or potentially equivalent to another atom). In other words, there must be a cause for the energy-input-as-cause; so, to call one output an effect and another its cause is to "solve" the problem by again posing the problem. Although it is objectionably mystical, the concept of an "actualizing probability" can be thought of as the actual cause, and "input" eliminated even as a secondary cause. To posit both is illogical and redundant. To explain this, I perform a thought experiment. Here we have three atoms, and any of their behaviors which have been caused by physical variables external to them (other atoms, cosmic rays) are excluded from the following consideration. If one emits a photon, a subsequent emission of a photon by one of the others may be attributed to the original photon-emission, as its effect. And it may be said that only the atom which originally emitted a photon must have been affected by a cause other than photon-energy. But if one atom emits one photon and then both others do, the original photon-emission could have caused only one, or half of two, of those emissions, so two, or three, of the three atoms were affected by a cause other than photon-energy. And if, as is unlikely but possible, all three simultaneously emit one photon each, then none of the emissions can have been due to the output of another, and all three have been affected by a cause other than photon-energy. Since that cause is sufficient to three simultaneous photo-emissions, it must be sufficient to two or one. It follows that the "causal chain reaction" of output/input is an illusion encouraged by our fixation on time as a condition for cause and effect. Our physicists realize this: their talk of "input" is a notational convenience. But most forget, as is shown by the distinction the majority of them make between "actual" particles as results of "input" and "virtual" particles as "spontaneous." In their present understanding, it is not that the output of one atom affects another atom, but that when an atom outputs, the "probability" which governs the behavior of the group has changed: the output merely symptomizes that.

In the previous section, I substituted for our physicists' mystical probability-as-a-natural-force, *ch'i*: super-subtle, demi-physical, configurative energy. I did so because probability is a mental construct, which therefore exists only in the scientist, not in the phenomena he is explaining. I also did so because such a concept, followed through, leads to belief in a God of one sort or another as the ultimate cause of the behaviors in question. If probability governs photon-emission, because it is a mental construct, ultimately human minds do. But if the physicist is not present, that is, under non-laboratory conditions, photon-emissions nevertheless occur, and assumedly they were occurring long before humans existed. Hence it is not *the physicist's* mind which must be thought ultimately responsible for photon-emission. Another, all-encompassing, mind then comes to mind, as the budding epidemic of religiosity among contemporary physicists testifies. But the reason for choosing *ch'i* over probability (ultimately, God-as-causal-mind) is not scientific. The only *mental* factor involved is aesthetic: the Western "solution" splits the world into a natural half and a supernatural half, whereas the Chinese one leaves it whole, which is more elegant.* The preceding matter cannot be decided by logic or empirical investigation, which brings us to the quantum jump in question.

At this "ground" level, the truth criteria of science are useless. The choice between understanding that *ch'i*—a natural force—or God—a supernatural force—causes inanimate-phenomenal change must be made on combined grounds of morality, instinct, and higher sensibility. I have already made the moral alternative clear. If cause is attributed to God, the whole of Nature, including man, becomes merely reactive, passive, impotent. Not only is the world which is constructed through such a worldview dull and negative, but its result is maximal irresponsibility on the part of influential people, who as a consequence of this worldview have very low standards for themselves and even lower standards for others. In contrast, if cause is attributed to an aspect of Nature, a configurative hyper-energy, then Nature, including man, becomes active and potent. Nature and man are intrinsically luminous and all of Nature is interconnected by its own invisible-yet-perceptible energy flows, as in a Van Gogh painting, a *positive*, self-assertive, world.

The choice between these two implications and worlds is up to each individual as a whole human being. What chooses *ch'i*, what interacts with the preceding moral-instinctive-high-sensory grounds, is *spirit* (*shen*). It is those who do not wish to enjoy or be fully responsible for the effects of their spirits who choose to understand that the ultimate cause is a supernatural property of a God—Spirit (or Dialectical-Material "Law"**) over which there is no human control, and

* Actually the religious argument is intellectually weaker than the Taoist-Confucian one, as I make clear in the next section. The difference, however, is not strong enough to be definitive.

** Not *called* supernatural but, because it is *un*-natural, being a going-from-one-extreme-to-another, and, because it "justifies" and encourages human weakness, it is essentially the same.

which can leave the world dark if it so wishes. One can even go so far as to say that the choice is political. To choose to understand that *ch'i* (both inanimate *ch'i* and Human spirit) governs inanimate phenomena is to elect a decentralized, basically local, "home-world, home-planet, *totally Human*" system of government. To understand that a God (or "Law") governs inanimate phenomena is to elect a centralized, "other world, alien-galactic," system of government over which there is minimal local (Human) control. Ecology means "household-knowledge": an ecology of the spirit reveals that spirit's source is the "home": "this" world, man. As God said to Cain: *Timshel,* Thou *Mayest*: You have the *spirit* to choose. I believe that God, in all his manifestations from Jehovah through the Marxists' inexorable Dialetic of history, wishes to retire. Our self-delusion, to date, has been to think that Western science has permitted him to do so. Only with totally Human, self-, government will that come about, and a necessary movement toward this end is to recognize that the world-force is not "probability" but *ch'i,* and that humans possess it in its highest form.

As I said in paraphrasing Confucius and Chuang-Tzu at the very beginning of this book, science is ultimately dependent not on the quality of the intellect but on the quality of the human being. "Before there can be truth there must be a True Human." Truth is not only discovered; it is also *deliberately produced.*

One can know instinctively that if there is inanimate *ch'i,* spirit (by which *I* mean *Human* spirit, with the understanding that, with the exception of exceptional individual animals, that is the only kind of spirit) is the highest form of *ch'i,* at the opposite end of the *ch'i*-spectrum from inanimate *ch'i.* Between the two is the region of basic biological *ch'i,* which, unlike inanimate *ch'i, is* subject to direct empirical verification. It can be felt, indirectly photographed, otherwise indirectly registered by instruments, mapped, and, on the basis of that mapping, verified by the predictive power, chiefly in medicine, of that mapping. It should come as no surprise that inanimate *ch'i* is not subject to logical-empirical proof, but can only be strongly indicated, whereas, as is now shown, animate *ch'i* is. Since we are animate beings our own centers are animate *ch'i,* whereas inanimate *ch'i* is peripheral to us. We truly resonate with the *ch'i* of animate external objects, but only partially resonate with that of inanimate ones. Let us move, then, toward some empirical evidence of the truth-creating human energies which our scientific paradigm has denied.

THE GERMS IN THE THEORY OF EVOLUTION

Biological, or vital, *ch'i* can be recognized from two perspectives: a perspective on the origination of life-forms as species, and a perspective on the evolution and continued vital functioning of each individual. I begin with the former, adding, to elicit the attention of contemporary philosophers of science, that the following will include a *scientific* alternative to the Theory of Evolution—something that is much desired by contemporary philosophers of science, as I have indicated in the first chapter of this book.

Avant-garde modern science is on the brink of recapitulating Chinese science, although absolutely prevented from doing so by the Mentalist paradigm which underlies it. The brilliant German biologist Ludwig von Bertalanffy, in his *Systems-Theory*, has pointed out that just as molecules have structures which are not predictable from the physical properties of their parts (atoms), cells and organisms have structures which are not predictable from the physical properties of their parts (molecules and cells, respectively). These structures, like those of molecules, he calls "emergent properties." And, just like our avant-garde physicists, he has responded to this recognition by positing a missing causal factor which is basically an idea. Where our physicists came forth with "probability," he came forth with "biological form, or organization." Organization itself, without physical substance, is a pattern in the human mind which arises as a result of empirical observation of organisms and intellectual reflection upon the results of that observation. It is called "relations" among the parts of the object. One cannot point at biological organization, relations, in the external aspect of the world, any more than one can point at "probabilities." But one can point at the physical aspect of organisms of which biological organization is recognized, at the so-called "parts" of organisms. Obviously, if biological organization, or configuration, exists in the external world, it must be ultimately indistinguishable from what it organizes. And it *does* exist there, because the "parts" are, indeed, organized, whether humans are observing them or not.

Again, *ch'i* is implied. Our own scientists have not come up with this concept because the Mentalist paradigm keeps the physical and metaphysical levels Absolute-Fragmentally distinct. Now they are closer to making that connection than ever before in the last 2500 years; the effect of that paradigm is wearing off. Conflict—theoretical, social—is the most basic characteristic of the Mentalist paradigm. It is highly entropic. What involves friction wears itself down, and, as friction reaches its peak, the energy—in this case Human spirit-*ch'i*—which was trapped inside is liberated.

Von Bertalanffy and those who either follow him in avant-garde biology or have covertly borrowed his insight probably have compunctions about following the implication of "emergent biological properties" through, because they think it leads to something in which they do not wish to believe—Vitalist Biology, which posits souls and a God behind them. The reason they do not like this implication, however, seems, from the remarks of those who have commented on their refusal to take the Vitalist position, to arise from a fear of ridicule by their scientific peers for being out-of-date, and the mistaken idea that Western science, unlike religion, is not mystical. Hence it would be more to their credit if they forthrightly took that position. The main point, though, is that Vitalism is not the only implication of the recognition of biological "emergent properties." Indeed, when things are thought and felt through, as just above, God-and-souls is only weakly implied, whereas the Tao understanding of the matter is strongly implied: namely, that there is a biological *ch'i*-organization, a demi-physical configurating force, in living things, which interacts with their substances to produce what Westerners schizophrenically call their (physical) substances, on one hand, and their (metaphysical) organization, on the other.*

As established in the previous chapter, whatever the nature of vital energy is, it must be understood that each individual is "primed" for living by some vital energy from its parents. Without it, the "machine" will not begin to "run." It needs that initial input to absorb and process energy-matter from outside itself into its own vital energy. Furthermore, as shown in animate Sextimagic forms, life-forms' processes, unlike those of machines, are not sequential. Human beings do

* Over twenty years ago Joseph Needham pointed out this "schizophrenia" in the second volume of his encyclopedic work. Here I simply follow through the implication, adding more recent Western theory and data.

not start to live; rather, parents continue to, in their offspring. One assumes, on the basis of solid evidence of ancient cosmic conditions in which life could not have existed, that life-forms once had beginning. Therefore, whatever the conditions of the origination of species are thought to have been, one is led back on a chain of reproductive "links" and energy-"primings" to the first individual or individuals in the generational lines which constitute each (and, if there was one, a very first) species. Having no parent or parents, where did the "priming" energy of that first organism or those first organisms come from? In other words, it is logically implied that before the first life-form(s) existed there was a life-form-*ch'i* which both organized substances into life-form(s) and primed that organized substance to intake and process more energy. Note that this does not exclude, for example, the proposition, entertained by most biologist-Evolutionists, that the first life-form resulted from an interaction of electricity (lightning) (which, we already understand, is configurated by *ch'i* in at least some cases) and organic molecules (which, we already understand, themselves result from molecule-organizing *ch'i* operating on atoms). It is just that it must be understood that the lightning in question was not mere electromagnetic waves but also biological configurating energy; and/or that the organic molecules in question were not mere organic molecules, but also biological configurating energy. This is just as in the case of the amino acids resulting from radiation plus ammonia. If our biologist-Evolutionists were thoroughly logical and avoided Mentalist mysticism, they would arrive, from the data they themselves regard as basic to this question, at the conclusion that life-forms are an effect of biological *ch'i*, and that historically this *ch'i* exists prior to the life-forms. Again: they do not get to this understanding because they fear ridicule for being Vitalists, on one hand, and because their paradigm forbids this Tao alternative, on the other.

It is worth showing how weak the Materialist alternative to the implicitly Vitalist position of the Systems-Theorists is. Called "Physical Reductionism," it proposes that actually the organization of atoms into molecules and of certain molecules into cells, and so on, *will be* predictable from the physical properties of the atom and the inanimate aspect of the cosmos as a whole once they are fully known. This position is weak not only because it is purely an item of faith, but because, thought through, it is really the Vitalist one. The Reductionists know that it is unrealistic to ask the atom to contain all the "information" required to organize all molecules and cells and organisms, especially in view of the fact that the same kinds of atoms form different molecular combinations. So they also posit that the (inanimate aspect of the) cosmos as a whole contains a set of conditions which governs the atoms' "evolutionary progresses." This set of conditions of course involves *change* in cosmic structure. For example, the distance between sun and earth becomes appropriate to the energy-level essential to life on earth. Such changes they attribute to what they call "chance." But the chances, or probability, of life-forming cosmic structural conditions must become 100 percent before life-forms are actually produced. So the model is exactly the same as that of quantum physics: "probability" governs the phenomena in question, and as the conditions for them come into being the probability changes by "actualizing itself." This "self-actualizing probability" is the God and souls of "closet" Vitalists expressed in quantum physical terms—just as God and souls are "closeted" behind our quantum physicists' "self-actualizing probability."

Of course, one can counter this argument by suggesting that the "chance" in question is not self-actualizing, not systematic, ordered. Rather, after the "Big Bang" (the cosmologist's correlate of the life-originating lightning bolt), the cosmic structure was so random and its variables so great in number that it was *only* a matter of time before they combined, in one tiny sector, as life-forms. The process of life-forming was totally random, but inevitable because of the enormous range of possible results. The problem with this theory is that for there to be *different* results—life in this solar system, none in that one —there must be *differences* in the variables of the "Big Bang." If the "Cosmic Egg" and its explosion (or, as I have suggested, its explosion-implosion, or contraction-expansion) were perfectly uniform, the results, everywhere, would be the same. The "Big Bang," then, had a definite form, an aspect of which was specific to the formation of life, which is to say that a very definite potential for life-forming was in it from the start. The "cards were stacked." As a metaphysical thing, that potential is "self-actualizing probability"; as a metaphysical-physical thing, it is biological *ch'i*.

There is an advantage to the Materialist argument. It is that it takes the whole universe into account as the condition for life. In this respect, Tao theory agrees with it. It is understood that biological *ch'i* (metaphorically, "seeds," "germs") exists even when the cosmos is in its phase of terminal-entropy-and-rebirth. The implication is that it "waits" for inanimate *ch'i* to configurate the substances with which it can interact to form organisms, or that actually inanimate *ch'i* is an aspect of biological *ch'i*: the aspect of it which interacts with (hypermatter and then) substances first, in the evolution-to-life-forming of the cosmos. The difference between these implications is not important. The point is that a *ch'i* which is specifically biological must be posited, unless one wishes, for lack of spirit, to posit God.

The only objection which might be made to the proposition that biological *ch'i* or (species-)*ch'is* exist among the basic conditions for cosmic evolution is that the structures of life-forms are thought to be much *more complex* than those of atoms or preatomic gases. Therefore, the conditions for atoms are prior to those for life-forms. This "logic" is motivated by two things. One is a wish to reduce the cosmos, through time, to a relatively simple set of variables, so that the variables are mathematically manageable. (For the same reason, biological evolutionists posit that one-celled life-forms preceded all others.) The other is the understanding that life-forms include atomic-molecular forms: the whole must be more complex than any of its parts. The first reason is not at all scientific except insofar as it concerns practicality in exercising a peculiar form of scientific method. By the same token, to posit several species-specific *ch'is* is just as logical as to posit one basic-life-form-*ch'i*.) The second reason ignores the fact that wholes necessarily are *not* more complex than their parts. As Chew points out, for a particle to exist it must vibrate and, therefore, have internal structure—parts. The implication is that everything has an infinite number of "parts." As Chuang-Tzu put it, "There is no end to smallness." Hence everything has the same "number" of parts. Korzybski, in nearly producing Tao Logic, pointed out that the Aristotelean law of inclusion, according to which the whole has a greater number of components than any of its parts, must be rejected. (He was ahead of his time, except for the fact that that same Aristotle, self-contradictorily, also reasoned that there must be an infinite number of parts!) Biological *ch'i*, then, inclusive of inanimate form, is not more complex than inanimate *ch'i*. Rather, it cannot exist without it, whereas the converse, apparently, is not true. Hence whenever, including at the organized universe's beginning, there is inanimate *ch'i*, there can be biological *ch'i*.

The inference of the existence of biological *ch'is*, prior to the existence of the organisms they form, is empirically evidenced, as follows. There are many forms of disease, one aspect of which is a high incidence of microorganisms ("germs") of one kind or another. Of these, many varieties are not found as parasites or symbionts in the *healthy* human body. For such varieties, it is posited by Western medical scientists that disease consists in "catching" them. There are an indefinitely great number of kinds of them which are not known to be present outside *sick* human (or animal) bodies. Of such diseases, it is said that they are contracted through human contact with humans (or animals) who have the disease, in an active or latent form. There is an obvious, gaping, hole in that explanation, as the Chinese doctor pointed out to me. (Doing so, he cured me of the Western-educationally caused disease of ignoring the obvious and not thinking things all the way through.) Since it is increasingly widespread, gonorrhea is a "good" example of such a supposedly "contagious" disease. Let us trace the disease along the Western-medical epidemiological line back to the first human or humans who passed it along. Where the gonococcus does not exist outside the sick human host, how did *the first* host get it?

Chinese medical theory provides the only logical answer: *by creating the conditions for its existence.* Obviously, such conditions can have but one aspect: the fittingness of the host to the microorganism. But such a condition is not *sufficient* to the existence of the microorganism. It is its environment, not the organism itself. *There must be another condition, and that is that the organism already exists in a kind of potential, or demi-physical form.* Its *ch'i* is already there, either in the healthy body or outside it (which doesn't much matter). Emphatically, this is not the Medieval European "Doctrine of Spontaneous Generation." It is precisely a theory of demi-physical organization-energy which acquires physical form (and may alter in size) under ecologically suitable conditions, whereupon the resulting organism is able to multiply. And it is a theory perfectly consistent with, all but necessarily implied by, the very basis of von Bertalanffy's Systems-Theory, which is the best biological sub-paradigm we have ever come up with.

There is ample clinical evidence for the Chinese medical theory.* To begin with the previous example of "disease-from-human-contagion": our medical scientists have identified four different types of gonococci, labeled types 1, 2, 3, and 4. Type 1 is obtained only from humans with acute gonorrhea, which implies that it is not present at

* The following data are from *Harrison's Principles of Internal Medicine* (the "doctor's bible").

the initial stages of the disease. Types 2, 3, and 4 tend not to survive when introduced into healthy subjects, which implies that they survive only in unhealthy subjects. Since these subjects' "unhealthiness" cannot be due to type 1 germs, it follows that germs are not a sufficient condition for gonorrhea in its initial stage. Rather, the germs are agents of its aggravation. The Chinese theory fits nicely: there is an alteration in the urogenital tissues of the subject which constitutes an environment attractive to and then constituting the substance and breeding grounds for germs of types 2, 3, and 4—but not type 1, because it is not found at the pre-acute stage. The effects of the non-1 types then constitute the substance and breeding grounds for *ch'i*-germs of type 1 in cases of ultimately acute gonorrhea from a single contact, and the substance-grounds for the germs themselves, as well, in cases of multiple contact.

There is an alternative, Western-style, explanation of single-contact gonorrhea—excluding the unexplained factor of susceptibility to initial germ infection. Type 1 germs are mutant descendants of one, two, or three of the other three types. This is as much as to say that three different sub-species of the shepherd species of dog, say, a collie, a Great Pyrenees, and a Saint Bernard, could produce, through a few generations of selective breeding, German shepherds. (Since this is a close approximation of the actual derivation of German shepherds, the hypothesis is credible.) But there is an additional datum, this time not a thought-experimental one but an actual laboratory finding, which has incredible implications if the same assumptions are made. When type 1 germs are subcultured (bred in laboratory host environments), without germs of type 2, 3, or 4, germs of types 2, 3, and 4 later are seen under the microscope. To assume, as a faithfully Western scientist must, that they are the mutant descendants of type 1 germs is tantamount to saying that within a few generations of selective breeding (a condition much stricter than the change from human to laboratory culture) German shepherds have become the ancestors of collies, Saint Bernards, and Great Pyrenees—which is preposterous. The German shepherd genes have been selected from among those of the other three types of dog and the genes distinctive of the other three types have been eliminated (which is why German shepherds breed German shepherds, and, even when bred with non-shepherd-like dogs, tend to parent shepherd-like dogs; their genes, being selected and extremely limited in "plasticity," are strong, stable). Further, the environment in which non-type-1 germs are adaptive (the laboratory subculture) is also an environment in which type 1 germs are adaptive. This is as much as to say that the German shepherds were able to produce the other three types of dog (as well as to reproduce themselves) *without* selective breeding, that is, by breeding them both for German-shepherd characteristics and for characteristics of the other three types—not in separate lines—but simultaneously, for each member of each generation. (The laboratory subculture is not differentiated into type 1 sectors and non-type-1 sectors (separate "lines" of selection)—it is perfectly homogeneous.)

One could counterargue part of what I have said by objecting to the analogy to dog breeding as faulty, because a dog is more organized than a germ and therefore more genetically stable. Indeed, it would be due to the genetic plasticity (multiple-alternative programming) of microorganisms that they can do a trick that dogs could not. Microorganisms may well be more plastic than higher organisms, but this objection ignores the fact that we have already posited that type 1 germs have a stable program selected from the programs of the other types. In addition, it might be objected that although higher organisms also evolutionarily mutate, it takes much longer for them to do so. This ignores the fact that increasingly our biologist-Evolutionists attribute plasticity to *higher* organisms, as well, in order to preserve the Theory of Evolution; and that consequently the relatively long time posited for the evolution of higher organisms, as of humans from an unknown form of primate, is less intrinsic to the Theory than before. To explain: originally it was believed that random mutations were caused by impacts of cosmic rays or nuclear radiations on DNA-codons. Hence, enormous time was required for the "chances" of adaptive and multifold mutations to be "actualized." But now Evolutionists recognize that the cosmic-ray theory is outrageously far-fetched. Hence they rely for their explanation chiefly on genetic plasticity and "drift" (selective recombinations of established genes). It follows that under new highly selective conditions significant mutation could occur within a relatively brief period. While we wait for the type 1 humans who live in Secaucus, New Jersey, or work in coal mines, or inhabit cult and Communist communes subsisting on experimental diets and being selectively bred to mutate into type 4 hominids or even primates, let us note, then, that the argument against my analogy to dog breeding cannot be firmly rejected on the grounds of contemporary Evolutionary theory.

In contrast to the gaping-holed and otherwise weak or untenable

Western germ theory, the Chinese one accounts for all the data and is common-sensical and (unless one is encumbered by ethnocentric prejudice) fully credible. First, it accounts for the germ-infection of the *first* person or persons who had gonorrhea, whereas the Western one totally ignores this matter because it cannot handle it. Second, it specifies the conditions under which a human is fully susceptible to contagion by germs of the non-1-type. Third, like the Western theory, it accounts for cases in which gonorrhea reaches the acute stage without an influx of type 1 germs subsequent to the original influx of germs, type 1 of which does not survive. But, instead of mutation, which cannot account for all the data in question, it posits what already *must* be posited in any case: that *ch'i*-germs interact with an altered tissue environment to produce full-fledged germs. Fourth, although mutation-theory reasonably accounts for the "appearance" in gonorrhea victims of type 1, it cannot reasonably account for the matter of non-1-types "appearing" in 1-type subcultures. So, the Chinese theory alone accounts for these "appearances," which may, therefore, be regarded as empirical documentation of the general Chinese theory of *ch'i*-germs.

There is direct evidence of pre-acute gonorrhea contracted without contact with a carrier, which, in addition, tends to confirm the Chinese explanation of its actual cause. Our epidemiologists are reasonably sure (and Chinese epidemiologists are quite sure) that a weak ("subclinical," non-1-type) form of gonorrhea "occurs in many patients with either acute or asymptomatic urogenital infections," "but the mechanisms responsible for dissemination are unknown."* In other words, *people who have not had sexual intercourse with others infected by gonococci may develop gonorrhea* when they have, or with, urogenital infections, as by *E. Coli.*** Nevertheless, "dissemination" is assumed, and the apparently scientific but totally vacant expressions "mechanisms for" and "are as yet unknown" are tacked on to (appear to) preserve the paradigm. The implications are that gonorrhea can be contracted without contact with gonococci and that urogenital infection is the, or an aspect of the, host environment for gonococci (gonococcus-*ch'i* and then gonococci).

According to Chinese medical theory, the original cause of gonorrhea is sexual intercourse by a woman with many different men. "Sexually liberated" people will find this notion unacceptable, since it

* *Harrison's Principles of Internal Medicine.*

** The hero of our recombinant DNA lab.

means that the life-style they promote is disease-causing. And it is predictable that some among them will accuse me of confusing morality with science. Let me admit in advance, then, that I am shamelessly "guilty" of what they might accuse me of. But before I elaborate on the necessity for the confusion and further jeopardize the civil rights of such people—not to mention the present market for ampicillin—let me point out that this Chinese theory *must* be essentially correct, for there *are*, or at least, there have been *initial* carriers of the disease.

According to Chinese medical theory, the details are as follows: When a man and a woman have sexual intercourse for the first time there is a very mild immune response in the vaginal tissues. Subsequent intercourse with this same partner will not exacerbate it. But a series of *different* immune responses to different males constitutes the host environment for the kind of urogenital infection which, in turn, combines with those immune responses to constitute the host environment for the gonococci. What makes the male fully susceptible to infection, then, is the state of his genital tissues during sexual intercourse.*

In the 1980's it is probably impossible, because of the increased frequency of pre- and extra-marital intercourse and because of an epidemic of dishonesty, to empirically test that Chinese-medical theory, but one can nevertheless perform a sound thought experiment. Here is a village in Ohio in 1950, long, long ago. With a few exceptions, the only teenagers having sexual intercourse in the village are married, and all the sexual intercourse in the village takes place between spouses. This is a religious community of face-to-face relationships. Anyone who violated the sexual morals of this community would be found out and experience a guilt more painful than the pleasure in question. There are a few exceptions, the majority of which are sexual unions between boys and girls who will marry each other, having had sexual liaisons with no more than two people other than their spouses, and the remainder of which are short-lived adulteries with only one partner. Twenty miles from this village is another just like it. All sexual unions and marriages occur within or between these two communities. Furthermore, none of the males who have traveled elsewhere have had

* As I proofread this two months after writing it and six years after learning it from my *shih-fu*, *The Medical Post* reported the finding that bacteria in semen is linked to cervical cancer. It is *promiscuous* females, specifically, who are almost unexceptionally its victims. Below I show how VD such as gonorrhea is a link, with help from modern medicine, to cervical cancer.

sexual intercourse with prostitutes except for seven men, five of whom did so in France during World War I and two of whom did so in Youngstown on a business trip. Only the latter two were married at the time. The former were cured of gonorrhea in France and the latter and one of their wives, who had contracted it from her husband, were cured by the local doctor. Now there is no gonorrhea in these communities.

But there is a teenaged girl in one of these communities who is subnormally socialized, on one hand, and, on the other, plain "hot"; and the heat is too much for several of her male classmates and their male friends to bear. A year later, several of the local boys have gonorrhea and the girl has gone off to Youngstown to be better rewarded than at home for her services. An epidemiologist from Youngstown records the incidence of gonorrhea reported by the local doctor, who is unable to trace its carrier beyond the girl. The epidemiologist concludes that the girl had had a liaison with an infected man from yet another community. The local doctor replies that he is sure she has not; he knows her parents very well, and his own daughter was her classmate, and as a matter of fact there has been no opportunity of that kind. Furthermore, he knows each of the boys and their parents very well. The majority of them are his fourth cousins or closer, and he is quite sure that none of them brought the disease in from outside the community, either. And he *would* know, because in this community "the clap" is regarded as such a filthy affliction, not only physically but morally, that no one who contracted it could fail to come to him with his problem at the earliest manifestation of its typical symptom, because they would find carrying it intolerable. The epidemiologist, despite his knowing that subclinical gonorrhea, at least, may be contracted without contact, assures the doctor that he must be mistaken. Perhaps the girl was taken advantage of by one of the local businessmen who had been to Youngstown and successfully threatened to keep quiet about it. Knowing better from the embarrassment on the faces of those businessmen when they came to be cured, the local doctor shrugs his shoulders.

To comment on a predictable response to the theory in question: that Southwest Pacific peoples, who tend to be sexually promiscuous prior to marriage, had no venereal disease prior to "white contact" is a Western-scientific (and masochistic if well-intentioned Humanist) folk belief. In turn, are we to believe that Captain Cook's sailors, on discovering certain Pacific Islands, made an epidemiological survey of the incidence of gonorrhea among the natives of the region before they accepted the sexual invitations of the local women?

Our thought experiment cannot be conclusive; but it is only ancillary to an argument which is. Gonococci do not exist outside the human body (except in laboratory subcultures derived from human ones). Hence, originally they come-into-full-existence within it under certain specifiable conditions.

Here is a last, Western-style alternative hypothesis. Humans are genetically programmed to produce gonococci, along with their own organic structures, when they are conceived. Let us examine this (ugly and "intuitively false") hypothesis. If the human body is genetically programmed for gonorrhea, the result must be demi-physical (because under normal social conditions the majority of people do not have gonococci in their bodies)—unless the human program also calls for gonococcus antibodies, which kill off gonococci but which, in a minority of genetically defective humans or humans whose defense mechanisms somehow have been damaged or overtaxed by fighting other germ infections, fail to completely eliminate them. This is a plausible hypothesis, except, as we will see, for the unrealistic demands it makes on human DNA-"programs."

Our medical scientists have concluded, on the basis of empirical observation through microscopes, that there are about a million different kinds of antibodies in the human body. To make an assumption generous toward this Western explanation, assume that merely 100 codons, that is, combinations of three DNA molecules which are regarded as "bits" of information instructing the formation of the human organism, are required to "program" each kind of antibody, each of which is a complex chemical structure which fits into a germ's molecular surface with the precision "of a key," as our scientists put it. It would follow that about 100 million DNA codons in the fertilized ovum are required for the "programming" of each kind of antibody. There are about 13,000 million codons in the fertilized ovum. The Western theory that DNA encodes the information for human antibodies therefore implies that 1 percent (rounded off to the nearest percent)of human DNA is devoted to programming for antibodies. Adding antibodies for gonorrhea and so on* makes it a solid 1 percent. Recent discoveries imply that only one third of the DNA in higher

* All other disease germs which are not present in all humans but which breed only in human bodies.

organisms is specific to any part of the anatomy. That raises the percentage in question to 3. Assumedly, an equal amount is required to produce the germs themselves, that is, germs which are not present normally but which breed only in human bodies, which means about half of them. That raises the percentage of DNA devoted to germ-and-antibody-production to 4½.

Because the immune system is estimated to account for much less than *1* percent of the body's information, our biologists abandoned the hypothesis that all the kinds of antibodies are DNA-programmed as soon as they estimated that there are a million kinds of them!"* Here, we must ask that over four times that amount of the DNA is devoted to the immune system. Our biologists propose, therefore, that all that is programmed-for are a few "precursor-cells" whose descendants, in an unexplained manner, mutate into the million varieties in question.

That this hypothesis is unexplained because it is really fantastic should be obvious. The mechanism that would select which of the randomly produced (if programmed-for, we are back at square one) mutants survive obviously involves, more than anything else, the germs in question, whose effect is to reduce the number of antibodies against them to a minimum. In turn, the remaining antibodies, being without enemies, would multiply without limit. Finally, being very similar and requiring the same environment, they would compete with the disease-fighting antibodies for living space and, of course, win!** In addition, it stands to reason that if the body environment for the selection of desirable mutant antibodies is specific enough to select for them, ultimately one has to posit that it must have been specifically programmed to be so! To credit this selection to the "environment" of the antibodies in question is, by Mentalist sleight of hand, to shift the attention away from the whole actually in question onto a part, so that the remainder is ignored.

There is further reason to adopt the Chinese *ch'i*-germ theory. It is *good* in the moral as well as intellectual sense. Its implication is that humans, unlike animals, have a *moral* dimension to their physical health. If they follow civilized sexual mores, they do not contract ve-

* See National Research Council in the *Bibliography*.

** One could alternatively propose that there is another limiting factor operative on both types of antibody. Let me simply say that that factor would centrally involve more genetic complexity in the DNA, ultimately much more than the amount already rejected as implausible by our own scientists.

nereal disease. If they do not, they sicken and die. Nature has arranged it so that human morality, which, as will be explained, is *socially* functional, is also *physically* functional. This is only to be expected, for reasons which, because they render religion unnecessary, the similar pre-"Enlightenment" Western-religious versions of this belief cannot include. (And *that* is one reason why religious morality fails, leaving Westerners victimized by a chaos that includes, for example, historically extraordinarily epidemic venereal disease. Religious morality doesn't make enough sense.) To wit, the moral:physical parallelism of humans is to be expected because man is a thoroughly cultural animal. Man is the only animal (or partly-or-wholly-animal being) which lacks enough instinct to assure its survival. Culture—basically, morals—functions in man the way instinct functions in animals (as Mencius, in 400 B.C., and the French sociologist Emile Durkheim, in the late 1800's, were the first to point out). And there is every reason to believe theories of such moral:physical parallelism. Sexually promiscuous monkeys, for example, do not develop venereal disease, whereas man is *meant* not to be promiscuous, so that his/her sexual tissues are more sensitive than those of promiscuous monkeys. Human society cannot function the way monkey society does: it requires more order so that the rest of culture, mainly technological culture, can be transmitted and applied. So humans are physically programmed to be deterred from sexual promiscuity to enable them to perpetuate the rest of their culture. And sure enough, as Freud observed, sexually promiscuous human societies, such as certain South Pacific ones, have subnormally potent technology, including self-defense technology, and consequently no longer exist. To believe the Chinese theory is not only more logical than to believe any alternative one; it is also better for humans to do so. (Again: true and false are ultimately indistinguishable from right and wrong.)*

The list of germs thought to infect humans only by human contagion, and for each of which it follows that they are either programmed-for by human DNA or originate from *ch'i*-germs under

* Our flirtation with sexual liberation is already being curbed on scientific grounds. In light of the recently discovered connection between cervical cancer and promiscuity, Dr. Aileen Clarke, head of epidemiology and statistics for the Ontario Cancer Treatment and Research Foundation, proffers advice to fellow females that sounds as maternal as it is medical: "Postpone intercourse until the late teens or twenties, and—when you start—don't sleep around." (See Francis, cited in the *Bibliography*.)

appropriate host conditions, is enormous. *E. Coli* is another example. Its natural environment, along with numerous others', is the intestines. When found in urogenital infection, it is supposed that it was transferred to the urinary tract, either through a perforation in the intestine, or via the bloodstream, or by humans extraordinarily negligent of basic hygiene or by infants expectedly ignorant thereof.* Given these three possibilities, all infections—the presence of the germ in its non-natural, non-intestinal, environment—can be reasonably well accounted for. But, of course, how *E. Coli* gets to the intestines in the first place is never explained. On closer inspection there is also a hole in the explanation of transfer from the intestines. Where there is "no demonstrable portal of entry" and the germs are therefore supposed to have been carried to the urogenital tract by the blood, it must be posited that the blood is sick. Otherwise everyone would be infected by them. But many patients have none of the diseases the explanation can rely on, namely, neoplastic and hematologic diseases, such as diabetes, cirrhosis, sickle-cell anemia, or the side-effects of Western medicines—radiation-"treatment," cytotoxic drugs, adrenal steroids, or antibiotics.** So, the Chinese theory nicely accounts for not only the natural presence of *E. Coli* in the intestines, but also for its infection of the body in cases where mechanical contagion is unlikely or impossible. That is, through activity unhealthy for the urogenital tract it becomes a host environment to which *E. Coli-ch'i* is attracted.

One should also wonder how many infections explained as due to lack of hygiene are really due to that. As stated earlier, *E. Coli* is commonly found in high concentrations in the urogenital tracts of people with venereal disease. This is also true of people, especially females, who do not (yet) have venereal disease, but who are sexually promiscuous. Is it realistic to propose that *each and every one* of these women either is extraordinarily negligent of basic hygiene or has had sexual intercourse with a man who is?

Where the strains of *E. Coli* implicated in infantile diarrhea, which usually occurs in hospitals, are *not* indigenous to the human gastrointestinal tract, what could their origin possibly be, other than *ch'i*-germs attracted to infantile gastrointestinal tracts which have been weakened, usually by the hospital environment (or probably by some-

* *Harrison's Principles.*
** *Ibid.*

thing associated with it, such as antibiotics)? Our epidemiologists, faced with this paradigm-jeopardizing fact, illogically propose that such strains "probably are disseminated within nurseries by asymptomatic infant carriers, mothers and nurses."* How, then, did these "carriers" become infected? The "carrier" is like the "inputting" atom, the victim of contagion like the "outputting" atom.

Further evidence of the *ch'i*-germ theory lies in the Western explanation for the presence in the body of micro-organisms which do not cause disease, but, rather, are essential to the living of the host. As we saw, the number of kinds of antibodies in the human host is so great that our biologists cannot posit that they are programmed-for by human DNA. For the same reason, they would not posit that benign intestinal parasites are programmed-for by human DNA. They have not considered this possibility, however, any more than they have considered the possibility of the gonococci being programmed-for by DNA, because they "explain" the presence of these micro-organisms by contagion (as with gonococci and *E. Coli* of the diarrhea-causing variety). But, as in the case of gonococci and that strain of *E. Coli*, the explanation does not withstand scientific examination. Let us take the example of cattle, to whom benign bacteria and protozoae are relatively important, for they are themselves part of the food of cattle, and are indispensable to digesting food from the outside.

Those micro-organisms digest grasses in the rumen (pre-stomach) into a form the abomasum (the equivalent of the non-ruminant stomach) can handle. It is supposed that they are transmitted to newborn cattle through the air and/or the newborn's licking of the mother's feces or the mother's licking of her feces and then of the newborn. But not all can be transmitted by air because some of the bacteria are strict anaerobes, are killed by oxygen. That means that the explanation relies strictly on fecal contact. Many of those micro-organisms are digested for protein; hence there is the possibility of all of them being digested before exiting in the feces—especially if the animal has an abnormal need for protein. Further, even granting that our biologists may be wrong about certain kinds of these micro-organisms being anaerobic, it is common practice in modern farming to remove the newborns from their mothers and all other cattle very shortly after birth. This makes the probability of transmission through contact much

* *Ibid.*

lower than the survival rate of those cattle is high. All told, the Western explanation stands on too thin ground. Of course, a contact explanation of the presence of similar micro-organisms in infants, who in modern society are not only separated immediately from their mothers but kept for several days in sterile conditions and often fed nothing other than sterilized, non-human, milk, is even thinner. Even in exceptionally unhygienic societies the likelihood that all infants have direct or indirect contact with adult feces is extremely low. Not to mention the fact that the theory is as insulting as it is unrealistic. If human infants depended on contact for their benign micro-organisms, most would die in early infancy!

Again, the *ch'i*-germ theory fits perfectly into the hole in the Western one. It is just that, in contrast with the foregoing examples, the host environment for the *ch'i*-germs in question is healthy. It might be added that at least some of these *ch'i*-germs must be inherent (but not, for reasons made clear, via DNA-programming), for if they were available to the body from outside itself, the sterile hospital environment would prevent the realization in infants of any digestive microbes, and kill them. In turn, the appearance of diarrhea-causing strains of *E. Coli*, especially in hospital-born infants, might be due to the lack of some of the digestive *ch'i*-germs whose place they take—that is, of varieties whose *ch'i*-germs are present in non-sterile, natural environments.

There is also evidence of the existence of *ch'i*-germs for disease germs which breed outside, as well as those which breed only inside, the human body. And it is well worth pointing this out because some germs presently thought to exist only in the human body may later be found to exist outside it as well. Our scientists do not, after all, have their microscopes trained on more than a tiny fraction of the environtent of humans, and even that fraction is not constantly observed. Furthermore, most of that fraction is observed under unnatural, laboratory, conditions which create data almost separated from the animate-inanimate whole supposedly in question.

With regard to such environmental germs, it should be understood that all that makes them germs, as distinguished from micro-organisms in general, is that they are agents of disease. So, their origination is just a special case of *the origination of micro-organisms in general,* of kinds of life-forms, which is the matter in question in this section.

Leprosy germs are an example of environmental germ. As our epidemiologists know, they breed in dirty areas, especially of cities, and humans susceptible to them may become their diseased hosts.*

Here is a clean small town near a river. It becomes a river port, a trade center, a big industrial-and-trade city. Near the docks live poor and demoralized families in squalid conditions. In the area are sleazy bars, whores, and pimps. In the material interstices of this dark and dirty area leprosy germs are found. How did they get there? Before the city evolved and deteriorated there were none. If they were brought by humans from outside the area, ultimately the same question must be asked about the origination of leprosy in *their* areas, so that factor is irrelevant to the explanation (just as the "output" of one atom, which is said to cause, as the "input" to another, that other's "output" cannot be the ultimate cause of that other's "output"). In general, we are considering any germ disease which is originally, or if not contagious only, contracted from the non-human environment and which is limited in the non-human environment to specific kinds of areas which may be created by changing natural or manufactured conditions. Such germs must come into existence after such changes, and therefore come into existence partially or wholly because of such changes.** Do germs of other kinds there before the change, pieces of newspaper, banana peels, flies' wings, the excretions of uninfected humans, and so on, magically combine into leprosy germs? The "proper" combination of such variables is the host environment for leprosy germs (and their precursors, if any). It follows, as it follows with respect to the human-and-laboratory environment for types 2, 3, and 4 gonococci, that a biological leprosy-*ch'i* exists, and is attracted to and interacts with these physical factors to become leprosy germs.

In other words, among countless other microbial *ch'is*, there are leprosy-*ch'i*-germs in the environment—demi-physical proto-germs which have everything required for the existence of what is found under the microscope in the medical laboratory save their relatively-physical substances and an environment which contains substances and energies which they convert into their own substances and energies, so as to be-

* I am indebted to my friend A.L. Saunders, M.D., for this datum, which in no way should imply that he supplied it with the end of making the present point.

** Due to the deterioration of cities worldwide and the wide spread of war which deteriorates everything, leprosy is making a big (unpublicized) comeback. (There are at present about 20,000,000 lepers.)

come fully physical and able to multiply. Here Aristotle, whose father was a physician and who, as I have said, was logically far superior to his successors,* deserves much credit. He reasoned that there must be "air-germs," defined essentially just as precedes save for the Polar-Complete dimension, to account for the origination of life-forms. His theory was rejected on extremely limited experimental grounds** and on the grounds that it was metaphysical by post-"Enlightenment" Western scientists, who at the time were Materialists. And now that they are becoming Idealists, they may revive it, thus keeping in step with the new Idealistic physics and the Vitalist implication of Systems-Theory.

Aristotle's theory is, indeed, objectionable because it is metaphysical. What is here proposed is a synthesis of the two extreme just-metaphysical-or-just-physical and therefore equally unrealistic theories which Western medical science alternates between. What I am proposing is the excluded middle of Western scientific theory, both of its alternating halves—for once, at once.

Although the Chinese theory fits in a moderate manner between the two contradictory extremes of its Western correlate, it is nevertheless, from our perspective, revolutionary and disturbing. So let me fortify it. With respect to the question of leprosy and other diseases which appear to be contracted from the non-human environment, the logical alternative to the Chinese theory is to posit that leprosy germs are simply life-forms normally present in the environment, like the majority of other life-forms, but in such diffuse populations that they cannot be found. It is just that they concentrate in certain kinds of filthy environments until the chances of contracting leprosy increase to the point where leprosy is in fact contracted, and isolating cultures becomes possible. There is a problem with this explanation. It equates the environment in general with a specific kind of filthy one in which leprosy germs breed, distinguishing the former as a less concentrated form of the latter. This is because, as Mary Douglas recognizes, filth is something that is out of place. The ecological niche of leprosy germs is not an intensification of, but a distortion of, a re-combination of ele-

* Save von Bertalanffy and his followers, whose theory of biological form as a natural "given" is consistent with, indeed, when thought through, implies, metaphysical (ideal) proto-germs.

** It was proven that fly maggots do not "spontaneously generate" in rotten meat. If screened from the environment, rotten meat will not come to be infested by fly maggots.

ments within, a healthy environment. It follows from this that leprosy germs are not present at all, except in potential, *ch'i,* form, in the normal environment, and that they come into existence when that environment is altered in a specific way.

That stands to reason: health is a balance within the harmonious manifold of Nature. Disease, as distinguished from natural death from aging (death is *not* a disease), would not exist if that balance were perfect. The same would follow for microbes which are agents of disease, with the exception of those which, like intestinal flora, are necessary to the survival of their hosts. The balance never is perfect, of course, but neither could it be such that *all* disease microbes are present at all times, for not all diseases are contracted at a constant rate, and some are *not contracted at all* for long periods, for example, the Black Plague. It follows that some microbial and viral agents of disease come into existence under certain conditions, and that they exist in potential form before and after such conditions occur.

The appearances of new life-forms, including Humans, must be considered from a similar perspective. The difference is that their potential forms, or configurating energies, must be stronger and more widely distributed in the environment than those of agents of disease, prior to their appearances. And the changes in the environment which constitute the conditions under which they fully corporealize and then multiply must not be due to alterations in the environment which are unhealthy for existing life-forms. Therefore they are due to an interaction between those organic patterns, or *ch'is,* themselves, and existing life-forms. Otherwise put: it is not that God caused the world to unfold along a line of higher and higher organisms so as to express Himself and be adored, but that the Human *ch'i* has acted upon that environment since its inception so as to create the conditions for its own corporealization within a manifold living whole which is entirely sacred and for which we are totally responsible. In this sense, Humans, not God or the too-improbable Chance of the Evolutionists that has replaced Him, created all life-forms. In this, I would suggest, lies the answer to the question of why the human body, especially in its foetal evolution, recapitulates the forms of so many lower animals. The Human *ch'i* is one with that of all other life-forms, is distinguishable from them in one, Polar, respect, but identical to them in another.

Before elaborating on the question of the coming-into-existence of higher organisms, I should add that our own medical scientists have recently begun to reexamine and increasingly to doubt the germ-theory of contagion, although they have by no means made a connection between this question and that of the origins of life. In his recent book, devoted to the question of influenza epidemics, W.I.B. Beveridge observes that direct human contagion cannot sufficiently account for flu epidemics because they often spread much faster than they could be sneezed, coughed, and contracted through human populations. He attempts to account for this discrepancy by positing that the viruses are carried by the wind in exhaled water droplets. They may survive in this environment for "up to several hours" *providing there is no sunlight.* He himself suggests doubt about this explanation, observing that a concentration of viruses is quickly diffused under such conditions: the chance of inhaling viruses exhaled by others at a great distance is extremely low. Flu sometimes breaks out on ships sailing long after the incubation period, far from shore. The chances that sailors will inhale viruses coughed out on shore are too low to take seriously. In any case, it is not only that flu spreads faster than the germ-theory "permits." As Louis Weinstein, another flu epidemiologist, observed, epidemics of flu do not spread outward from a source; the Hong Kong Flu epidemic of 1968, for example, appeared *simultaneously* all over the world.*

Even if a superb, chance, orchestration of direct human contagion, winds, and darkness might account for roughly simultaneous appearances, however, the problem of how the disease originates in the first place in each instance of an epidemic remains unanswered. I am by no means denying that there is virus or germ contagion. Rather I am agreeing with present growing recognition that it is an insufficient explanation, and noting that the *basic* problem, of origin, is virtually ignored.

The flu viruses survive only in some warm-blooded animals, including man. If they are *always* present, in some form, in part, parts, or all of this environment, one must posit either that in the case of epidemics human resistance to them, planet-wide, is lowered, or that they become more virulent, or both. The only clue we have to the answer is a strong one: flu co-occurs with exposure to a *specific variety* of wet, and usually cold, wind, which is instinctively perceived by humans as unhealthy—"*bad*" wind. Our epidemiologists are perfectly aware of this and can even predict outbreaks of flu by the season and the weather; it is just that they deny a causal connection. Granting a connection,

* See Inglis, cited in the *Bibliography.*

there are four logically possible interpretations of it. (1) The viruses are always present, as (a) demi-physical or (b) fully physical organisms in the host, and the simple physical property of cooling, by wind, either lowers resistance to the viruses or increases their virulence. (2) The viruses are not present in the healthy host and are carried, in (a) demi-physical or (b) fully physical form, by such winds. (3) The viruses are always present, as (a) demi-physical or (b) fully physical organisms in the host, and a non-physical property of the wind either lowers resistance to them or increases their virulence. We already know that (2) (b) accounts for flu only exceptionally, so it may be disregarded here. A mysterious experimental result obtained in this connection indicates almost conclusively that the first option is false, thus leaving us with a theory that involves a non- or demi-physical variable, either a wind-*ch'i* or a *ch'i*-virus. Humans contract flu chiefly in winter, suggesting that a climatic and weather factor, namely the physical cooling effect of wind, is the independent variable. Beveridge reports that mice kept in laboratories year-round at 72 degrees and 50 percent humidity mysteriously contracted flu roughly one-and-a-half times as frequently in winter as in summer. The experimenters "did not identify the remaining factor operating." Obviously, some aspect of Nature, associated with climate, penetrated the laboratory, unless mice are programmed for lower flu resistance during the winter. Beveridge notes that there is no evidence of this in mice or in humans. The agent in question must have been non-physical because it was immune to concrete walls, heated air, sterilizing filters, and glass. And given the strong correlation between wet, usually also cold, wind and flu, and the absence of any other correlation, one would infer that the agent is *in* such wind.

According to Chinese theory, as in Sun Sze Mo's *Chien-Chin Yao-Fang* (eighth century A.D.), it is the second of the remaining two options that is correct: the specifically identifiable kind of wind in question—instinctively perceived by humans—is the natural environment of flu (*kan-mao*) microbes, which enter human hosts and there become stronger (*chang*) and multiply. This "becoming stronger" refers to the corporealization of *ch'i*-germs. Whether his choice between those two options is correct or not, a non-physical variable which is formative of the diease is strongly indicated on the basis of present evidence. It happens that Chinese herbal medicines, whose prescription is based on the preceding theory, cure all varieties of flu by creating an environment within the body which is opposite to the properties of and reverses the effects of wet-cold wind, the posited ecological niche of *ch'i*-flu-viruses.

Again, I am persuaded to credit human common sense with greater truth value than instrument-and-laboratory-dependent Western science. The term "influenza" is from the Italian: "in-flux."

Avant-garde Western speculation on epidemics follows lines quite consistent with the preceding Chinese theory and derives from much earlier objections to the germ theory. Reporting on the new doubt of the germ theory of contagion, in the November 80 issue of *Omni*, Brian Inglis observes that half a century ago Hans Zinsser, an epidemiologist, hypothesized that "such outbreaks were the consequence of some interaction, an as yet unexplained biological relationship, among the microbe, the environment, and man."

If, like the Chinese, we envision the world as a whole and regard disease as a change in the *whole*—and therefore an "effect" of *ch'i*—which is undesirable to humans, we will be able to address this matter realistically and effectively. If we do not, we will continue to wander down an increasing number of blind alleys. We rushed down such an alley when the germ theory of contagion was originally subscribed to as the sole explanation, and until recently, evidence of our error has been ignored. As Inglis observes, Max von Pettenkofer and his colleagues, by swallowing teeming cultures of cholera bacilli, successfully challenged the germ theory when it was first advanced by Koch. All contracted only mild diarrhea.

When I sought enlightenment from my *shih-fu* on the discrepancies in the Western germ theory, one of his observations was that "There is no end to smallness" (Chuang-Tzu) is as true of the *biological* as of the inanimate world. Hence, as Western microscopic techniques improve, smaller and smaller disease agents have been and will be discovered until the distinction between matter and energy, physical and metaphysical, will have to be dispensed with. Two years later I read in the December 1980 issue of *Scientific American* that a *micro*-micro-organic, and (where viruses are equally molecular and organic) *more* molecular than organic, agent of disease, the "viroid," had recently been discovered, and that it may be prior, as a disease cause, to viruses and bacteria. (I then again updated this text, accordingly.) It is my hope that, as we work in our peculiar, theoretically weak and experimentally intense, way toward discovering disease-*ch'is*, we will not fail to take advantage of a medical system which, on the basis of *ch'i*-theory, cures many diseases which we cannot.

Western awareness of the inadequacy of germ theory has been suppressed to preserve the paradigm, but is now rapidly emerging. Inglis reports that shortly after World War II, tests at the Common Cold Research Unit in England "showed, disconcertingly, that in controlled conditions people rarely catch cold from one another, no matter how intimate their relationship." Tending to confirm my theory of hospital environments as conditions for germ-*ch'is* is this last observation: "Even more baffling are the eruptions of epidemic neuromyasthenia in many countries. One of its favorite targets is hospitals. No surprise in that, except that it is commonly the nurses and doctors who collapse with tremors and twitches, while the patients remain immune."

The pioneer of ecological medical science, Réné Dubos, in an article, "Second Thoughts on the Germ Theory" (*Scientific American*, 1955), wrote:

> During the first phase of the germ theory the property of virulence was regarded as lying solely within the microbes themselves. Now virulence is coming to be thought of as ecological. Whether man lives in equilibrium with microbes or becomes their victim depends upon the circumstances under which he encounters them.

Now, "circumstances" are precisely conditions other than the variables in question, hence, neither the microbes nor the human body as defined by Westerners such as Dubos. These circumstances are conceived, ecologically, as *relations* between human bodies, their environments, and the microbes. Relations are abstract concepts. It follows, von Bertalanffy-wise, that relations cannot be the operative variable. Rather, it is something like them, but which, unlike them, is not a mere idea in the minds of scientists. So it is something demi-physical, which is a tendency, or which has "intention." The psyche, as *ch'i*, and physical-disease *ch'is* fit this bill.

In some cases "relations" do refer to concrete conditions, such as dietary deprivation, wherein microbes present in the body become virulent by default of metabolic resistance. But, in turn, what is meant by "resistance"? In Western terms: either antibodies or the normal chemical structure of cells. But Dubos is sure that at least in some cases it is not low or inappropriate antibody production which is responsible for susceptibility to microbial disease. As he observes,

> . . . it is well to keep in mind a fact so simple that it is never

> talked about—namely, that the tissues of man and animals contain everything required for the life of most microbes. This is well shown by the ability of tissue cells to support the growth of bacteria and viruses in the test tube.

That takes care of the hypothesis that resistance equals normal cell structure. And so, all of it takes care of the hypothesis that resistance is simply and purely *physical*. Therefore, it is either meta-physical (as in "relations" in the abstract sense), or it is demi-physical. Likewise, *lack* of resistance must, in at least some cases, be a meta-physical or demi-physical variable, or condition. Of course, a germ-*ch'i* and lack of resistance are the same thing seen from two different perspectives. And accordingly, Chinese medical therapy chiefly seeks to restore to normal or make temporarily super-normal the resistance of the patient, rather than to denaturalize and complicate the matter by introducing a variable in addition to the patient and the microbe: a chemical which poisons the microbe, and which, if administered without regard for the rest of the clinical picture, as in the West, has side-effects—that is, transforms the disease into another, more complicated, one.

Evidence of the coming-into-existence of organisms from *ch'i*-germs is limited to diseases, because the living world is stable through the ecological interdependence of all species. There is no room for new species (save, perhaps, a species of a higher order than humans) unless one is eliminated and its environmental niche is fundamentally altered. All we can expect to observe is the coming-into-existence of organisms whose environment is faults in the living field. Hence, disease data, such as precede, are the most powerful evidence which could be sought. In turn, wherever faults, or disruptions of the ecological harmony, appear, one would expect, according to the Chinese theory, to find the appearance of organisms. There is another sector, then, in which such evidence should be found; and if it is, the Chinese theory is even more powerfully confirmed, because it *predicted* it. This sector is the correlate of the human body: the planetary body, which over-mechanized and chemicalized farming and industry have disturbed. And the correlate of disease germs in this sector is the non-microorganic type of being called the "pest." Such evidence, in terms which are as conclusive as could be expected, is all around us. Here is an example, taken from the scientific magazine *Discover* (October, 1980):

> Like a swarm of hungry tourists, Mediterranean fruit flies are

back again, this time eating their way through more than 500 square miles of backyard fruit and vegetable gardens in California. . . . Agricultural specialists hoped they had solved the Medfly problem once and for all. Four years ago, after a large-scale infestation in California, they released millions of male Medflies that had been sterilized by radiation and sent them off to mate with their unsuspecting consorts. Normally, the female lays as many as 300 eggs in her two-month lifetime, but during the 1976 season, the mating rites produced no offspring. The situation looked settled—until now. Somehow Medflies—including virile males—have managed to appear again. Officials speculate that the pests stowed away on fruit that was smuggled in from Central America.

CHILDREN OF GOD, CHILDREN OF AMOEBAS, AND CHILDREN OF MEN

In each of the preceding cases there is a question unanswered (or very poorly answered) by Western science: Where a specific environment may be free of a given micro-organism (or organism) and then become its environment, where did the organism come from, or how did it come-into-existence there? And we saw that theories of contagion skirt the question in just the circular way that the Western explanation for an atomic photon-emission does: by attributing the phenomenon, as an effect, to another of exactly the same kind. We then examined alternative Western-style theories and found them either quite untenable or unacceptably weak. And we saw that the Chinese *ch'i*-germ theory not only realistically accounts for the data, but also indicates why Western theory fails to. I also illustrated how it explains *why* disease germs arise in given host environments: for such arisings, it specifies the conditions which are the actual causes of the diseases in question. Finally, we saw that the Chinese *ch'i*-germ-plus-environmental-conditions theory is consistent with our best biological sub-paradigm, Systems-Theory (Ecology)—a sophisticated version of the paradigm set forth by Aristotle, 2500 years ago. And it transcends it in that it is not religious-mystical: it does not posit Spirit or Form without Matter or Substance. In Chinese terms, it does not posit a yin-less yang.

After many years reflecting I cannot find any reason to make a conclusion other than the one made above: the Chinese theory is scientific and the Western one is not. (Meanwhile, Westerners are increasingly doubtful about the germ theory.) I am open to the ideas of anyone who might show me where I have erred in reaching that conclusion. Until that might happen, I will continue to regard the Chinese theory at least as superior to any Western one, and maximally as the truth, plain and simple. With that understanding, I move with confidence, and much more certainty than any Western theory provides, to a scientific alternative both to the Theory of Evolution (the Materialist half of the matter) and the Theory of Divine Creation (the Idealist half of the matter). That is, for every life-form (roughly speaking, every species) there is a life-form-*ch'i* which (with its proper environment) is the specific condition for its origination. And these *ch'i*-germs are "givens" in Nature, are always present in the universe, even at its cyclic termination-rebirth, being no less basic than atomic and subatomic energymatter-"particles" and their structures, which as we have seen along von Bertalanffian lines, must also be regarded as *ch'is*.

The only problem the human mind should have with the Chinese theory is in attempting to imagine the actual process of coming-into-full (substantial)-existence of a *higher* organism. It is easy enough to imagine—indeed, many biologists have *observed*—the coming-into-existence of micro-organisms. There is nothing specific in the observed environment, then there is a speck which enlarges and acquires a complex, life-, form. Ultimately, the "speck" in question is a subtle energy-matter interaction which, as explained, begins with hyper-energy and matter-that-is-almost-without-form: There is a transformation from the half-existent to the fully existent which, as shown, can be tied to a unified scientific theory. This transformation is easy to imagine with reference to the appearance of micro-organisms because the "lines" of the phenomenon are fuzzy, and they are fuzzy because the organisms in question are tiny. But big and small are relative to the human observer, who is the measure of all things. It follows that it is easy to imagine the coming-into-existence of a type 4 gonococcus but extremely difficult to imagine that of an elephant or a human because the latter organisms are, to us, big. But objectively (where objectivity is also a human measure), from the perspective of something much bigger than a human—specifically, the universe as a whole—it is easy to imagine the coming-into-existence of an elephant or a human. It is like observing the coming-into-existence of a micro-organism. Not hypnotically focused upon it because it is "large," one perceives its environment as well as it, and it is only a tiny aspect of its environment. The overall phenomenon consists of a tiny (that is, *swift* as well as spatially diminutive) change in the whole. And where each life-form is in fact not a perfectly individual part, but an interdependent aspect of, the whole, such is the actual case.

The preceding imagery, it should be realized, addresses the mystery of creation in a way to which Western savants are quite unaccustomed unless they have both experienced the awesome identifications of the ego with the world which spiritual adepts and some LSD trippers report, and performed the exercise of trying to visually comprehend Einsteinian relativity, which involves conceiving contractions and expansions of the time-flow and variation in the intensity of "there-ness," or mass. In parallel with the movement through which I concluded, above, that in a sense Humans, not God, created all life-forms, it involves a human reappropriation of God's perspective, which, in this view, is *inside* as well as outside the world, simultaneously. That this exercise in cognition is difficult should not be mistaken for an indication that its direction is false. The truth about something may be simple, but its attainment is not necessarily easy. As the reader who has absorbed the text thus far realizes, the preceding imagery has a rational and empirically grounded basis. It is far from a conclusive one, to be sure, but its basis is sounder, I think, than that of the Theory of Evolution. In turn, I know that the imagery I have elected is inadequate, although I do not know to what extent that is my fault and to what extent that is due to the nature of the problem, which is, after all, the ultimate mystery. Our seeking to know our own origination is itself a consequence of that origination, and therefore involves transporting something that exists into a zone where it does not, and involves a part in knowing the whole. That the result requires twists of the mind is inevitable. The alternative of letting God do this for us is simply a retreat. And the alternative of "scientifically" theorizing that it is all due to Chance is no different, save in that it seeks to describe and explain the effects of Chance in detail, as the evolution of species, one from another. Since credence in the Theory of Evolution depends on that explanation, I now compare it to the Chinese one.

In attempting to image the whole, the Chinese theory agrees with the Western one—in its sophisticated, ecological, version. Higher organisms do not emerge in a primitive environment. Before they emerge, lower ones must already have emerged, for they are necessary conditions for the higher ones. The difference is that the Western theory—the Theory of Evolution—posits that higher life-forms derive from lower ones, by mutation-and-environmental-selection, whereas the Chinese theory, in this respect like Divine Creationism, regards the proposition that one species can transform itself into another as a preposterous violation of common sense, which, given the "give-ness" of life-form *ch'is*, is totally unmotivated in any case. As explained, the motivation for positing inter-species transformations is to reduce the matter to a simple form—a *Fragment*—so as to make it amenable to Western-scientific-style experimentation and theorizing. Being *Complete*, Chinese theory has no need to propose inter-species mutation. Without that motivation, our own scientists would drop that proposition like a hot potato, because the Theory of Evolution, whose purpose is to relate the "first blob" (the micro-Cosmic Egg) to all other life-forms, is fraught with difficulties which they themselves recognize. Let's have a brief look at them.

Because the theory that species adaptively mutate due to impacts of cosmic or mineral rays on genes is so far-fetched and unrealistic and because genetic plasticity has been discovered, genetic plasticity and genetic drift (recombinations of genes along new lines through sexual reproduction) are increasingly relied upon to account for mutation. The concept mutation, accordingly, has been weakened into something more like self-selection, a complement to environmental selection. Genetic plasticity of an extent which permits changes drastic enough to fit inter-species mutation has never been indicated by empirical observation. But plasticity to the extent of inter-subspecies mutation has been, as for example on the part of staphylococci which become penicillin-resistant. Not only are such phenomena limited to extreme, human-caused conditions, but they are achieved only by micro-organisms, especially bacteria, which are the lowest of micro-organisms. Further, such phenomena may very well be more akin to mutation by mutagenic chemical action, namely, of Western medicines and pollutants, on DNA than selection from a plastic genetic program. But let us simply understand that the lowest of micro-organisms appear to be sufficiently plastic to be subject to inter-species mutation.

If that is so, nevertheless it is *only* such organisms which exhibit this capacity.

There is reason to expect only the lowest of life-forms to even dimly exhibit the Evolutionary plasticity our scientists pray for. If higher organisms derive from lower ones by selections and permutations of pieces of their programs (and note that permutations themselves are parts of those programs), then the higher the organism, the less plastic its program—like our German shepherds derived from our other European shepherds. In other words, the more "evolution," the less potential for evolution. Yet our Evolutionists want us to believe that as drastic a mutation as the primate-to-human one could have occurred, whereas micro-organisms, in fits of creative pique, only manage to produce relatively simple mutations.

But let us grant that all organisms, regardless of the degree to which they are organized, are extremely plastic, so that it seems conceivable that all multi-celled organisms directly or indirectly derive from one or more single-celled organisms.

If all life-forms evolutionarily derive, directly or indirectly, from amoeba-like, single-celled organisms, those single-celled organisms must contain within themselves the genetic potential for all life-forms on the planet! As we have seen, our scientists reject even the proposition that human DNA has the potential to encode-and-build all the kinds of antibody in the human body. Our scientists implicitly are clinging to a theory which they themselves should declare totally unrealistic, outrageously fantastic. What's more, if amoeba-like single-celled organisms had such enormous potential, would it not follow that the odd amoeba would change in just one fundamental way for us under the mutation-encouraging conditions of a biology-laboratory culture? In this light, recombinant DNA experiments are an angry reaction against Nature for not having cooperated in this regard, and the Theory of Evolution is a mythical charter designed to legitimize this obscenely dangerous reprisal.

Meanwhile, evidence to the effect that the *first* organisms were not necessarily maximally simple and that life *began* as a complex, interdependent whole, which is what the Chinese theory predicts, is beginning to accumulate. Just this year it was announced that five different types of biological cells have been discovered that existed a mere billion years after the earth formed; it had previously been expected that any organism existing that early would have been much simpler.

What are the chances of *multiple* "chance" originations of life? Likewise, human skeletons older than those of some of their supposed proto-hominid ancestors are being discovered almost every year, and our "family tree" is constantly being desperately rejuggled to prevent it from looking less like a tree than like a ball of yarn that the kitty got at.

Much can be said about the absurdity of the now secondary and complementary theory of mutation—mutation by natural chemical or radioactive accidental alteration of genes. Although it is unacceptable, it is necessary to the desperate preservation of the Theory of Evolution. I'll be brief. The chances of such a mutation being adaptive to the environment of the organism—for a dumb particle to have an impact as smart as Nature's manifold harmony—are conservatively estimated, by the Evolutionist Julian Huxley, for example, to be about 1 out of 1000. Given that normally the organism is *already* adapted, one might argue that actually they are almost zero, but I'll adopt this common Western-biological estimate. Now, the way our Evolutionists look at this—Fragmentally, of course—provides for some plausibility. Let us say that over an unimaginably long period, twenty new single-gene (or single gene-combination) traits are acquired by the descendants of a given species to form a new species. The implication is that minimally 19,980 other unsuccessful varieties of mutant of this species have emerged and failed to survive. This is conceivable. But when one recognizes that, on the average, the proportion of unsuccessful mutations must be the same for each and every species, the implication is that on the average only a tiny minority of life-forms are adaptive at any time; the great majority, through widespread ecological war, are heading for extinction! Otherwise put, Nature, which as our Ecologists appreciate, is an incredibly complex harmony of interdependence, is pictured as a failing experiment conducted by a militant and lunatic gambler, who, like the God it began with, works in ways that are not only mysterious but intolerable.

Henry M. Morris, a very bright Creationist who makes more sense than our Evolutionists, but not enough sense because his objective is to prove the Bible's validity, not to discover truth, has made the following, valid, points in his book *Scientific Creationism*. Morris points out that the Second Law of Thermodynamics, by which entropy increases, predicts the opposite of what the Theory of Evolution does: increasing *dis*-organization. Hence, even for the level of organization of the totality of life-forms to remain the same, there must be another, *specifically animate*, organizing, factor. More specifically, DNA can only be replicated with the help of certain enzymes which, in turn, can only be produced at the direction of DNA. As Homer Jacobson put it in *American Scientist* (1955), "Did the code and the means of translating it appear simultaneously in evolution? It seems almost incredible that any such coincidence could have occurred, given the extraordinary complexities of both sides and the requirement that they be coordinated accurately for survival." Morris then examines this matter of chances, first noting that, ironically, apparently they are not good enough for Evolutionists themselves, who tend to campaign zealously for removing mutation-producing radiations from the environment. Why, after all, not speed things up, get supermen into existence, and solve all our problems? Then he explains that according to information-theory the number of steps needed to build the first protein molecule in 1500.* The probability of this being achieved by chance, given a generous 50 percent chance of positive mutation at every step, is one out of 10^{45} (10 succeeded by 45 zeros). The number is more than astronomical. "Even if we were to assume that the complete set of trials up to the point of failure (or 1500 in the event of success) could be accomplished in a billionth of a second, and even if we assume there are 10^{80} systems attempting these trials (10^{80} equals the total number of particles in the universe), and that they keep trying for 30 billion years (10^{18} seconds), there could still be only the following number of attempts to achieve such a replicating molecule in all the universe in all time: . . . 10^{107}." Again, plainly an organizing variable whose existence is denied by the Evolutionists must exist. The Creationists cannot be blamed for concluding that God is implied; in doing so, they are being infinitely more logical than the Evolutionists, whose disregard of such probabilistic calculations is totally irrational.

Morris then turns to a gaping hole in the Theory of Evolution implied by the geological record. To wit, as the highly reputed paleontologist (and Evolutionist) George Gaylord Simpson has observed, ". . . as every paleontologist knows, . . . *most* new species, genera and families, and . . . nearly all the categories above the level of families [that is, the most general types of life-forms], appear in the record suddenly and

* From atoms. Note that even if we assume that certain external earthly factors lower this number, those factors, in turn, are also attributed to chance—ultimately, the "cards" not stacked when the Cosmic Egg Big-Banged.

are not led up to by known, gradual, completely continuous transitional sequences." As T. Neville George has observed, "Granted an evolutionary origin of the main groups of animals, and not an act of special creation, the absence of any record whatsoever of a single member of any of the phyla in the Precambrian rock remains as inexplicable on orthodox grounds as it was to Darwin." Morris writes, "Invertebrates have soft inner parts and hard outer shells; vertebrates have soft outer parts and hard inner parts—skeletons. How did the one evolve into the other? There is no evidence at all."* Simpson writes, "This regular absence of transitional forms is not confined to mammals, but is an almost universal phenomenon . . ."

Another datum previously thought to support the Theory of Evolution and which must now be discarded is "atavistic organs": organs in higher life-forms which were thought to have no function in those organisms and to be mere testimony to the evolutionary pedigree of those organisms. Now that our Mentalist surgeon-maimers have Narcissistically imposed their worldview via scalpels, removing such organs as tonsils, appendices, and so on, to "improve" the human being into a streamlined model, full living has been recognized, again, without apology for any "inconvenience" caused. In turn, the extensive organic similarities among animals testify to the fact that there are common denominators in the animal form in general. Not needing to posit that these different animal forms evolved one from the other, and with evidence to the effect that they did not, one is then obliged to posit, if out to preserve the Theory, that the great similarity among different animal organizations is a matter of chance—a fantastic proposition.

Finally, Morris makes the following devastatingly common-sensical point: "But even if variation, or recombination, really could produce something truly novel, for natural selection to act on, this novelty would almost certainly be quickly eliminated. A new structural or organic feature which would confer a real advantage in the struggle for existence—say a wing, for a previously earth-bound animal, or an eye, for a hitherto sightless animal—would be useless or even harmful until fully developed. There would be no reason at all for natural selection to favor an incipient wing or incipient eye or any other incip-

* Actually, there is a theory which circumvents this, on the (dubious) assumption that the *larva* of an invertebrate became its adult form and then mutated into a vertebrate. There is also the alternative of vertebrates and invertebrates being independently evolved from a prior, shell-less-skeleton-less, life-form.

ient feature.* Yet, somehow, if the evolution model is valid, wings have evolved four different times (in insects, flying reptiles, birds, and bats) and eyes have evolved independently at least three times . . . Charles Darwin said that the thought of the eye, and how it could possibly be produced by natural selection, made him ill." Put otherwise, the Theory of Evolution calls for faces even mothers couldn't love, or which would elicit *negative* selective responses from all normal members of their generation.

Much more, and in much more sophisticated terms, can be and has been said for and against the Theory of Evolution, but there is no need of that here. Since such arguments use either the Theory of Evolution or Creationism as their basis—two variants of the Binary Con against each other—they are necessarily inconclusive. In contrast, by comparing both those theories to the Chinese one it is possible to establish in a fairly simple manner that the Chinese one is superior to either of them and incorporates the better aspects of both.

Like our religious theory, and consistently with Systems Theory, the Chinese theory posits the existence of energy-forms ("souls," or life-form-*ch'is* Absolute-Fragmentalized) prior to the full existence of the organisms themselves. Like our Evolutionist theory, the Chinese one posits that it took a very, very long time for the present diversity of species to establish itself, and it recognizes change by mutation, but to a much more limited extent than in Evolutionist theory. To my knowledge, the Chinese savants never worked out to the detailed extent that our scientists have the order in which different species must have emerged, but for at least 2500 years they have understood the basic order—the only aspect of the matter of which our scientists are quite sure, in any case. That is, that life-forms began as pond scum, that there must be plants before there are animals, that among animals there must be mammals before there are humans, and that humans must have been the last life-forms to emerge. Hence, there are definite stages, just as the fossil record, in "violation" of the Theory of Evolution, shows. This is playfully put in a passage of Chuang-Tzu's *Nan-Hua Ching*, a passage which has been quoted by some Sinologists as possible evidence of an ancient and fantastic Chinese Theory of Evolution. (Rather, it makes evident the ancient Chinese Ecological

* Although that is true, there are incipient or atavistic anatomical features, such as the almost-toe above a horse's hoof. But such relatively minor changes, which are here regarded as *intra*-species changes, are accepted by the limited, Chinese, theory of evolution, explained below.

theory of evolving conditions for higher and higher life-forms.) What has misled such translators is the use of the term *sheng*, whose non-technical sense is "to produce," "to give birth to," "to live." In Chinese science, unless living or birth is literally referred to, the meaning is precisely "is the direct condition for the existence of"—as in the famous Cycle of the Five "Elements," of which it is said, for example, that water "produces" (is the direct condition for) wood. This is obvious to anyone who understands Chinese science, or Taoism, for one of its central truths is that for any given thing's existence the existence of all other things is essential. The only qualification is that certain things can and must *fully* exist, not only as *ch'is* but also as physical bodies, before others can because those other things require their substances for their own full realizations. A further complication with that passage is that it uses ancient code names for most of the life-forms it refers to—a practice also widespread in the medical texts, whose purpose is to insure against the abuse of powerful knowledge by the morally unqualified.* Here are extracts from it, revealing said order-of-coming-into-existence:

> The seeds are subtle. On the water they form a scum. At the juncture of water and land they form lichen and oysters. Living on mound-hills they form plantains. Given rich soil they form [the grass called] crow's-feet. The roots of the crow's-feet become [are the food of] grubs and its leaves become [are the food of] butterflies [which] . . . become [are the food of] a bird . . . Leopards are the [limiting] direct condition for the existence of horses and horses are the direct condition for the existence of the human [civilization].**

The final component of the Chinese theory, which is both similar and dissimilar to the correlative Western one, is its definition of kinds of life-form and of the limits of mutation. Kinds of life-form at the general level of felines, for example, which I have non-technically called "species" so as not to enter the rather arbitrary intricacies of Western biological taxonomy, are defined first and foremost with reference to their *functions*—that is, their energetic, or *ch'i*-, qualities,

* A concern which, in this age of recombinant DNA, anti-Human social engineering, and impending Ultra-Violet Catastrophe, has little meaning.

** All great civilizations use horses; as the supporting cast in our cinematic cowboy mythology they have been underestimated with respect to their surrogate, combustion-engine vehicles.

including their personalities; and their modes of survival and uses to other life-forms. In contrast, the Western classification is first and foremost morphological (having to do with shape). The Chinese system is an energetic action-system; the Western one is a thing-system; the Chinese system deals with the organism as a living thing; the Western system deals with the fixed appearance of the organism. Ecology is beginning to lead Western classification in the Chinese direction, and its need is recognized because relatively objective and sophisticated morphological reclassifications have brought about a recognition that there is often as much morphological sameness as difference between what, with reference to function, are obviously different life-forms. (The Polar-Complete continuum in the animal kingdom is becoming obvious, and, because the data are Fragmentally superficial, is posing a problem.)*

To give an extreme example; as my *shih-fu* once put it:

> Your scientists are quite right in understanding that certain forms of life have become extinct. What they don't understand is that their functions, because they are an essential part of the established whole, remain. These functions are then filled by new life-forms or, in their absence, inanimate moving things. For example, humans killed off the most formidable predators—giant bears and [saber-toothed] tigers, and your scientists believe they [their functions] no longer exist. What then do they think automobiles are?

Not only is the high carbohydrate consumption and massive life-killing function of the giant predators satisfied by automobiles driven by humans in a state of somnambulent intellectual idiocy equivalent to predatory animal intelligence, but automobiles even closely resemble those predators—especially the felines. This is particularly true of the super-cars of the 1950's and 1960's, when a grill was a huge fanged maw and a car really had a tail. Dinosaurs also fit the preceding functions and not for nothing have the largest of our cars been compared to them.

The preceding example is as enlightening as it is amusing, for it illustrates a general feature of evolution as it has actually occurred and as Chinese theory has it. As humans (around much longer than our scientists presently believe) have increasingly expropriated the energetic

* Aristotle recognized it—and then ignored it.

functions of animals, highly energetic life-forms have reciprocally shrunk: evolutionary change has chiefly to do with *size* and secondary shape characteristics. The saber-toothed tiger, without his saberteeth, is still around, but smaller—as the tiger; the mammoth and mastodon are still around—as elephants; the dinosaur is still around—as the crocodile and the lizard; the eohippus is still around, larger because man uses his energy—as the horse, his extremities changed from paws to the huge single nail called a hoof. Likewise, animal-engineer-functions have been expropriated by men and the high-energy functions of their animal agents have been expropriated by axe, scythe, and shovel wielders, by back hoes, bulldozers, and steam shovels. Morris writes: "The most remarkable feature about fossil insects as are known is that they are very similar to those living now. In many cases, however, they are much larger than their modern relatives. There are giant dragonflies, giant cockroaches, giant ants, and so on. But their form is no different in essence from that of modern insects." (Nature has literally made room for us; think what it would be like if it hadn't.) As tends to confirm the Chinese theory, the largest extant insects are found precisely in areas where human material technology is least advanced, and neither their localization nor the extinction of their even larger ancestors can be adequately accounted for by the nature of climate and available food. The giant cockroaches would do quite well in any city dump in the temperate zone.

In accord with the relatively deep, functional, understanding of the characters of life-forms, the whole of the established natural order of life-forms may be recognized. None of the function-classes within it changes once it exists, for, as our ecologists have begun to recognize, the balance of Nature, like the DNA-structure of any organism, is self-maintaining. If one agent of a function is removed, another agent takes its place, and the coming-into-existence and going-out-of-existence, and/or reciprocal changes in size or potency, which evidence such changes are governed by interacting life-form-*ch'is*. The species-limited morphological changes, likewise, are more plausibly attributed to such interactions than to mutation.

Recently, our evolutionists have changed their theory precisely in the direction of the Chinese one, and have recognized data that they cannot account for in their terms. Evolution is now distinguished by the majority of theoreticians into "micro-evolution"—change within species—and "macro-evolution"—change from one species into another. This is not only because (as the Creationists have long pointed out) the emergences of new species are sudden, but also because the direction of most change has been found *not* to be toward higher life-forms. In other words, in the view of the recent majority of theoreticians the emergence of new species can no longer be predicated on gradual changes within species. This leaves the Western theory in a vacuum: it now has no explanation acceptable even to evolutionists, for the emergence of new species. Accordingly, those who recognize this refer vaguely to "some different mechanism"—"even a gross random mutation in a single generation."* The desperation of "random," in light of the preceding chapter, is plain: the God of "Chance" is being covertly imputed a highly organized and detailed purpose and an awesomely efficient way of achieving it. Their paradigm has failed, and they have returned to thinly disguised religion.

In turn, there is new—or newly recognized—evidence of species-*ch'is*. This, I pounced on with excitement, for again I had found a precise confirmation of an idea totally independent of my expression of it in the original version of this book. It has been found that some life-forms, for example, chestnuts, have changed in *substance* (chemical form), but not *biological form*. It follows that the substantial and formal aspects of these life-forms are equally basic: the biological form is to an extent independent of, and outlasts, the substance. Von Bertalanffy is quite right: it has existence of its own. What, then, is this "biological form"? In the Western, Mentalist, terms which von Bertalanffy neither rejects nor affirms (because he is an inch away from the Chinese conception), it is a metaphysical thing, a God-caused soul; and a theory based on such a concept inevitably is mystical. In Tao-terms it is species-*ch'i*, a concept consistent with a scientific paradigm which, unlike our own, is empirically sound.

I conclude by quoting Ludwig von Bertalanffy, inventor of a scientific sub-paradigm that is more subversive of our established one than has yet been realized. His is one of the most respected and seriously taken big ideas of this century.

> Here we are dealing with fundamental problems which, I believe, are "swept under the carpet" in the present biological creed. Today's synthetic theory of evolution considers evolution to be the result of chance mutations, after a well-known simile (Beadle, 1963) by "typing errors" in the reduplication of the

* *Newsweek*, November 3, 1980.

genetic code, which are directed by selection, i.e., the survival of those populations or genotypes that produce the highest number of offspring under existing external conditions. Similarly, the origin of life is explained by the chance appearance of organic compounds (amino acids, nucleic acids, enzymes, ATP, etc.) in a primeval ocean which, by way of selection, formed reproducing units, viruslike forms, protoorganisms, cells, etc.

In contrast to this it should be pointed out that selection, competition and "survival of the fittest" already *presupposes* the existence of self-maintaining systems; they therefore cannot be the *result* of selection. At present we know no physical law which would prescribe that, in a "soup" of organic compounds, open systems, self-maintaining in a state of highest improbability, are formed. And even if such systems are accepted as being "given," there is no law in physics stating that their evolution, on the whole, would proceed in the direction of increasing organization, i.e., improbability. . . . Production of local conditions of higher order (and improbability) is physically possible only if "organizational forces" of some kind enter the scene; this is the case in the formation of crystals, where "organizational forces" are represented by valencies, lattice force, etc. Such organizational forces, however, are explicitly denied when the genome is considered as an accumulation of "typing errors."

. . . Presently the genetic code represents the *vocabulary* of the hereditary substance, i.e., the nucleotide triplets which "spell" the amino acids of the proteins of an organism. Obviously, there must also exist a *grammar* of the code: the latter cannot be . . . a chance series of unrelated words . . . Without such "grammar" the code could at best be a pile of proteins, but not an organized organism. . . . I therefore believe that the presently generally accepted "synthetic theory of evolution" is at best a partial truth, not a complete theory.

The "organizational forces" in question are biological *ch'is*. For 2500 years human beings in the West have been condemned to thinking of themselves as the children of non-human beings—either of God or of pond scum—extreme opposites typical of the extremist Mentalist paradigm. The effect, of course, is to encourage less than human conduct on the part of humans: "I didn't *mean* to rape that girl: God in His infinite wisdom filled me with lust and put her into my hands.

Perhaps it was His way of showing me the Light. And now that I am threatened with imprisonment I want you all to know that I have been reborn as a result." A flock of Protestant ministers flaps in to escort the accused from the courtroom. "What can you expect from the descendants of mere pond scum? Compassion? Self-discipline? Respect? Dignity? Send him to a shrink," declares the judge, as he sinks back into his pond. It is time we reappropriated the vital force and became the children—and the parents—of *humans*. What is good concerns social principles, and it is in this way that "hard" and social science are actually totally interdependent, so that correct hard-scientific theory leads to social justice, so that a valid hard-scientific sub-paradigm leads to a good sociopolitical one.

CHAPTER 6

Luminous Beings

I see man as a ball of lightning. A living organism is nothing but a giant liquid crystal, a semiconductor composed of an intricate system of conductors of various stages of conductibility. Hello. . . . Can you hear me?
—VIKTOR INYUSHIN, avant-garde Soviet researcher, via an electromagnetic conducting device called the telephone

A four-cornered vessel without any corners—What a strange four-cornered vessel! What a strange four-cornered vessel!
—K'UNG-TZU (Confucius)

THE SOFT-MACHINE UNWOUND

At the beginning of the last chapter I said that Western scientists will not only never understand "how acupuncture really works," but never understand how *anything* "really works" until they recognize the need for the concept of *ch'i* in science. Then I showed that there can be no causal explanation—at least, no *rational* causal explanation —in physics or biology without that concept. If I did my job well, understanding the necessity of that concept to genuine medicine will now come easily. I will begin by pointing at the evidence of physiological *ch'i*, or vital force, between the short-ranged and broken lines of Western science. Then I will explain "how acupuncture really works," and solve the mystery of cancer.

Chapter Five showed that species-specific *ch'is* must be posited to account for the organization of inanimate substances into life-forms. Being a necessary component of those organizations, and those organizations being persisting essential features of life-forms, it follows that a life-form-*ch'i*, as physiological *ch'i*, is continually present after the origination of an organism or line of organisms. Further, to account for any newly conceived individual's processing of energy-matter, one must posit a given, "priming" vital energy. And that energy is not (only) the chemical energy commuted to the foetus from the parents, because it must have been available to the first organism of the species in question, which did not have any parents. Through sexual intercourse, parents extend their own *ch'is* into a priming life-form-*ch'i* passed onto their progeny. In short, the life-form-originating *ch'i* and the ordinary physiological *ch'i* of all living individuals are one and the same thing viewed from different perspectives.

As the foetus develops in the mother's womb, brain cells are produced in the lower part of its body, and they migrate to precise destinations, according to type, to form the brain. The number of types is not known, but it appears that brain cells (like all other cells in the body) belong to functional groups. These groups, as might be predicted on the basis of general Polar-Completeness, and as modern research increasingly confirms (Nauta and Feirtag), appear to totally overlap in function. There is no brain function without the participation of all groups. For example, brain cells that empirical testing shows to be concerned with sensation are also empirically shown to be concerned with muscle movement, and all brain-cells, with the exceptions of some highly specialized ones such as optic-nerve-receptors, seem to cooperate in governing sensing and movement in all specific locales in the body. (This is much as each color is all colors.) But in turn, each neuron also appears to have a focal specialty. (This is much as each color is chiefly itself.) It therefore appears that an attempt to segregate the brain cells into a number of functional groups is much like the problem of distinguishing colors: it is a function of the degree of fineness desired by the investigator. At one extreme, virtually each individual neuron might be regarded as having a unique function. At the other, all the brain cells might be regarded as a single functional group. But we want to determine the size of brain cell groups of an order analogous to primary colors. We are looking for their most obvious border-"lines," which like those of colors, would surround a chosen focus (a cell, like a color). What then naturally offers itself as a standard is the number of contacts a brain cell has with other brain cells. At the most, there are about 1000, and we will use that number because we want to make the groups as large as possible, so as to simplify our picture of the organism and thus give Western explanation of the phenomenon of brain cell migration the best chance possible to make sense.

There are 10 billion (10^{10}) to a trillion (10^{12}) brain cells. To estimate the number of primary groups of brain cells, that is, those that are identical, then, we should divide by 1000. Again, I take a lowest number (here, 10^{10}) to give the Western theory the benefit of all doubt. In this instance, my generosity, it should be noted, is 990-billion-fold (10^{12} minus 10^{10}). This gives 10 million (10^7). Ten million, then, is the number of specific destinations for specific types of brain cells, very conservatively estimated. How the brain cell "knows" it is taking the right path as it migrates is a mystery to our scientists. But common sense dictates that if it is done physically (by DNA-programmed chemical responses), the cell must be programmed to respond positively or negatively to its environment as it migrates. The first cells to migrate need no information, because they are relative to no other types: they are coordinates for the orientations of others. The last cells to migrate need to be able to recognize every other type of cell and to respond accordingly. The complexity of the paths taken by cells, and the exactitude of their destinations, surpass traveling by the shortest route from a living room in an apartment on the tenth floor of a high-rise on the northwest side of Manhattan to a kitchen in an apartment on the fortieth floor of a high-rise on the southwest side of Manhattan, while destruction and reconstruction of half the buildings on Manhattan is going on at a feverish rate. To again make a conservative estimate, a directional response would be 1 of 20 possible different directions in three-dimensional space. A final cell, then, must contain up to 10 million (for the different types of brain cells) times 20 (for appropriate directions in response to each kind) bits of information, or 200 million bits. Let us say that it need contain only a hundredth of that (2 million bits), because it migrates through an area containing, say, only one one-hundredth of the different kinds of cells, its chances of running into different kinds of cells decreasing as a function of the directional decisions it has made. The average cell, then, contains half that information: 1 million bits. There being 10 million different groups of such cells, the implication is that for directional migration of brain cells alone, the DNA of the fertilized ovum must contain 10 million times 1 million, or 10 trillion (10^{12}) codons for this function alone. If merely 100 other bits are required to program each type of brain cell for its form and all its other functions, the number of DNA-bits for brain cells alone is 10^{14}. But there are only 1.3 times 10^{10} codons in all the DNA in a fertilized ovum*—1/10,000 of that.

So, an estimate generous toward a Western, DNA-, explanation of brain cell formation and migration plainly implies that a non-physical configurative factor, a definition which *ch'i* fits perfectly, is involved. In light of the preceding chapter this is just what one would expect.

Specifically, the most basic lines of *ch'i*-flow in the developed human body, according to Chinese medical theory, run from just the place where the brain cells migrate from (what becomes the base of the spine) upward to the head, meeting, from the front and rear of the body, at the anterior lower region of the brain (behind the nose), and this general configuration is consistent with the formation and final shape of the human brain.

Let us do another rough but legitimate and illuminating computation. The human body contains about 60 trillion (6 times 10^{13}) cells. About half of them form organs. Each of these organs has a definite complex shape which implies a program for the grouping of cells which makes it up, that is, each cell has a definite position within the organ, or, to be more generous toward Western DNA-theory, every thousand cells make up a group which do. To continue to be generous, let us again say that only 100 bits of information are necessary to each cell finding its organ and location within it and to the formation of a given kind of cell in all other respects. So, we have 3 times 10^{13} cells divided by 1000 into 3 times 10^{10} group kinds of cell, for each group of which there are 100 bits, or codons, in the fertilized ovum. That makes 3 times 10^{12} bits, again a number greatly in excess of the

* One standard text gives maximally .3 times 10^9 and, as is typical but increasingly questioned (see below), states, on mysterious grounds, that the maximal number of genes necessary to the formation of vertebrates is merely 10^6, which implies maximally 10^7 "bits of information"—a plainly unrealistically low number in light of this text.

total number of codons in the fertilized ovum. Plainly, the organic formation of the human body—and much more so its complex functioning—cannot be programmed for by DNA. There must be a non-physical configurative variable.

This being recognized by a minority of realistic biologists, Paul Weiss, for one, author of *Principles of Development*, suggests an alternative to the model of a little DNA-factory churning out all parts of the organism on the basis of a plan which plainly could not be contained within the walls of the factory. He proposes that the developing organism is a field of which the parts are not discrete elements but tensions in the field. Logically enough, a field is posited to account for the coordination of what have been thought of, mechanistically, as building blocks. With this, a Chinese-style Polar-Complete continuum of almost-parts which do not exist apart from the whole enters the Western paradigm with subversive, and much needed, results. If one part of the field changes, then all other parts are immediately aware of it. This general structure, then, accounts for the holes left by DNA-programming-theory. All we have to do is fill the hole that Weiss, in his turn, has left. Being vague in this regard, he seems to picture the field as being purely physical, made up of genes, and he says that the role we attribute to genes is actually played by "guiding cues of the dynamic field structure to the total complex." He therefore distinguishes between the field and its structure, and adds the unidentified factor of "guiding cues" to explain its non-random behavior. The structure, as already pointed out and as von Bertalanffy recognizes, is not the physical components themselves and it is an emergent property, which, therefore, must be thought of as a non- or demi-physical thing. The concept of guiding cues, of course, is nothing other than a concept of a *determining* function of that structure. The need for *ch'i*, to fill the hole in Weiss' picture, is obvious: the field in question consists of the genes and their products, cells, which migrate away from the DNA-factory to form the body and which have no physical connection to, or chemical means to communicate with, the factory, or most of each other, once they do. There must, therefore, be a connecting-coordinating variable of a higher order—just as in the case of photons heading in opposite directions toward polarizing filters at different angles.

Along the same line, both Weiss and W.M. Elsasser, author of *The Chief Abstractions in Biology*, recognize that, as I have shown statistically, there is extensive evidence to the effect that morphology (for example, organ shapes) is not transmitted by DNA.

Likewise, the "program" itself must contain a higher coordinating variable which is being ignored by those advancing a mere biochemical explanation. Von Bertalanffy calls it a missing "grammar," comparing the DNA codons to the mere vocabulary that the "grammar" manipulates. To go a little further: there must also be a "meta-grammar." This is what decides which "sentences," properly constructed according to the "grammar" and constituted by "words" which together make sense, should be generated at what points in the development of the germ cell into a complete life-form, so as, for example, not to produce kidney cells and send them into the lung area or to produce mature germ cells in a foetus. This meta-grammar must have *intention* and a total proto-image of the body at every stage. It is like the event in the mind just before a sentence is thought: the *it* that is "put into words." That *it* also chooses at any given moment sentences which are appropriate in relation to all sentences generated and all sentences to be generated. It is the trainer of trains of thought. The *it* I have just defined by analogy to the highest aspect of the biological program precisely fits the definition of *ch'i,* and so, therefore, does that aspect of the program. (Not only is the grammar-vocabulary model inadequate to explaining biological development, it is also inadequate to explaining thought and speech.) Similarly, as L.G. Barth (in Apter, cited below) has observed, if every reaction (sentence) in a cell requires an enzyme (grammar), and enzymes are produced by cell reactions, then either there is an infinity of enzyme-reaction cause-effects or there is a self-catalyzing enzyme—or a self-reacting reaction. One cannot escape the implication of *intention* in the biochemicals —*ch'i.*

Cybernetics is the hope of those who wish to refute the strong indications of the role of a vital, patterning, energy in the formation of living things, and the tentative explanations it offers are hopelessly inadequate. This is chiefly because cybernetics *simulates* the data it seeks to explain, with models which are far too much simpler than the living data they are pretended to represent. For example, Michael Apter, author of *Cybernetics and Development,* grants that, if the development of the body from the germ cell is language-like, the DNA would have to contain more information—perhaps even an infinite amount—than there are features in the developed body. Failing to make any computations such as these, he simply grants that that amount of information could not be contained by the DNA in a germ cell, and proposes to identify the factor, additional to the DNA, which directs the evolution of the body from the germ cell. He draws on Magoroh Maruyama's topological (a variety of geometric) experiments to propose that the structures of the cells which the germ cells produce are that factor. To explain by analogy: the structures, or shapes, of the circular joints and straight bars in a Tinker-Toy set determine the total possible set of ways a number of them could be joined, so that their simple shapes partly determine the configurations of quite complex structures. Further, once one has begun to construct something with these pieces, the number and shapes of possible complete structures that can now be built is limited and in a sense predetermined by the first, partial construction. The joints and bars of the Tinker-Toy set, then, are the body's cells, and the construction of the toy is the relations among those pieces during construction and at completion of the construction, which are due to a simple set of rules (insert which bar into which hole in which context) and the shapes of the pieces.

Of course, this is essentially the perspective I took when I considered the actual data of the foetal evolution of the brain, to show mathematically that even if the ongoing configuration of the cells is taken into account as a choice-delimiting factor, the results could not be attributed to DNA-programming plus that factor. So Apter's model is plainly quite unrealistic, although undeniably it is correct in positing that the structure of cells partly determine their organizations in the body. What's more, even if cell structure explained *all,* one would have to recognize that their structures result in an outcome, the total configuration and functions of the body, which cannot be regarded as a mere accident. *There would have to be a "plan" which predicts that outcome from cell structures*—a preconceived image of the Tinker-Toy before its construction.

Notwithstanding that—and it is something he could know by simple calculation if he does not know it already—Apter jumps to an extreme conclusion consistent with his model: germ cells are "templates" which *"reproduce themselves, and as a side-effect, each one has the potentiality of developing into a larger and more complex version of itself"* (emphasis his). Living adult bodies are a mere accident, a side-effect, of the self-replication of the germ cell. Of course, the "potentiality" that Apter ascribes to these germ cells is precisely the mystery which his model fails to account for, so the concept "side-effect" is merely a means of connoting that the mystery is a minor problem. But in any case, the germ cell does not, in fact, reproduce itself: rather, as shown, it directly and indirectly produces more different *kinds* of cells (minimally 60 billion) than it contains codons (13 billion), where codons are responsible only for one bit of information each. And it is the organism made up of these kinds of cells, organically patterned, which, in its turn, produces new germ cells, at sexual maturity. Apter's "side-effect" covertly is also his "*cause*-of-side-effect," just as the side-effects of our scientific paradigm as a whole are, in fact, actually "templated" self-replications of our paradigm as a whole.

Cybernetics applied to living phenomena can only account for their inanimate aspects, and accordingly its ultimate contribution, like that of our basically inanimate paradigm as a whole, is to demean human beings. If *we* are mere side-effects, how can we object to our *science*'s side-effects? The cybernetic explanation of vital configuration is as ignoble as it is scientifically inadequate. If fosters the attitude exemplified recently by one of North America's foremost TV-scientists, who stated that because of computers' "super-human" capacities to process information, he accepts the computers of the future as a higher product of *biological* evolution than human beings. Picturing themselves as "detached minds," in the 2500-year-old image of the architects of our paradigm, such scientists are condemned to regard computers as their superiors, all the while ignoring that their bodies "processed more information" as they developed in their mothers' wombs, than their conscious minds could ever contain, in a way that a computer cannot do at all. Unable to know anything through their bodies, of which one aspect is their senses, they are totally insensitive to the chilling horror of a world in which humans, as political subjects, workers, and medical patients, are treated according to computer logic as inferior forms of computers.

To leave those who see themselves and all others as mere side-effects: rational biologists such as Weiss, Elsasser, and von Bertalanffy all point directly at biological *ch'i*, but not one of them goes so far as to assert the existence of such a variable. Why? Because they are afraid of being accused as Apter does his opponents of Vitalism and because they are not aware of the non-religious alternative here provided. Again, we are faced with a choice which has existed for 2500 years: to agree with Aristotle that there are souls, which are metaphysical and ultimately traced to a supernatural cause, or to non-mystically posit that there are *ch'is*, which are a demi-physical, given, aspect of

Nature. There is no third way out of this and that is why our thinkers on this matter have come up with nothing basically new for 2500 years.

Now let us move toward the truth about acupuncture, by examining some facts about what Western biologists and medical scientists assume is the governor of all life processes: the nervous system.

There are minimally (see above) 15 billion nerve cells (neurons) in the human body. Supposedly, messages are sent to the brain and received by 5 billion sensory or motor non-central neurons. The remaining 10 billion neurons in the brain interpret all incoming data, decide what to do about it, and send out appropriate commands. All of this message sending, or encoding and de-coding, is thought to be accomplished by electrical impulses set up by chemical reactions in and between neurons. With a few exceptions, messages sent to and received by the brain must pass, if only the nervous system is involved in this phenomenon, through the spinal cord to the lower brain. The implication is that the spinal cord and lower brain channel all the electro-chemical information to and from each group of non-central neurons. Where non-central neurons impinge on 100 others on the average, 50 million groups, conservatively estimated, are involved. (Visual neurons, for example, are each unique, belong to no groups; there are 2 million of them.) Let us say that each group can transmit and receive only 2 messages each: "stimulated, very stimulated." The "channel" must therefore be capable of handling 100 million (10^8) different bits of information. The diameter of the spinal cord is about 2 centimeters, so its cross-sectional area is a function of 4 square centimeters. The diameter of an axon (the message-outputter of a neuron) is about 2 microns, or ten-thousandths of a centimeter, so its cross-sectional area is a function of 4 over 10^8 square centimeters. Hence 10^8 axons, one for each bit of input information, can be packed into the spinal cord, which is enough to handle the information.

So far, so good. But as soon as we get to the "net" of lower brain cells receiving the input, the "electro-chemical circuit" model breaks down. These lower brain cells spread over the spinal axons, each impinging, on the average, with about a thousand of them. And it is physically impossible for them to distinguish, in their outputs, the input of one axon from that of another. As soon as the reason for this is understood, it is then also understood that, in the first place, the information coming to the spinal axons (or the few neurons directly connected to the brain) is also not encoded—makes no electro-chemical sense—in any case. Each neuron receives electro-chemical input from 100 (non-central) to 1000 (central) other neurons through branches called dendrites. When the number of inputs, and their frequencies, are sufficiently high, the neuron then transmits the *sum-total* input through an output-branch called an axon, *of which it has only one.* The implication is clear: the output of a neuron can in no way encode the input which led to it. When it "fires," it is simply saying "I have been sufficiently stimulated, to a degree indicated by the frequencies of my waves, by an unspecified number of unidentified neurons, to fire"; and when it doesn't fire, it is simply saying "I have not been sufficiently stimulated, by an unspecified number of unidentified neurons, to fire." The "channel," above, could not process 10^8 bits because they have not been sent, in the first place.

In short, there is no way that a chain of neurons running from a sensing-neuron to the brain can encode information specifying the location or nature of the original input. And reciprocally, there is no way that a chain of neurons running from the brain to a motor neuron can encode information specifying the nature of the message and its intended destination. Obviously, the nervous system is a *primitive* physical substratum for informational energies, or forces, just as the particles in an atom are a physical substratum for the informational force which organizes an atom and governs its fluctuations.

That the nervous system is not a precise electro-chemical circuit capable of processing precise information, but rather is a physical *field* through which precise information moves, is indicated by recent studies of the brain. For example, it has been found, while attempting to determine the "mechanism" of thinking, that brain cells secrete fluids over different areas at different times, and that these fields of fluid directly affect electro-chemical discharging on the part of cells within their spheres. To the investigators, who wish to see the brain as a physical machine, this phenomenon is analogous to flooding multimillion-dollar computer circuitry with water. They conclude that, because such secretions are so crude, so unselective, involving all neurons within a sphere determined by mere osmosis, they must govern nothing more complex than "changes of mood." How, then, do we manage to think rationally, when these secretions drastically alter the brainwave patterns? Those investigators should be given 1000-microgram doses of LSD 25, which has an effect very similar to those secretions,*

* But, unlike them, dangerous: my recommendation is strictly rhetorical.

and then be asked to explain why their intelligence quotients have doubled. Their intelligences doubled, they might be able to actually answer the question: the conductivity of the *primitive field* through which informational energy, *ch'i*, moves has been doubled.

The fact aside that brain cells have only 1 output per 1000 inputs, so that they could not possibly exchange precise information electro-chemically, consider the number of codons which would be required to program the brain for such exchanges. Again, the conservatively estimated number of groups of same kinds of brain neuron would be 10 million (10^7). As stated, according to empirical findings, each group participates in and is essential to the functions of each other. This implies that each group is programmed to identify information from each group, and to interpret it according to its own specialty. For example, a group specializing in coordination in walking or running must be able to recognize a message transmitted from a group specializing in hearing—as of a car coming from behind—and translate the message into a specific balance of the limbs and trunk permitting the individual to turn his head without losing balance. So, assuming that each group can send *only 1* message, each group has a program which is a specialized version of the totality of 10^7 programs, which means that the DNA responsible for the super-computer in question contains 10^{14} bits, or codons. But the totality of DNA in the fertilized ovum contains only 1.3 times 10^{10} codons.

Calculations such as the above make obvious that the electro-chemical activity of the nervous system, taken *alone*, is not the actual transmission and coordination of specific information required for the constantly changing and totally intercoordinated vital functioning of the body. Rather, it is a system of electro-chemical homeostasis (self-corrective balancing) for *crudely* harmonizing vital functions, without specification of those functions, or decision-making about them. When, for example, one stubs and cuts a toe, the sensor nerves in the area are excited and in their turn electro-chemically stimulate the neurons they impinge upon, and so on. As explained, the physical aspect of the nervous system provides neither for *direction* to this expanding wave of electro-chemical activity nor for transmission of information as to its point of origin. The electro-chemical activity simply fans out, weakening as it does, like a wave expanding outward in all directions through water into which a pebble has been dropped. One neuron turns on 100 which turn on 10,000, and so on. Eventually a long neuron inside the spinal column, or a long neuron attached to one, picks up the impulse, from one sector of the "wave-front." It has no way of "knowing" that it came from the specific toe in question; the best it might do is "know" that it came from below the waist, and on which side of the body, by virtue of its own location. The same goes for the brain to which this message is directly transmitted by that long neuron. This is obvious from examination of the nervous system: there are only a few, long, neurons which connect specific points in the body to the spinal column or the brain, and there are ultimately billions of different possible inputs to each of them, without any physical mechanism for distinguishing one input from another. The best the brain can do is respond by channeling the pain-signal-energy back down into the body, one effect of which would be the stimulation of the hypothalamus and liver, responsible for secreting hormones which promote healing. The necessary and much more specific information-processing: "the toe has been stubbed and cut, pass this on to the brain; manufacture hormones and antibodies and send them to the toe" must be effected by a subtle (yang) variable other than, but based on and operating with, the gross (yin) nervous system and its electro-chemical activity—a configurative energy or force, *ch'i*. As we will soon see, this is precisely what is directly implied by empirical evidence about acupuncture.

Our neuroscientists, in typically Fragmental fashion, study the sensory and motor neurons at one end of their imaginary "chain" and the brain cells at the other, and conveniently exclude the middle, for its structure necessarily implies that their researches at either end could not possibly solve the problem in question. Further evidence that the brain, taken alone, is not a computer, but rather is a central station and generator-amplifier for neural energies and the physiological *ch'i* which configurates and directs them, may be recognized in the results of Paul Pietsch's famous experiments with salamanders. Pietsch surgically scrambled salamander medullas (which are most of a salamander's brain) in every conceivable way short of totally destroying them and failed to stop the salamanders from behaving normally! Plainly, the function of the salamander medulla is largely independent of its physical structure, and a non-physical, or demi-physical, variable is operative in centrally governing the salamander's physiology. Likewise, many humans have had large parts of their brains removed without marked, and sometimes with no, loss of "brain"-function. This is not to say that the brain is dispensable, not by any means. It is to say that *alone* it is not responsible for the physiological, emotional, and intellectual computing required by the human organism. It is the physical basis for that computing. It is for that reason that brain injury *does* sometimes impair, and oxygen-deprivation of the brain always distorts, physiological functioning: the *ch'i* is without a neural base and electro-chemical medium, or its medium has been disturbed.

K. S. Lashley and P. J. Van Heerden and then Karl Pribham have come up with a theory which purports to account in physical terms for the brain's functioning but which, objectively understood, merely serves as a plausible specification of the electro-chemical field upon which the *ch'i* must operate. They compare the brain to a hologram, which is a photographic plate that records a three-dimensional light picture through light-wave-interferences. When a laser beam is projected onto the plate, the image of the original input is reproduced three-dimensionally in space. Like the groups of brain cells, any piece of the hologram contains the entire image. The brain, then, is like a holographic plate and the interference patterns of the electrical waves its neurons generate and transmit chemically to each other is like the hologram image. The information storage in the brain is attributed to electrical waves constantly produced by each neuron, independently of the greater activity of interaction among neurons which constitutes the interference pattern, the total output of the record which is the sum of the "knowledge" in each neuron part. Simply put, this theory is to say that the information and information processing of the brain is not chemical but electrical—is an electrical field. It in no way explains how information might be decoded as to point of origin and content, or encoded and directed to specific points.* All it does is eliminate the problem inhering in the neuro-chemical explanation by positing that rather than a chemical network, there is a continuous energetic field. The solution is just like Weiss' field model of foetal development. It moves from inadequate parts to an adequate whole. When any aspect of the field is altered, all other aspects of the field reflect that alteration, so that, as it were, the whole, but no part of it,

* Mathematical translational formulae are posited, but this is to attribute physicality to a mental construct—just as in physics' probability-as-a-natural-force. It is to imply *ch'i* in a mystical way.

is aware of the alteration. This is essentially the same homeostatic model I introduced above to illustrate the probable "mechanism" of the entire nervous system.

What Lashley, Van Heerden, and Pribham have produced, then, is a model of brain consciousness and associative memory: a whole responsive to change in any of its parts, one aspect of which might saliently respond, perhaps by resonance (sympathetic vibration), to an alteration elsewhere in the whole which is like the interference pattern in its own area. The central problem with this model (its isolation from the rest of the nervous system aside) is easily recognized by extending its analogy to a hologram. To get the image from a hologram, light must be beamed at it. What is the correlate of that light in the brain? If we speak of brain responses to sensations from outside or inside the organism, it would be an energetic input from nerves outside the brain. But what of the brain's ultimate capacity, thought? When one "turns one's mind to something," in their terms one is beaming light onto an area of the brain in which memory of that something, and all relevant data, is concentrated. Of course, that input, and much more so its direction, is totally unaccounted for by their theory. In short, their analogy to a hologram virtually calls for a variable other than electrical interference patterns, which *governs* those interference patterns. And, because they have not only asked the brain cells to electro-chemically do all they are known to but more (to set up constant as well as fluctuating wave patterns), there is nothing left to serve as a candidate for "generator of governing input." In any case, that variable would have to have intention, will, purpose—a property which no realist could attribute to even the most sophisticated interference pattern, for that pattern is a mere result of billions of independent and therefore maximally stupid chemical electrical potentials.

The preceding argument is a sophisticated version of the old religious-Idealist one used to establish the existence of the spirit, or soul. In the old days it was observed that, as everybody knows, there is something inside them which does the thinking, and which is aware of that fact, and which can be aware of being aware of that fact, and which therefore is obviously not machinelike, but quite deliberate, and not physical, because it is infinitely regressive. Here, because there is independent evidence of it (above, and below), instead of a metaphysical soul which is therefore attributed to a Great Spirit (God), I posit a demi-physical energy which is an aspect of Nature.

HOW ACUPUNCTURE REALLY WORKS

Despite Western data exactly to the opposite effect, the following assertion is found in the most official medium of Western science, the report of the National Research Council, *Science and Technology: A Five-Year Outlook* (1979):

> It is very unlikely that any vital forces remain to be discovered in the major mechanisms still buried in neurobiology or developmental biology.

Note how, as exhibits the basically Narcissist character of Western science, the empirical data are confused with theories about them: the "mechanisms" (if any) are "buried" in the organisms in question, not in "neurobiology or developmental biology," which are theories about those organisms. The assertion is continued, to say something which, in light of the preceding section and chapter, rings oddly.

> Although there is no unique vital force in living organisms, there is a unique molecular basis for their organization which is not found in the inorganic world: the molecular storage of information programming the development and functioning of the organism.

The truth is suggested by the above writer's very phraseology, "molecular *basis* for their *organization*": it is, precisely, nothing more than a basis, and the organization in question is conveniently left unidentified. He might well have written:

> Although there is no unique vital force in organisms, there is a unique neural basis for their harmonious and complex functioning—metabolism, cell regeneration, disease resistance, and so on—which is not found in the inorganic world: the electro-chemical transmission of information through the nervous system.

(The nervous system is also not found in the *vegetable* world, one might add, but plants nevertheless have vital functions which differ from animals' chiefly only in that they are slower! And that is what nervous systems are basically all about—dynamicity, speed—which is what distinguishes animals from plants more than anything else.)

Western attempts to explain "how acupuncture really works" are as stubbornly ignorant of facts established by the very Westerners who make the attempts as are the preceding official assertions of Western (or more exactly, American) biology.* And at the points where they fail they reveal with equal force that the Chinese theory in question is, *if* not absolutely correct, far superior to any Western one.

There are two varieties of Western explanation of acupuncture. One comes from the Capitalist world and is typified by reliance on unimaginative and narrow-scoped experimentation and extremely conservative theory. The other comes from the Communist world, specifically, the Soviet Union, and is typified by open-minded and creative experimentation and theory but strict, life-denying, Materialism, as is obliged by Marxist dogma. The Capitalist, or American Medical Association-protecting, variety is most unpromising, almost has the quality of burlesque. The Communist, or spirit-suppressing, theory is so close to the Chinese one (its suppression aside) that the evidence behind it constitutes unambiguous proof of the latter's validity. One need merely remove its obscurantist dogma of Materialism (which does not inhere in the experimental data). Let us examine the A.M.A.-protecting one first.

The most acclaimed attempt is the Canadian psychologist Ronald Melzack's, summarized in *Psychology Today*** as "How Acupuncture Works: A Sophisticated Western Theory Takes the Mystery Out." Melzack does not deal with acupuncture as it is used in traditional medicine almost exclusively to cure specific illnesses in specific ways—for example, to cause the body to absorb and eliminate brain tumors and to stop in seconds massive brain hemorrhages. Rather, he deals with the least-used and most primitive effect of acupuncture, analgesia-anesthesis (pain-killing and numbing), upon which Western medical doctors, due to Communist Chinese demonstrations of its su-

* It should be noted, to the credit of German scientists, that they long ago adopted von Bertalanffy's biological paradigm, which, as explained, is only one, Polar-Complete, step away from the Chinese one. It is no accident that Chinese biology and medicine is being taken more seriously in Germany, and by diffusion, the rest of Europe, than in North America.

** Also see Melzack's more detailed article, cited in the *Bibliography*.

periority to Western anesthesia in surgery, are fixated.* However, even when acupuncture is imaginatively reduced to a mere pain-killing technique which leaves A.M.A. types feeling comfortable, Melzack's de-mystification fails. It is worth briefly examining, because it does address something true of acupuncture in general: the control, from one point on the body, of a distant area of the body which has no Western-known functional relation to the point of control.

First, Melzack grossly misrepresents some Chinese medical theory involving yin and yang, then ritually declares the Chinese theory "unsatisfactory" because "It falls too far outside our own scientific approach to medicine." Another example of the ethnocentric irrationality that characterizes the Western-medical reaction to Chinese medicine. He then purports to solve the mystery, chiefly on the basis of his and Patrick Walls' "gate-control" theory of pain, which is opposed to the accepted "specificity theory," as follows:

> The patient's faith in the procedure as a result of long cultural experience, together with the explicit suggestion that the patient will feel no pain, greatly diminishes his anxiety. Mild analgesic drugs further relax the patient, who is about to have long needles stuck into him and twirled around, or electrically charged. The patient's predisposition makes it possible for him not to feel the stimulation as pain. The nerve impulses produced by twirling the needles, or sending electrical pulses through them, activate parts of the brainstem that block pain signals coming from the site of the surgery. In other words, the gate closes. The signals never reach the parts of the brain involved in pain perception and response, and the surgeon is free to begin his work.

"The patient's faith . . . as the result of long cultural experience" must be amputated from his argument because *ch'i*-probing ("acupuncture") works on North Americans—often, skeptical ones, to boot—as well as Chinese. Indeed, without qualification "the patient's

* The "show-biz" mixture of traditional Chinese and modern Western medicine with which the Communist Chinese amaze Western physicians obscures the fact that many diseases it deals with surgically, such as brain tumors, can be (and should be) cured by acupuncture. They are giving Western physicians as little as possible, and what they really want: not superior medicine, but a technique which will permit them to preserve their own, ultimately surgical, unnecessarily injurious, medicine and to obscure the rich therapeutic range of traditional Chinese medicine.

faith," regardless of cultural experience, must be amputated from his argument. Young children could hardly have faith in the procedure when they have no conception of any procedure at all. And, as the executive editor of *Prevention*, an American health magazine, Mark Bricklin, has observed:

> The problem with that reasoning is that doctors have been able to perform operations on *animals* using only acupuncture techniques for anaesthesia. It is doubtful, to say the least, that an animal is going to hold still and not cry out in pain because it has somehow been "psyched" into believing that it won't feel anything....

"Mild analgesic drugs further relax the patient . . ."—well, none whom *traditional* Chinese doctors treat, because they do not use them.*

The core of Melzack's theory must also be removed. First, why, unless *directed* to do so, should "parts of the brainstem" or any other component in the nervous system block pain signals when it is the function of the nerves to transmit them? There *is* a "gate," but plainly it closes by virtue of a factor which overrides the nervous system altogether. Second, according to his theory, the entire body might be anesthetized, whereas only *specific* loci are. There is no attempt whatever to explain why certain points affect certain areas and others do not. Third, there is no way to explain in terms of the structure of the nervous system why certain points anesthetize certain, usually distant, areas, and there is no known specific connection between any of these points and the "gate."

In any case, a "pain-gate" would bear no relation whatever to, indeed, would *obstruct*, the *healing* which acupuncture is almost unexceptionally used for. Western-medical findings bear the latter, Chinese-medical, principle out: when neuro-chemical paths are obstructed, the muscle tissues at their termini degenerate. This phenomenon remains "unexplained." (Reciprocally, as Vladimiric Polezhaev discovered in the 1940's, the stumps of amputated frog's legs regenerated when irritated by needles. A technique entailing *pain* stimulated healing of the most dramatic kind.)

Since Melzack's article appeared in 1973, it has been discovered

* And in their demonstration operations, Communist Chinese physicians do not always use analgesics.

that a pain-killing chemical, "endorphin" (*internal* + *morphine*), can be released by the brain, and, along Melzack-like lines, researchers at the University of Virginia hypothesize that acupuncture somehow triggers its release. This theory presupposes that the neuro-chemical system makes it possible for endorphin to block pain perception at *specific* loci. But, first, how does the brain know where the pain locus is, when the nervous system, alone, cannot transmit such information? Second, how can pain perception be blocked in the brain without shutting almost the whole brain off, because almost all brain cells are involved in any perception? Note that this theory, also, bears no relation whatever to therapeutic, as opposed to deadening, effects.

Dr. Robert Becker, chief of orthopedic surgery at the Veterans Administration Medical Center in Syracuse, New York, has made discoveries atypical of the "A.M.A.-calming" type, and close to the Soviet type, which led to a theory much closer to the Chinese one, which, to his credit, is far more innovative than those of the A.M.A.-calming variety. I quote from Stephen Dewar's article in *The Canadian*, "A Second Nervous System?"*

> "Just consider healing for a minute," [Becker] says. "When you cut yourself, some system in the body monitors the damage and turns on the healing process. Then as the wound heals, the process changes and finally shuts off. That can't simply be the chemical actions of DNA in the cells: there has to be a system controlling it." In short, a second nervous system, which Becker believes is located in the perineural cells and which operates like a direct-current analog computer. (Don't let the language throw you: DC-analog computer is just an automatic control system that turns different functions up or down or on or off by varying DC charges. A thermostat is a DC-analog device and so is the volume control on a radio.)
>
> The existence of a second nervous system would help explain an awesome range of phenomena. For example, Becker has shown that the sensation of pain and consciousness itself seem to depend on the nature and level of DC currents in the perineural cells of the brain. He has shown that a reversal of the electrical current flowing through the brain causes a loss of consciousness. . . .

* The article summarizes Becker's "Boosting Our Healing Potential," cited in the *Bibliography*. It also appeared in full in *Reader's Digest*, November 1979.

> A similar process seems to be involved when hypnosis is used to block pain. Becker's measurements show that in a hypnotic trance the electrical potential drops or even reverses in the nerve sheath. . . .
>
> Becker has found that some painkilling drugs have electrical effects. Novocain, for example, drops the electrical potentials in the nerve sheath. As the drug wears off, the potentials begin to rise again . . . (and) the patient feels pain once more.
>
> Acupuncture, too, is illuminated by the theory of a second nervous system. Becker reasoned that such a system would need amplifier points to boost the signals every so often, just as you need an extra amplifier if you run overlong speaker wires from your stereo. When he went looking for these amplifier points he not only found them, he also discovered they corresponded to most traditional acupuncture points. He theorizes that the insertion of a needle into these points may interfere with the flow of direct current and consequently reduce pain.
>
> . . . (S)ays Becker, "such a system also provides a potential explanation for the success rate of modern medicine. We know that about 50% of modern medicine's success happens simply because the patient *believes* the doctor can help. It looks as if *that* faith helps some patients to generate the appropriate electrical potentials and heal themselves."

In general, Becker replaces *ch'i* with electricity, whereas, as I stated earlier, the Chinese theory includes in its energetic map of the body electricity as well as *ch'i.* As soon as one understands why electricity alone cannot account for the phenomena in question and alters Becker's theory accordingly, one has produced a rough, Western correlate of the Chinese theory.

First, the association between electrical resistances, on one hand, and pain sensations and alterations in consciousness, on the other, does not necessarily imply that electricity is the controlling variable: it may simply be an aspect of what is controlled. Second, and as shows that it is *not* electricity which is the controller, it has been found that acupuncture partially restores motor control to victims of multiple sclerosis, which degenerates nothing other than the perineural tissues which Becker posits constitute the physical basis of his "second nervous system"—and precisely in the area of Melzack's "gate." Here, then, we have a direct test of his hypothesis, and it disconfirms it. Mark

Bricklin, in his *Natural Healing,* cites Dr. Arthur Kaslow, a Californian M.D., who reports

> that the first MS case he treated was a man so incapacitated that he couldn't wash or dress or feed himself. "After ten treatments at my office and a series of treatments at home, the patient was able to take care of these needs without help. He's thrown his cane away," Dr. Kaslow said. . . .
>
> Like others, Dr. Kaslow cannot explain how or why acupuncture "works" to improve the symptoms of multiple sclerosis. But he does have a theory that MS is not purely a problem of demyelination of nerve fibers. If it were, he pointed out, the destroyed myelin sheaths would make it impossible for any procedure to eliminate symptoms and yet the procedures he uses do just that, to a greater or lesser extent. Furthermore, he argues, MS patients typically go through periods of remission of natural symptoms—again something that couldn't happen if the destroyed myelin sheaths were the sole explanation of the disease.

Two assumptions can be made about the remission of MS symptoms. One is that the myelin sheaths remain destroyed and that an alternative pathway to and from the brain is somehow established. The other is that the sheaths are regenerated. Kaslow assumes the former, because without treatment the symptoms may naturally remit (cease) and then resume. Along that line, another physician, B.F. Hart, "theorizes that this therapy stimulates alternative pathways in the central nervous system, so that message impulses to the arms, legs, etc. bypass the nerve fibers whose myelin sheaths have been damaged or destroyed."* It may well be that alternative pathways are somehow established, *but they could not be nerves,* because MS happens to specialize in the upper spinal cord, which is the channel to the brain—and the location of Melzack's "gate"—and the arms and legs are not directly connected by other, non-spinal, long-axoned neurons to the brain. It is more likely, therefore, that the spinal cord continues to serve as the pathway. In either case, it must be *ch'i,* not electricity, or *ch'i* "pulling" electricity along, which courses to and from the brain, either as a consequence of acupuncture or when MS symptoms naturally remit.

* *Ibid.*

In turn, for cases of permanent remission resulting from acupuncture, there is good reason to posit that the myelin sheaths have been caused to regenerate. At first sight, the healing might be attributed to electricity alone. The healing function of electricity is an established fact, and has led to a specialty called "electro-biology."* Dr. Andrew Bassett has placed electrical coils on the flesh over non-united bone fractures and discovered that the electricity causes the soft tissue at the fracture to produce calcium to unite the bone. In one case a fracture, unhealed for nine years, yielded to this treatment in nine months. (Such regeneration does not normally occur.) Dr. Stephen Smith of the University of Kentucky has used similar methods to cause frogs' legs to regenerate after amputation—a phenomenon that is impossible according to "established" biology.

In the case of the supposed regeneration of the MS-destroyed myelin sheaths, then, one can posit that acupuncture enervates healthy myelin cells below the destroyed ones, super-charging them electrically to multiply and flow into the damaged area. And we can attribute to the DNA-structure of the healthy cells the specificity of this response to electricity. But in the cases of regenerated human bone and whole frogs' legs, we cannot attribute healing to electricity and DNA alone. In the former case, *non*-bone cells produce bone cells and in the latter case several different kinds of cells, in perfect coordination, constitute a limb form other than their parents'. They must have been specifically directed to do so. That is, to make this conceivable, there is a *ch'i* in the damaged area (a "ghost bone," a "ghost limb") which is strengthened by *ch'i* brought from elsewhere by acupuncture and then, possibly through the energetic medium of electricity, it configurates the form of new cells and the form of their organization—much as a gonococcus-*ch'i* uses the unhealthy tissues of its host to constitute a gonococcus germ.

Hence, mere electricity and self-replicating DNA programs may be the case in the supposed regeneration of myelin sheaths, but in many other similar instances, such as bone regeneration (also accomplished by acupuncture) that *cannot* be the case: a *ch'i* is necessarily implied. Our biologists must *know* it but do not want to admit it. If DNA-information in cells adjacent to a damaged area contained all the infor-

* The following data are taken from the TV news show *Prime Time Saturday*, 19 April 1980.

mation necessary to generating appropriate cells and organizing them, each group of cells of same kind would have to contain the information for at least one other kind of cell, for the form of the organ or limb they constitute, and for knowing when an adjacent area needs repair. If such information consisted of only 100 bits and each group of same kind consisted of 1000 cells (an overly generous estimate on both counts), the number of bits in the fertilized ovum would be 60 trillion (the number of cells in the body) divided by 1000 (60 billion) times 200, or 12 times 10^{12} bits, and there are only 1.3 times 10^{10} bits in the fertilized ovum. (Of course, the same calculation implies that even the normal, foetal-developmental, formation of the whole organism also could not possibly be accomplished by DNA: the number of bits would be 6 times 10^{12}.)

Whether the theory is based on a gated neural circuit, a neurochemical governance of hormones, or a second nervous system surrounding the first, it is plainly disconfirmed by the fact, long known, that pain may be felt by humans—and dogs—in areas whose connections to the brain via the spinal cord have been severed. Where the spinal cord is the obvious central *neural* channel, the only neurological way out is to posit that alternative pathways through the secondary neural network are involved. As explained, this refers to a mere net which has neither direction nor a cellular structure which permits precise transmission of information. Each theory is to the effect that an intelligible telephone conversation can be held when the central line is down, through a web transmitting several billion conversations simultaneously to each of several billion receivers. It follows that if an alternative neural pathway is involved there is a non-electro-chemical variable which guides nervous impulses along it and which contains the information in question. The phenomenon is appropriately called "pain referral," but what might be doing the referring is conveniently never mentioned.

Becker's discovery of a near-perfect correlation between zones of high electrical activity and acupuncture points is the place where acupuncture explanations from the Capitalist world overlap with those of the Communist world. Becker was not the first to establish the preceding correlation. Much earlier, in 1963, the North Korean Dr. Kim Bong-Han found a *perfect* correlation.* What the North Koreans and

* Bricklin, *op. cit.*

Communist Chinese have experimentally discovered since then is not known, but it is known that avant-garde Soviet scientists have arrived at fascinating conclusions along this electrical line, which, when rationally interpreted, confirm the traditional Chinese theory.

It was Semyon Davidovitch Kirlian, the photographer of life-form coronas, who invented the first Soviet acupuncture point finder-verifier, an electrical amplifier connected to a light bulb, called an "acupointer." It has become an accepted diagnostic tool in Soviet hospitals but is used only by one to a few innovative and open-minded heretics in North America.* Soviet scientists have also found that skin differs in temperature and sound at the acupuncture points. The reason that readings of electrical activity at acupuncture points may be used diagnostically, in reverse of traditional Chinese practice, is that each point is the focus of a sphere of *ch'i* attached to one, or a definite set of, specific physiological functions. For example, the "gate of the liver," a point on the radial surface of the foream, is, literally, the body's external access to the liver, and almost any liver disease can be ameliorated or cured by inserting a needle at this point.** Consequently, a sick liver might be symptomized by abnormal electrical activity at the "liver-gate." This Western variation on traditional diagnosis, and the treatment by acupuncture based on it, is, however, both primitive by comparison to traditional diagnosis and therapeutically inadequate, for reasons already indicated. Symptoms of diseases are usually very complex, involving several organs, and reciprocally, each symptom, taken alone, may be of any one of several dysfunctions; and the organ function which is the source of the disease, and/or which is the strategic focus for therapy of the disease, cannot be known through such one-dimensional diagnosis. As a consequence, direct address to an organ with an electrical symptom may worsen, not improve, the patient's condition. Accordingly, almost always a specific combination of points (which varies during the course of the disease) is indicated in true acupuncture therapy. With those qualifications, then, it may be understood that a correlation between acupuncture points, represented by electricity, and organ-locations-and-diseases has been systematically established by Soviet—and apparently some heretical

* I have that by word of mouth; one, unidentified, like myself is a graduate of Yale and (unlike myself) is a practicing M.D. in California.

** This is not to invite amateur experimentation; non-side-affecting healing is not so simple.

American—medical scientists. Now let us examine the resulting Soviet theory, and then pick it apart to reveal the valid Chinese theory underlying it.

As reported in a book misleadingly entitled *The New Soviet Psychic Discoveries* by Henry Gris and William Dick, Viktor Mikhailovich Inyushin, a research director at the top-priority research institute in Alma-Ata, has gone so far as to recognize that simple electricity is not in question: the energy is also *configurative*. And he bases this observation not merely on logical inference from related empirical data, such as preceded, but *direct* experimentation. Understandably, he does not specify the experiments, but he generally indicates their nature. His basic variable was the Kirlian "aura" and the independent variables were illnesses, alterations in electromagnetic input, moisture, oxygen, diet, and temperatures around and of living subjects. Positing a material "biological plasma," Inyushin states:

> Biological plasma, in contrast to inorganic plasma, is a structurally organized system. In it, the chaotic heat motion of particles has been reduced to a minimum. That is, the entropy of the system is minimal. Moreover, biological plasma in its thermodynamically unbalanced state is notable for its considerable stability in varying temperatures and other environmental conditions.

Obligatorily a Materialist, Inyushin hypothesizes that biological plasma is "an integral system of elementary charged particles" which "is a dominant factor in all biodynamic relationships within these organisms." He therefore reasons that "it is bound to be luminescent under certain conditions." And he adds, "However, concrete proof of the presence of quantities of free electrons in living systems is still forthcoming. Therefore, the presence of a fourth state of matter in live organisms remains so far as a hypothesis." In short, he has conclusive evidence of a bio-configurative force, one with precisely the non-entropic property of what, from the Sextimagic perspective of Chapter Four, was called hyper-energy. And he has a Materialist hypothesis essentially to the effect that the force is an electron-ic substance, which, so far, he has been unable to prove. This is as much to say that he has empirically discovered a configurative *demi-physical* vital force. If the substance is free electrons or any other kind of particles—say, the subtler neutrinos—nevertheless it has an order, a pattern, which is not predictable from the features of its substance, and which therefore is an emergent property. As pointed out along von Bertalanffian lines, such order cannot be merely physical. Hence, whether Inyushin will obtain luminosity or not is irrelevant to the matter. (It is predictable, by the way, that he will, or, without realizing it, already has, for *ch'i* does have a physical *aspect* and human spirit is, at the subtlest of levels, luminous.) He has already established, then, that the variable in question is metaphysical or demi-physical.

Now here is a crucial point. Inyushin is able to adhere to his Materialist and ultimately Mechanist hypothesis only by recourse to the same conservative beliefs of our own "explainers" of acupuncture—beliefs which I have shown must be recognized as false. Inyushin regards "variations that find reflection in the brightness, color and dynamics of the microchannels" (in the acupuncture points' auras) as "physiochemical reactions that bring about changes in electric parameters of the cells as well as in work functions of the electrons." The circularity of this reasoning should be obvious: what are those "physiochemical reactions" reactions to? Among the stimuli mentioned by Inyushin are changes in the environment around the body and in the nutritional content in the body. But the medium in question, between those stimuli and the cells, is nothing other than the nervous system, which as I have demonstrated cannot alone, nor only in conjunction with cell-"programs," be the governor of metabolism or cell regeneration. In short, Inyushin is right back where he started, victimized by the Marxist-Materialist variant of the Binary Con. His own data to that effect, sophisticated correlates of lie detector results, are that another cause of said variations "was found to be emotional—namely, emotional reactions and oxygen consumption changes brought on by them." If emotions cause physiochemical reactions in cells, then they cannot themselves be the physiochemical reactions of cells. (Variation in oxygen consumption is itself an effect of such physiochemical alteration—here, of the cells of the diaphragm, so it is strictly the emotions, not also oxygen, which are the causal factor.) It follows, in the absence of any other physical variables possibly responsible, that they are a form of *ch'i*.

Of course, it could be argued that emotions are the electro-chemical reactions of cells other than those they affect, but again, because emotions are or are responses to changes mediated by the nervous system, we are faced with the need to posit a variable other than the nervous system, which governs the nervous system. This is more obvious in cases such as the raising of a burn welt by a subject hypnotized to believe she has been burned. Where the nervous system cannot even be attributed the capacity to direct a simple signal from the brain to a specific motor nerve, and cell-DNA cannot be attributed the capacity to generate an organic form, how can we ask it and cell-DNA to direct the formation of a burn welt at a specific, exact location, and of a specific size and shape? A "four-cornered vessel without any corners," that is, an energetic and physical body without configuration, without an ordering variable, is, indeed, a strangely inadequate model of the living body.*

Further Western data show that there could be no question of the "biological plasma" being merely electromagnetic or merely an output of cells determined by their chemical structures. At the same time, these data confirm the Chinese *ch'i*-map in striking detail. Kirlian photographs have revealed "phantom" parts of organisms where the physical parts have been removed. The most famous example is that of a leaf of which one of the sections is absent: the precise form of that section appears in the photograph as a light-configuration. Since there is no physical substratum, and since it is inconceivable that the surrounding physical area could be radiating an electromagnetic field of such precise, organic, form, it follows that a configurative force other than electromagnetism (acting on the electromagnetic field input by the photographer), and which does not derive from the cells of the organism, is present.** That the physical body is governed by a *ch'i*-body which has the same form as it does is precisely what Chinese medical theory states.

But there is more. According to that theory, in addition there is a *ch'i* which surrounds the body and which acts as a primary, invisible skin between it and the environment. That is, if we could see the human body-*ch'i* (as Carlos Castaneda appears to have learned to) we

* In classical Chinese texts the human being is sometimes compared to a ritual vessel. "Corners" is standardly used metaphorically to refer to configuration of a definite kind, or limiting factors, such as ritual rules. At the same time it is understood that true configurations, being demi-physical (like the "lines" between adjacent colors) are imperceptible and indeterminate at the physical level. Hence the "vessel" both has and does not have "corners." Confucius' exclamation is classically Taoistic, Polar-Complete.

**It is contested by some that the precision of such "phantom parts" is accidental, because they do not always appear. On the same basis it could be argued that the photographic-and-subject conditions vary so that only sometimes is the image captured. But my point is not made only on the basis of such evidence.

would see a luminous egg and, inside it, a human form. This energy-shell is the human organism's primary defense against diseases of external origin (such as influenza and leprosy), and it acts as a homeostatic medium, balancing the temperature-moisture and all other metabolic conditions inside the body against those outside of it. This is almost precisely what Inyushin's researchers have discovered through systematic empirical experimentation, involving what is called a "Kirlian scanner," which allows observation and filming of Kirlian activity (electrical manifestations of *ch'i*-flows) at any part of the human body.

> Apparently the active points of the skin actually serve as instruments for an exchange of electric magnetic energy with the surrounding atmosphere. The experiments of our scientists have shown that there is a connection between the electrical ability of man and his general state of health. With sick people, the character of certain points of the skin changes radically. It is those people who best react to changes of the electrical magnetic fields [as in weather-changes, sunrises and sunsets] of Earth.

The Chinese theory also specifies, to put it roughly, that points on the yang area of the body chiefly output and secondarily input *ch'i*, and points on the yin area of body chiefly input and secondarily output *ch'i*. The resulting image of the energy-shell is of horizontal flows of *ch'i* emanating outward on either side of the spine and circling around the body on either side to enter it at the front—the configuration of an electromagnetic energy-flow around a magnet. Flows around the limbs and extremities have the same configuration. This is just the configuration of a Kirlian corona around a finger. Tending to confirm the picture of *ch'i*-flows around the trunk is the following discovery reported by Inyushin: "In some points, the illumination is higher, and in others much lower. One can assume that some points contain a large number of negative charges, and others, positive charges." The Soviet avant-garde scientists, by photographing cross-sections of the energy-shell flows, are amassing empirical evidence of the yin-yang configuration of the external *ch'i* of the human body.

The non-electromagnetic property of *ch'i* may be directly sensed through acupuncture needles (or to use the original and more accurate Chinese term, *pien*: probes), and this gives us a key as to how the Chinese medical-savants, without Kirlian scanners, mapped the body-*ch'i*, in the first place. There is a pressure that *ch'i* exerts outward from within the living body. If one inserts acupuncture needles into one living and one dead animal or human body of same size and shape, in the same place, one finds a palpable difference: a lack of resistance against the needle from the dead body. Likewise, the resistance of a robust *or happy* patient is greater than that of a weak *or depressed* one of the same size. The Western explanations which might be provided for this phenomenon confirm the Chinese theory, and leave off at a point beyond which only Chinese medical theory can provide an explanation.

The difference in blood distribution in the living and dead bodies is a neutral factor. Because the volumes of blood are the same, on the average the dead flesh is as turgid as the living. A fairly plausible Western explanation, then, would be that the momentum of blood coursing at right angles to the needle and the contractions (if there are any) of muscles around the needle account for the greater resistance of the living body (or of the more robust of two living bodies). But there is a further datum which necessarily implies that *ch'i* has *direct* effects, independent of blood momentum and muscle contraction, on the needle. The resistance to the needle by a living body is greater in the areas which, according to the Chinese *ch'i*-"map" of the body, have *yang-ch'i*, as opposed to those which have *yin-ch'i*. Indeed, one can feel, through the needle, a distinctively "spirited" pushing-out from the yang areas. I refer the reader to Porkert for anatomical designations of those areas. On the whole, they have neither higher blood pressure nor heavier musculature than do the yin areas. What is responsible for the difference, then, is neither anatomical nor physiological (in the Western, physical, sense). There remains the hypothesis that electromagnetism is the operative factor. It could not be, because the measured force of electromagnetism at the points is only a tiny fraction of the force required for the differences in the pressures in question.

The unrealistic and reductive nature of the theory that the vital configurative force is mere electromagnetism is obvious when one considers the organic patterning of the human physiology revealed by the efficacy of acupuncture and the Soviet confirmations of the *ch'i*-map. There is simply no way to account for the specificity of the point-organic-functional connections in terms of a cellular or organic chemical substratum. How, for example, could the liver generate a sphere of electromagnetism whose apex is at a point on the forearm, or conversely? How, for example, when I cured a patient in ten minutes of swollen lymph glands under the jaw, was a field generated between those glands and points near the patient's *knees*? The physiological fields in question number in the thousands and are of all shapes and extend in all directions. Even if the cells were appropriately charging, could electromagnetic fields maintain these shapes and directions while being overlapped in this manner? They would cancel each other out. The energetic aspect of the living body is not a mere "ball of lightning," a "vessel" without "corners."

Here is yet another hole in the Western explanation. If the vital force were electromagnetic, why, then, do compasses not respond to human bodies, on one hand, and what is the form of energy operative in dowsing for water, on the other? Plainly, when the illuminating but also data-obscuring (because it involves artificially introduced electricity) electromagnetic field of Kirlian photography is absent, the human organism interacts with water in the earth, if not also air, through a form of energy which is totally unknown to Western science. Dowsing, an excellent indication of this, is such a mystery to Western science, on one hand, and so destructive to its grossly electromagnetic-energetic physical sub-paradigm, on the other, that it is simply ignored with a rigor akin to religious fervor. It is a fact that human beings—with the exceptions, apparently, chiefly of people so highly "educated" that they are physically as well as mentally alienated from Nature—can successfully dowse for water with an apple branch. Where I live, in the country, this activity is as common as sunlight. Plainly, there is a flow of unknown energy between the water in the earth, via an *organic* medium, and the human organism. And it is obvious enough that the poles in question are the water-*ch'i* of the water and the water-*ch'i* of the human, that is, the configurative vital energy which governs the water in the human body—water being what the physical aspect of the human body (and the apple branch) primarily is.

That the healing vital force which mediates between the body and its environment is not electromagnetism is also indicated by the nature of the material which conducts it. The best acupuncture probes are made of platinum ("white gold"), which is alloyed with steel only to increase the probe's strength and flexibility. The higher the platinum content, the more powerful the effect.* White gold is not the metal which would maximize an electromagnetic field. Al-

* And the more dangerous is the amateur use of acupuncture. Western researchers and Hip doctors: beware.

though it is the best conductor of electricity, it is non-magnetic. Its distinctive conductive characteristic is something else. I have already mentioned it in demonstrating the predictive generality of the Sextimage: gold especially conducts green light, which is to say minor yin (the yin inside yang) energymatter, the state which characterizes living energymatter, rebirth, regeneration (whose symbol is wood). What it conducts between the body and its environment is, in Western-physical terms, much more energetic and less material, much longer and slower of wave, than are electrons or positrons. Inputting electricity, the Soviet and Communist Chinese medical scientists (inputting electricity is called the "Shanghai method") are making acupuncture relatively physical, gross, and robbing it of its natural subtlety and flexibility—its *vitality*. Tending to confirm this analysis is Inyushin's punch line: the development of laser-"acupuncture," which is reported to be more effective than the Shanghai method. It stands to reason that it would be, for a much subtler form of energymatter—light*—is being used. But it is still gross by comparison to the traditional method, for only a thin sheet of gold conducts light: something much subtler than light must be conducted through a golden probe. Green light is only its gross symptom, a correlate of it. Finally, and here, again, the infinitely greater subtlety and naturality of the Tao-paradigm, if Nature is to be accurately understood, must replace any version of the Mentalist one: just as minor yin is a simultaneously opposite concept, so is the conduction of acupuncture. As Inyushin has discovered, the flows at points are dual: chiefly positive (outward) and secondarily negative (inward), or the reciprocal. Electromagnetic flows do not have such a configuration (they are uni-directional). Hence Kirlian photography records effects of only half the actual *ch'i*-flow around an organism, and the result is exactly as would be predicted: yang without yin and yin without yang (in Ektochromatic color-terms, red without purple and purple without red). Put otherwise, and as shown in more general terms with reference to animate Sextimagery, what is animate is maximally Polar-Complete. An electromagnetic model for acupuncture falls far short of the truth and, accordingly, attempts to modify it into a simple electromagnetic input which will have undesirable side-effects.

* And by it, heat, the chief agent in acupuncture's cohort, moxibustion. (The two are conjoined: *chen-chiu*:acupuncture-moxibustion.) Roughly speaking, they have opposite effects; the Soviets are totally confusing them.

Note that all the preceding arguments against Western interpretation and the empirical data they are based on reflect the general Quadrimagic principle, which our paradigm directly opposes: "yang is active and yin is receptive, yang is relatively subtle and yin is relatively gross." The neural, chemical, and electrical or electron-ic variables used in Western explanation are relatively gross: the body (*shui*) is the receptive substratum of an active-subtle yang complement, *ch'i.* That the empirical evidence to that effect is denied simply exemplifies the power of the Binary Con over the Western mind. In this instance, it is the Material half which is relied upon, for the (mis)understanding is that its only alternative, the other half, is a religious doctrine based on God and souls. Working with this (or the other) half, all Western causal explanations, inevitably, are circular. "Emotions cause physico-chemical reactions which are emotions" (Inyushin, above) and "The configurative vital force is electron-ic yet these electron-flows are configurated." Likewise: "God is ultimate, non-moving cause yet he depends upon humans for His existence because cause requires (in this case, human mental and emotional) movement"—the circularity in Aristotle's inferential description of God as total cause. As I said, Western science is basically religion.

The rationale is plain: Since acupuncture works without electromagnetic input, and since that totally subverts our basically physical-inanimate scientific paradigm, henceforth "legitimate" acupuncture will be done only by Western or Westernized physicians (or engineers) who input electromagnetic energy through the needles.

It is worth recognizing the ultimately political context of that issue. I said earlier that to elect, against God or Marxist "Natural Law," the understanding that *ch'i* is what powers everything, including ourselves, is ultimately a political decision, in favor of "local self-government." Here the connection is obvious. When diagnosis relies upon expensive and sophisticated machinery such as Kirlian scanners and treatment entails being attached to specialized laser-beaming generators in state-approved hospitals, not only is the anti-popular, machinal, and centralized character of Western medicine maintained; it is exacerbated. Because our paradigm is one designed for central control by a minority of people, whose responsibility to people is minimized by reliance on machines and externalized machine-like theories, it is no accident that the "scientific development" of acupuncture is taking the lines presently exhibited in the Soviet Union. It is also no accident that the Chinese term for "to heal," *chih,* is also the term for "to govern," and that it is said that "A good scholar-official also makes a good doctor."

How, then, does acupuncture *really* work? By directly manipulating the cosmic-and-human demi-physical configurative force called *ch'i*, which has been expropriated from Westerners by the architects and (for the most part innocent, victimized) vehicles of the Mentalist paradigm, thus depriving Western peoples of a concept which is necessary not only to genuine, Human healing but to medicine that can be practiced anywhere, which is therefore truly democratic.

THE TRUTH ABOUT CANCER

An examination of cancer, as mysterious to Western medical scientists as our need for Chinese medical science-art is great, is perhaps the most direct and interesting way to get from the level of the individual living body, via its psyche, to the level of the living social body and civilization—the promised outline of *shang i*:high medicine which it is the central purpose of this book to offer. Let me say at the outset that the closest thing in the West to a valid explanation of the present cancer epidemic is found not in Western medical science but in a novel by Alexander Solzhenitsyn, *Cancer Ward*, in which he suggests that the socio-political structure of Soviet society is like and therefore somehow causal of cancer. As I have indicated, the socio-political structure of *all* of modern society may vary from one society to another but it is basically the same, being based on the same beliefs and attitudes that the Mentalist paradigm is. Cancer is a disease chiefly caused by modern society. The missing variable is human psychical-physical *ch'i*.

The Canadian medical anthropologist E. Alan Morinis can provide us with the first of two steps required to understand the actual nature of cancer. As Morinis explained at the Canadian Ethnology Society Meetings in Montreal in March, 1980, cancer cells are distinguished from normal cells by their totally anarchic (anti-systemic) self-interest: they multiply without regard for the number and location of their kind which is required by the organism as a whole. Reciprocally, for an unknown reason the normal cells surrounding cancer cells—ultimately the body as a whole—fail to homeostatically correct this proliferation. Hence, cancer is a disease of the entire body which is characterized by cellular disorganization: what in the context of this book may be called a Fragmental perspective on the part of body cells. Morinis' original (and admirable) contribution is the jump he then makes to recognize not only a similarity but also a *connection* between the configuration of cancer and the configuration of modern society. Taking the regular association between increasing industrial pollution and the rising cancer rate in modern society, in conjunction with empirical findings that these pollutants are mutagenic (mutate cells) to indicate that industrial pollution is the cause of cancer, he observes that it is the cancer-cell-like self-interest and self-aggrandizing goal of the people behind our unlimited industrialism which is the ultimate cause of cancer. For those reasons, cancer can be neither cured nor prevented by modern medicine, which seeks, according to its germ theory, to find a cancer germ or cancer virus.

As Morinis recognizes, his theory can account neither for the fact that people whose exposure to pollutants and poisons has been nearly nil may get cancer nor for the fact that people exposed to the same industrial pollutants or poisons in the same amounts do not all get cancer. At least some cancer is caused by something other than industrial pollution. And there is a second factor which governs resistance to industrial pollutants.

According to Chinese medical theory, there are two types of cancer cause. One is physical and the other is social-psychical. The physical cause is poisoning, and this includes not only mutagenic industrial chemicals and nuclear radiation but also antibiotics. As Morinis points out, there is a correlation between the decline of germ-caused epidemics and the rise of the use of antibiotics, on one hand, and the rise in the cancer rate, on the other. According to my *shih-fu,* antibiotics disintegrate—indeed, as recent Western research has confirmed, explode—disease germs, and this violent process may mutagenically scarify neighboring cells, which then become cancerous, stealing energymatter from neighboring cells to proliferate and, reciprocally, weaken neighboring cells.* It is in this way, as Morinis says, modern science has substituted epidemic cancer for epidemic germ disease. (Not only does some of the effluent of a pharmaceutical factory cause

* And here is the promised connection between promiscuity and VD, on one hand, and, via Western medicine, cervical cancer, on the other.

cancer, so does some of the medicine it manufactures. There are side-effects at *both* edges of that sword.)

But again, not everyone who has taken antibiotics or been exposed to industrial pollution has or will get cancer. And people who have not been exposed to either industrial pollution or antibiotics have contracted cancer. The operative factor in both cases is psychical-physical *ch'i*, that is, a disease of it which, as Morinis unknowingly anticipates, is similar to the self-interested configuration of both cancer at the cellular level and the anti-social attitudes and goals of unlimited-growth-industrialists and their cohorts. Specifically, cancer-causing *ch'i* is characterized by an attitude, be it a matter of choice or a result of social conditions, or, as is most likely, a combination of the two, which is obstructive of the natural social-psychical flow of human life. (The technical term is non-*t'ung*, where *t'ung* is the term, introduced earlier, for through-going, as in Complete scientific thought, logical consistency, unexpected theoretical consistency with empirical data, and animateness of hyper-energy, or *ch'i*.)

An actual example: one of the patients I observed Chinese medicine cure of cancer had married a young man whom her parents profoundly disapproved of. After she had given him a child, he deserted her on the grounds that she had become sexually distasteful to him. Bereft of her parents' loving and valid wishes for her welfare, cut off from the life-flow provided by her husband, and being an emotionally sensitive woman in a thoroughly life-obstructing attitude, she developed cancer at one of the physical loci of her sexuality, her breasts. At the cellular level cancer is cell anarchy, and as we have seen, anatomical-physiological *ch'i* must be responsible for the non-anarchy, the organic systematicity, of the living body. Hence the nature of a *ch'i*-caused disease is like the nature of the disturbance of the *ch'i*.

The two types of cancer causes, physical and social-psychical, are ultimately one and the same. The so-called values behind our industrially polluting society and our unnatural, side-effecting medicine, and the attitude (psychical condition) of a cancer victim are both varieties of life-obstructing *ch'i*.

It would be enlightening, I believe, to determine what the incidence of gravely frustrated or abused love is among a statistically significant number of women with gynecological cancer and without a prior history of veneral disease, as compared to sample of women with neither disease.

As I have suggested earlier, modern society is distinguished from non-modern society, including the one in our own, Western, past, by its extreme unnaturalness, which is the most basic kind of anarchy, whose ultimate source is the Narcissistic self-interest of the architects of our scientific (including social-scientific) paradigm. Hence, as social-psychical reality is increasingly constructed according to that worldview, and therefore increasingly *mutated* by it, more and more people are put into life-obstructing attitudes, which affects their *ch'i*, first in its psychical and then in its physical aspects. Western society is based on and increasingly realizes a denial of Human Nature, with the result of decreasing self-government from the physical through the psychical to the social levels, and increasing central government of all normal psychical and social functions. Since central government of those naturally localized functions cannot work, evidenced by the increasing instability of the Capitalist and Communist polities and economies, anarchy results. From the perspective of the average individual, the result is *stress*. For example, to economically force men to work for faceless corporate executives, and women to work for men not of their kin or community, and to consign their children to day-care centers are denials of Human Nature, in that they depersonalize work and weaken families and communities. I speak of average people, not the exceptional minority whose talents may float them into service of greater communities-societies, which is Human Natural for them. These denials obstruct the life-flow of a couple's economic cooperation (the local-self-governing economic function of the family), the life-flow of a couple's coordinate decision-making (the local self-governing political function of the family), and the life-flow between parents and children (the local self-governing educational, Humanizing, function of the family), all of which contribute to or even directly result in interpersonal and intra-personal stress. Such conditions disorganize the *ch'i* to produce, among other possible "psychosomatic" diseases, cancer, and, when sufficiently grave, make the individual relatively susceptible to the cancerous potential of industrial pollutants and antibiotic cell scars. In turn, deliberately behaving in accord with Human Nature, which in modern society often must take the form of resisting unnatural life styling, is cancer preventive.*

* According to Chinese medical theory, stress endured to good purpose does *not* cause illness, because such stress is in harmony with the *ch'i* of suppressed natural society: one is *not* alone in such a condition; one is a member of a close community, albeit a potential one.

The fit of that Chinese medical sub-theory with all the indicated cancer data may be taken as evidence of the existence of psychical-social *ch'i* and the need for something totally lacking from our scientific repertoire: high medicine, that is, societal-psychical medicine which takes into account a societal-psychical-physical continuum of *ch'i*, whose chief symptom is emotion.

Richard Totman in his book called *Social Causes of Illness* documents that people who are socially-psychically injured become physically ill much more often than those who are not—or who are less so. And he establishes that cancer is one of the illnesses which such people more frequently contract. Tending to confirm the Chinese-medical explanation of social-psychically caused cancer are recent Western studies of the physical correlates of emotional stress.* It has been found that not stress alone, but the social-psychical conditions on it and the way it is handled, specifically correlate with physical disease. For example, workers whose jobs are stressful but who like their jobs become physically ill at a lower rate than do those whose jobs are stressful and who dislike their jobs. What most people dislike about their jobs (and most people dislike their jobs) is that there is insufficient security in them and, as Paul Goodman has pointed out, that they have little or no control over what is done, when, and for what purpose. All those dislikes boil down to centralized economic conditions that have replaced the natural local-self-governing ones of traditional society, where everyone was either the boss or had a personal relation to the boss. Rats, it has been discovered, are bothered less by electric shocks if they are allowed to control the timing of the shocks. How much more must it be true, then, that stress undergone by *humans* for ends they control must not be illness-producing, and that stress undergone for ends out of their control must be illness-causing.

Those who handle stress the best are those who have *strong family ties*, and those who handle it worst are those who do not, or whose family ties have been broken. Divorcees, for example, have more heart disease, strokes, and infectious diseases such as TB than do married people. Until recently, the high incidence of hypertension among black American males (who have a high incidence, as compared to white American males, of being regarded as dispensable by their welfare-supported women, as is encouraged by governmental welfare

* The following data are from Adler and Gosnell, cited in the *Bibliography*.

laws*) was viciously attributed to "genetic factors." James Lynch, scientific director of psychosomatic clinics at the University of Maryland, rightly suspects that it is due to "social alienation." In his book, *The Broken Heart: The Medical Consequences of Loneliness*, he establishes that single men have a higher physical illness rate than do married men. The death rate due to hypertension is almost triple for divorced men as compared to married men, and the cancer rate is more than double.

The major correlates of loneliness, or what is here called "life-flow obstruction," are well-known. Their cause-effect is anarchical social-psychical *ch'i*: forced retirement, death of spouse, divorce, departure of children from home, the reduced role of the family, and loss of a sense of community—both of the latter being correlates of the centralizing of formerly local social-political-economic governance.

Durkheim established that loneliness—what he called "the egoistic condition"—fosters suicides. The suicide rate of the young is skyrocketing: they are the chief victims of centralization.

Where the most binding relations focus in sexual reproduction, the core of modern social alienation is the disorganization of such relations.

Correlations between the decline of sexual morality supportive of those relations and rising rates of cancer, then, are expectable.

The uterine cancer rate is much higher in sexually promiscuous women than in monogamous women.** This is consistent both with the antibiotic explanation (their VD rate and consequently their antibiotic-using rate is also higher) and the obstructed *ch'i* explanation.

Cancer, then, is one of the possible results of stress, and the specific condition for it is life-flow-obstructing stress, the accurate conception of which taps human aesthetic, more than logical, capacities.

Chinese medicine further specifies both the energy-shell of *ch'i* at the periphery of the body and the psychical *ch'i* internal to the body, which is the relatively independent variable in cancer etiology and "psychosomatic" disease in general. The reader may have noticed that in the Sextimagic correlations listed earlier each of the six yin-yangs characterizes two organs except for reverting yin and minor yang, which characterize only one. The two missing organs are *ch'i*-func-

* See George Gilder's *Sexual Suicide* for the details.
** See Diane Francis, cited in the *Bibliography*.

tions. I did not introduce them earlier because the evidence that the chief aspect of human physiology is effected by *ch'i* first had to be presented, and it had to be made clear that even the physical organs, such as the heart, are not, in fact, only physical. In accord with that understanding, Dr. Porkert uses the term "orbs" instead of the term "organs" to distinguish Chinese anatomy and physiology from the grossly and Absolute-Fragmentally physical version to which Western medical scientists ascribe. Organs are *functions* (*yung*) as much as they are formed substances.

The minor yang demi-physical orb is called the *san-chiao*, which has been literally and therefore incorrectly translated as "the three burners," or "the three coctic spaces," regarded as a product of Chinese fantasy or of a Pythagorean-like need for numerical symmetry (they were invented, the Sinological story goes, to supply one more organ for each of the reverting yin and minor yang coordinates). Actually, *san-chiao* refers to three major divisions, upper, middle, and lower, of the energy-shell, and its extension into and throughout the body. Present space does not permit showing that, like the other organ-associated *ch'i*-functions, there is direct empirical evidence of the *san-chiao*. Suffice it to say that once Chinese diagnostic technique is mastered, their symptoms can be felt, that therapy based on diagnosis of disease of the *san-chiao* is effective, and that pain symptoms for which Western doctors can find no specific location are often symptoms of diseases of the *san-chiao*. The reverting yin *ch'i*-organ, one which, being extremely yin, is more physical than the *san-chiao*, is called the *hsin-pao-luo, literally*, "the pericardial meridian" (as an "acupuncture-meridian," or central line of *ch'i*-flow). One of its central functions is to help the heart perform and, as befits its relatively metaphysical nature, one of those functions is to regulate and generate *emotions*, which, and as Inyushin has demonstrated, directly affect the conductivity of all acupuncture points, and which are the relatively independent variable in so-called psychosomatic diseases. Thus, the *san-chiao* and the *hsin-pao-luo* form a relatively-non-physical functional yoke which has the same Sextimagic values as does the *especially animate* yoke of the liver and the gallbladder: extreme yin and minor yang.

Now I turn to direct empirical evidence, from the Capitalist sphere, of obstructed *ch'i* being the cause of cancer, or the condition predisposing a poisoned body to develop cancer. J. Michael Bishop, medical researcher at University of California at San Francisco, has isolated the gene of what he calls a "virus" which causes cells to become malignant. To his surprise he found that the gene in question is virtually the same as that of the cell it invades. It follows that, just as Chinese medical theory has it, this "virus" may very well be a degenerated normal cell (and/or a micro-organism resulting from the materialization of cancer-*ch'i* emitted by cells destroyed by poison). In any case, the cancer-causing gene being virtually identical to those of a normal cell, it follows that a non-DNA factor is the (de-)configurative cause.* There is a non-physical lack of integrity, which accurately refers both to cancer cells and to the basically Mentalist attitudes behind the anarchic growth of Western industry, and the breakdown of family and community that accompanies it, since the nineteenth century. Attitudes and physical bodies are one, and the connecting variable, ignorance of which makes them *appear* to be separate, is *ch'i*.

As my *shih-fu* put it:

> There is no such thing as clean spirit and a dirty natural environment or a clean natural environment and dirty spirit. Simply, there are the perceptible or imperceptible aspects of a clean or dirty *whole*. For all to become dirty, it is only necessary that one area of it become dirty.

Cancerous spirits cause cancer. Malignant ideas result in poisoned air and water, which produce physically sick, and, often, therefore, angry people, who produce destructive ideas that contagiously spread to all people and twist their spirits in a destructive direction, thus completing the ecological loop. A crucial example is the devaluation of children epidemic in the West, which is "adaptive" to an environment in which, if one cares for children to the degree that is natural, one must have the strength to tolerate constant worry about their psychical and physical health. How can contemporary parents guaran-

* For a carcinogenic virus to cause a normal cell to become cancerous it would have to be programmed to program a normal cell to proliferate. Hence it would have a gene which a normal cell lacks: one that programs the virus to eat a non-proliferative gene in the host or one that instructs it to proliferate and (if one viral invasion is sufficient to more than one cancerous generation) to reproduce that instruction in each generation. Hence, either the carcinogenic gene exists but has not in fact been found, or the carcinogen is in fact not a gene. "Virtually the same" suggests, of course, that the normal/cancerous difference is *sub*-genetic, *sub*-molecular, and where it could not be atomic, it is non-physico-chemical—*energetic*. In his article, Bishop notes that cancer-causing genes do seem to derive from cellular ones. The mystery is how they are *patterned* into cancer-causing ones.

tee their children's futures? They no longer belong to them, they belong to "society," which means a social dis-organization whose ultimate guides are, to use the Chinese-medical expression, "deer posing as horses," some knowingly, most not. Out of control of its own children's futures, society then further sickens social-psychically, and graver physical diseases are produced via the medium of enturbulated *ch'i*, and people are made more susceptible to externally contracted germ diseases. The chief cause of disease is an unnatural scientific paradigm, which derives from unnatural attitudes, or the failure to employ a natural scientific paradigm. Genuine and high medicine addresses itself not to the Narcissist-fantasist Utopian, or Heaven-on-Earthly, goal of eliminating disease and death, but the really possible goal of minimizing them and gracing life with the highest possible quality for the maximal number of people. The Tao-paradigm is basically *common*-sensical and based on *common* human feelings, and its realization does not put people into life-obstructing attitudes.

In heart-*ch'i*-wrenching contrast, the hottest item in Western cancer research (outside the Soviet Union) is interferon, a biochemical which can be manufactured in sufficient quantities for cancer treatment by *E. Coli*, properly mutated by recombinant-DNA vivisection. Apparently it neutralizes the "genetic" variable which causes cancer cells to proliferate. I leave it to our medical researchers, who, because they do not seek to *prevent* cancer and because they had reports from China that Chinese medicine can cure cancer, are not without blame, to take responsibility for the predictable vitality-interfering side-effects of interferon treatment.

Let us turn to the Soviet medical scientists. The super-centralized and therefore extremely anarchical social-political system which Inyushin's Alma-Ata Group serves is not taken into account in their research, yet it must be the chief cause of cancer in the Soviet Union—a society more centralized and, reciprocally, destructive of natural family and community structure, than our own. This major oversight aside, avant-garde Soviet cancer research is much closer to and, again, confirms that of genuine medicine. The Alma-Ata Group is confident that ultimately it will be able to use the Kirlian scanner to diagnose the *onset* of cancer, which they correctly define as a form of "deterioration in body and *mind*." This means (if they are telling the truth, and there is good reason to believe that they are) that they have empirically established associations between cancer, emotional states, and the electrical manifestations of *ch'i*-flows at acupuncture points. Inyushin has obtained positive results for cancer patients using laser-"acupuncture," which he declares to be far superior to both the traditional and the Shanghai methods. With high probability, arising from the competition in medical research between Communist China and the Soviet Union, that no highly competent traditional Chinese physician has ever candidly performed for the Alma-Ata Group, Inyushin is in no position to make the first part of his evaluation. But that he has found that light-input is more therapeutic than electrical input is consistent with the traditional *ch'i*-theory. As explained, light is less physical than electricty and therefore closer to the *ch'i* which governs it and which is the actual vital force. Hence his results tend to confirm the *ch'i*-theory of cancer.

Let us hope that those imaginative, if stubbornly Mentalist, Soviet medical scientists will not be politically obliged to define "mental deterioration" as "ideological disagreement with central-governmental policies," so as to select those with the healthiest spirits in the Soviet Union as victims of the side-effects of the treatments Inyushin anticipates, namely—as implies the actual inferiority to traditional methods of laser-"acupuncture," laser-acupuncture, buttressed by cancer-causing radiation and chemotherapy.

CHAPTER

7

Shang-I: Higher Medicine

Talent neglected or misguided, investigations into the nature of things not completed, what is right understood but not acted upon, and the lack of energy to rectify what is wrong—these are the things which pain my heart, which I exist to remedy.

—K'UNG-TZU (Confucius)

The lower doctor heals the illness; the median doctor heals the whole person; the higher doctor heals human society.

—SUN SZE-MO (a physician of the eighth century A.D.)

This is the shortest and simplest, but also the most meaningful chapter in the book. What is high is most alive, and what is most alive is most subtly structured and is everywhere at once, so this chapter has no sections, and it wanders from one topic to another, spanning a range between movie criticism and the feeling of a bitch for her pups, all the while having to do unremittingly with one specific topic: the source of all health.

This book has traced, through an array of logical trains and data from every aspect of scientific investigation, a Complete circle. It began with the proposition that the Western, Absolute-Fragmental, or Mentalist, scientific paradigm is false and side-affecting, indeed, disease-causing, and it identified its source as a physical disease which may be called intellectual and emotional Narcissism, whose chief characteristic is a disdain for Nature, common sense, and human beings other than oneself and those of one's (peculiar) kind. The book closed the circle with a demonstration that the chief and ultimate cause of diseases, of any kind, is, precisely, that Mentalist disease. It follows that health at all levels fundamentally depends on establishing a Polar-Complete *social*-scientific paradigm, one which has the same structure, or energetic configuration, as is found at all the levels of reality which have been dealt with, namely a Polar-Complete structure. This is precisely what the image of Human society that is based on *Jen*:Human-ness is. So here my task is not to discover or invent but to translate, and that means to translate "from scratch," because no aspect of Chinese knowledge has been so distorted by Western China-Specialists as has that Confucian one. Indeed, the entire previous part of this book may be regarded as nothing other than a preparation for correct understanding of the Confucian social-scientific paradigm. But a few more preparatory things should be said. What I will say here is by no means an exhaustive exposition. Rather, it gets at the essence of the Human social-scientific paradigm as part of the Polar-Complete scientific whole.

The best way I have found, in my teaching, to cut through the Absolute-Fragmental wall which separates modern (including Chinese) imagination from a natural social-scientific paradigm is to show, first, that we had but no longer have democracy and, second, that the two social models which monopolize present consciousness are nothing other than Absolute-Fragmental, opposed, abstractions from the natural human behaviors which existed almost everywhere, before those Binary versions of "Democracy" were made real and were used to decimate traditional society everywhere. These consciousness-monopolizing models are called Capitalist Democracy ("The Right") and Socialist Democracy ("The Left"). The former characterizes the military-technologically overpowerful countries of Western Europe and America and the countries influenced chiefly by them, and the latter characterizes the military-technologically overpowerful countries of Eastern Europe and Asia, and the countries influenced chiefly by them. The so-called Third World (relatively traditional) countries are understood to have no choice other than between the two. The Capitalist-Democratic countries are understood to have a choice between remaining Capitalist-Democratic or switching to the Socialist-Democratic path to the Heaven-on-Earth of Communism. The Socialist-Democratic countries are understood to have a choice between remaining Socialist-Democratic (and, some constantly receding day becoming Communist) or "regressing" (as though most had ever been there) to Capitalist Pseudo-Democracy. Those with mixed political economies are torn between the two. In other words, the whole world is condemned to one or the other form of "Democracy," or, as in the case for non-"Democratic" peoples such as the Chinese, Vietnamese, Cambodians, and American Indians, vigorous policing and psychical-physical exhaustion through "participation in Democratic People-Meetings" or starvation, torture, and genocide. In short, the "truth" of the inevitability of Binary choice is established not according to either of the criteria of truth but by irrational and usually coercive means.

Meanwhile, as a matter of fact, our Knowledge-Specialists and the highest-positioned politicians whom the elite among them work for are the people who least believe that either system is actually democratic, and who are most aware that both systems are cracking to pieces at their seams, the explosive force being nothing other than the social-physical disease-resistance of the people whom they have Evolutionarily conducted into a "higher social state."

The problem is the centralization, and its fragmentalizing of personal relationships, or distancing from common people. Our bureaucrats have so many problems from so many communities, about which they understand so little, that, according to a consensus established among them, they do nothing about them, or they aggravate them, by passing laws which state their dim understandings of the natures of those problems or by setting up more bureaucratic organs—for example, daycare-center-creating and monitoring—which further remove the common people from control over their own lives and further tax them. The bureaucrat is faced with one hundred problems from one hundred different communities, each problem somewhat different from the other, with which he is so unfamiliar that he could not possibly address one of them in a complete or Human manner. The laws he passes confuse these problems into one, and create side-effects. For example, because there are street gang murders in a big city, innocent city dwellers and hunters in the hinterlands are deprived of their rifles, thus (where the street gangs circumvent gun-control laws) creating a population more defenseless to street gangs than before. In contrast, if there were responsible local self-government (the conditions for which are outlined below) in the place of over-extended central government, given the same one hundred problems each community would have only one problem to deal with, and it would do so along tailored and personal lines real and heart-felt to its members. Intercommunity problems would then be addressed by representatives of each community in question, who would consult personally with the members of their own communities; and so on up the line, so that all responsibilities are shouldered at the most local levels. A result of this maximally Human and efficient structure is that the central government has only specifically central duties to perform—the defense of the whole, moral-educational and medical services* to that whole, and nurturing the harmonious self-governing of that whole.

Let me at once exemplify both the naturalness of the Confucian social paradigm and the malignant attitude of the architects of that betrayal of the original Euro-American democratic paradigm, contemporary so-called "Democracy." A few years ago a movie called *Walking Tall* was made. Based on fact, it told how a small-town American named Buford Pusser became the sheriff of his town in order to drive out of it a Mafia-run gambling racket and whorehouse that had recently been established there. His battle with these gangsters, pimps, and whores entails enormous physical and emotional suffering, including near-fatal injury to himself and the murder of his beloved wife. Pusser wins. The popularity of this movie proved so extraordinary that it has been followed by two sequels. And what accounted for its popularity, established by the word-of-mouth of its viewers, was the audience's perception of Buford Pusser as a hero fighting for good against evil. The mass media's reaction to this phenomenon represents the mentality behind our scientific paradigm. It was as follows: shock, dismay at the popularity of the film, mocking of its hero and the people who loved it, and ostensibly moral objections to its "violence." According to *TV Guide* (August 23–29, 1980), a trade paper justly characterized the *Walking Tall* movies as a redneck violence series.

* To refer to an opportunity long passed: such as suppressing the invention of the gun and executing its inventor on behalf of millions of potential murder and out-of-scale-war victims.

Failing to see any connection between the saga of Buford Pusser and the racism of the South, where "rednecks" are found, I must conclude that what is distasteful to such reviewers is nothing other than the heroism of Buford Pusser and the audience's admiration of him. Let it be noted that the popularity of these films is Northern as well as Southern, not to mention Oriental. The first one was well received in Taiwan, where I saw it, as betrays that its appeal is based on something which transcends the most definite cultural boundaries and which, therefore, is natural.

Now let us determine exactly what is so objectionable about the "redneckness" of the majority of American people, in the eyes of such reviewers. What it is in the terms defined by the social paradigm which such reviewers serve. First, Buford Pusser fought evil conceived as gambling and prostitution, which is evil for the reasons that gambling and prostitutes encourage men to deprive their families of money, which it is their right to deploy toward their survival and the increased material welfare of the next generation; that prostitutes spread venereal diseases, which may then be transmitted to their clients' wives, exacerbating unhealthy marital conditions (which of course must have existed a priori); that unmarried men visiting prostitutes acquire the marriage-jeopardizing habit of divorcing sex from love; and so on. All of this boils down to prostitution and gambling being a disease which takes root in unhealthy familial conditions and aggravates them, and that a community's acceptance of prostitution and gambling makes the implicit but clearly understood statement to all members of the community, and especially the impressionable young, that unhealthy families and the destruction of families are acceptable. Hence the distaste of such reviewers is a distaste for healthy families and communities and the people who belong, or wished they belonged, to them.

In addition, the "official" objection to "violence" cannot be to violence without qualification, for the violence of the legal system of our Progressive Social Planners is ample. Not only are those imprisoned degraded, beaten, and homosexually raped, but usually the perpetrators of those acts are given sentences unequal to the gravity of their crimes, which does psychological injury to those they have injured. Chances are (as sociological research has established) that once they are released from prison they will, with knowledge and attitudes derived in prison, continue committing violent crimes, but more professionally. The "official" objection, therefore, is to "violence" exercised justly and by natural, local authorities, which is a form of localized, community-based self-government based on the respect and affection of members of the community for each other—*Jen*: Human-ness. Such localized self-government, although it is part of the in many ways beautiful, and in comparison to the Socialist system qualitatively superior, original *democratic* plan for Europe and America, is abhorrent to the Social Planners who have exploited innocent but nevertheless real faults in that original plan to create a centralized governmental system which increasingly deprives local social units of the power to govern. Buford Pusser's second offense, in short, was "taking the law into his own hands"—as though the law could be other than people's natural rights, including that of self-defense.

The objection to *Walking Tall*, predictable from the social-"scientific" paradigm for Western evolution and the "redneck" response to it, is the thrust against centralized and therefore actually anarchical "government" and toward local self-government based on family-values, which is essential to health at all levels and which is one of the most fundamental features of *natural* society. Basically, what is "objectionable" is the Human-ness of the film and the audience. It is the fundamentality and healthy naturalness of genuine democracy, of local self-government based on Human-ness, which, alone, accounts for the (also unpredicted) success of the late E.F. Schumacher's *Small Is Beautiful*, whose message, based on an economic perspective characterized as "Buddhist," and which could have been characterized as "Human," is that modern society must de-centralize, be based on local self-governing units so that central government has a very light hand and exercises only the duties which a specifically central government is required for.

That concept of light-handed central governing, which Confucius devoted much of his effort to assure, is called in Chinese *wu-wei*, which may be translated as "minimally interfering central governing, made possible by the satisfaction of most governing functions by local groups in the most natural manner." Such local self-government, or genuine democracy, is like the harmonious interfunctioning of the bodily organs, each of which performs its special task of its own accord, but to the benefit of the whole, and such central governing is like the whole-bodily *ch'i* which assures harmony and defends the whole from external attack. What is today called "Democracy" is its exact opposite. It is like one organ presuming to determine fully the functions of all others—the brain which our medical scientists falsely imagine totally governs human physiology. And its effect is that of interferon: inhibiting the vital process of the whole in order to cosmetically adjust one of its parts.

Now that I have illustrated with a specific (and randomly elected) example the naturalness of Confucianism, and the unnaturalness of its opponent, let me move to the level at which healthy society can be generally described, by, as promised, showing that the Binary "Right" and "Left" are opposed abstractions from normal human behavior, less-than-halves of a natural whole which is called "Democratic" and is anti-democratic.

The supposed virtue of the Right is that it fosters free enterprise, which is good because it allows natural rewarding of excellence—imagination, service to the community, hard working, frugality. It also allows natural rewarding of inferiority—no service or disservice to the community, laziness, squandering. And, all other things being equal, free enterprise *is* good: it exploits the principle of *enlightened self-interest*. That is, those who can excel do so out of self-interest, ultimately interest in the future of their children, and the side-effect is the benefiting of the whole community through the products of their excellence, and, under traditional, normal, conditions, through their generous sharing of their wealth with those who have made it possible for them to amass it. In contrast, the supposed virtue of the Left is the opposite of self-interest: altruism. Its social-systemic vehicle is the opposite of free enterprise: socialism, whereby the means of production are owned and controlled by a central government (or hierarchy thereof) on behalf of the people, according to the principle that the distribution of wealth shall be homogeneous among all.

The Right, then, opposes the Left on the grounds that the Left deprives people of the natural rewards for their social contributions: those who excel are not rewarded for it, and those who fail or who disserve others are rewarded for it. The result is total demoralization and the deterioration of all products, or production. This theory is quite sound, as is evidenced, by the high alcoholism, divorce and suicide rates in Communist countries and by the fact that the Communist countries depend (because of their failures of production and emphasis on war machines) upon the Capitalist ones for food to feed their people so that they will not rebel against the demoralization brought down upon them by their central governments. The Left, in turn, opposes the Right on the grounds that the Right creates an economic class system in which the rich exploit the poor, often along ethnic or racial lines which intensify the injustice, thus demoralizing all but the upper class, with the result of the deterioration of all products. This theory is also quite sound, as is evidenced by the increasingly low quality of our products (cars, alas!), the high alcoholism, divorce and suicide rates in Capitalist countries, especially highly industrialized ones, and the fact that Capitalist countries increasingly depend, for support of their expensive economic system, on the purchasing power of Communist countries and the low-cost, human-labor-intensive goods that such countries can provide thanks to their centralized power to coerce the common people to work for much less than they deserve. And meanwhile, the most rich and powerful class in human history, that of elite Communists, is formed and entrenches itself.

Note that, like all Absolute-Fragmental so-called opposites, the Left and the Right each include features of their opposite and are in fact so totally interdependent and mutually supportive that they are actually Polar-Complete aspects of a same thing. The Tao is universal; all that can vary—and here I quote Confucius—is whether man's reaction to it furthers life or results in negative side-effects.

What the Right lacks, what the Left lacks, and what both sides lack are all obvious. The Right does not promote *enlightened* self-interest because it fails to assure that the products of the successful are good and because it fails to motivate a distribution of excess wealth —wealth beyond the level of relative material comfort and material security. It leaves society wide open to a perversion of free enterprise, called Capitalism, which fosters economic and consequently political injustice.

In turn, what the Left lacks is natural justice, or adjustment, which free enterprise makes possible, and, more generally, the natural local organization and self-government which, via coordinated self-interest within small groups, is the fertile soil of free enterprise. It distributes goods equally within but not between the classes it forms; destroys creative spirit or ideologically confines it to state-designated goals and means; fosters bribery to counter the disinterest of all occupational classes of people, such as doctors and nurses, in all other occupational classes of people. And, most basically, it subordinates parents to the state, depriving them of the authority essential to extending themselves as persons through their children so as to give their children maximal care and self-esteem and to achieve for themselves the social immortality which humans naturally seek (and which, as explained, Christianity and Communism work hand in hand to misdirect people away from with Heaven or Heaven-on-Earth).

The ultimate effects of the faults on both sides are essentially the same, just as are the inComplete explanations of both scientific sides of the Binary Con, those of the Left being worse for society as a whole because its central-governmental system has more coercive power than the Right's, so that its effects are even more anarchical, natural-order-disturbing. This is why people outswim gunboats through shark-infested waters to get out of Communist countries into Capitalist ones, whereas people unexceptionally do not do so to get out of Capitalist countries into Communist ones.

The Left and the Right are two hands without a body in their middle, and the right one slaps our faces as we turn left and the left one slaps our faces as we turn right.*

The body which the two slapping hands both lack is Human-ness, which I will now exemplify by reference, as promised, to natural societies from the social systems of which our Social Planners have extracted the less-than-halves called Right and Left. I am going to use a non-Chinese example for every feature in question so as to emphasize that Human-ness (*Jen*) is widespread, natural, and human, not specifically Chinese.

What we are looking for, specifically with reference to the half-merits of The Right and The Left and what they lack, are traditional social systems which combine, not as two separate systems but as Polar-Complete aspects of same systems, self-interest and altruism. This is too easy to find, because it is the usual arrangement when all of human history in all places on the planet, not only Western civilization, is taken into account. But I will use an excellent example, in which the typical case is graced with beauty and exceptional spirit—the Abkhasian people of the Southern Caucasus Mountains. Cultural-anthropologically they are also a good example because they have distinct social classes and thus are potentially subject to Marxist

* Being what in their minds are good Christians, most people in the non-Chinese Left and Right territories continue to fall for this. Jesus was not a masochist. "Turning the other cheek" refers to the attitude of a benign and powerful man, such as he was, toward a malicious weak one, whose slap is negligible. The common people are benign and weak and their rulers are powerful and, when not malicious, incompetent; the people are in no position to turn the other cheek.

(Leftist) analysis, and because they are both shepherds and farmers, thus combining the two principal economic modes. And I refer to pure Abkhasian culture, which no longer exists because the Soviets have conquered the Abkhasians and obliged them—although, because of the superior integrity of Abkhasian culture, admittedly to a lesser degree than they have been able to oblige Russians—to pervert their culture to fit into and serve the Soviet socialist system.*

The local organizations the Abkhasians employ to amass material wealth in the form of vegetable and animal food and shelter are the family and clans made up of families—kinship organizations. Relatively successful families, according to the natural and lawful system of free enterprise, acquire greater material wealth, and employ relatively unsuccessful families. This basic owner-laborer class-relationship is what in Right societies is a basis of conflict and injustice and what, according to Marxism, inevitably leads to conflict and injustice. The last is simply untrue, and that reveals that Marxism (the extreme version of the Left), because it is based on that assertion, has nothing whatever to offer by way of improvement over Capitalism (the extreme version of the Right). Capitalism is not, as is claimed by Marxists, a necessary consequence of free enterprise and is not, as is claimed by Capitalists, essential to the protection of free enterprise. What proves those points is the fact that the Abkhasian system's free enterprise is, like free enterprise in almost all traditional societies, directed toward altruistic ends—in this case, ingeniously, so that neither is there oppression by one class of the other nor is there the accumulation of goods for the sake of accumulating more goods (Capitalism). The basic means of accomplishing this is that families of the upper class give at least one of their children, in infancy, to be raised by families of the lower class.** The result is that nobles have kinship-love for members of the lower class. No church sermon or set of laws can compel self-interested rich people to treat common people with respect and justice; but no force on earth could compel a person to treat with disrespect or injustice parents who raised him with love, nor would such a person permit anyone else to do so. Such use of the *human heart*, instead of the impotent and hypocritical use of *social systems*, is precisely what is meant by *Jen*:Human-ness.

* See Benet, cited in the *Bibliography*.

** A politically enlightened practice which survives in Euro-America only, unfortunately, in the form of a common theme in fairy tales.

In addition, the Abkhasian upper class distributes much of its wealth to the lower class through the medium of feasts with gift-giving, which, unlike Christian charity, does not announce itself to be charity and therefore does not humiliate the recipient of good will. ("The charitable person is not charitable," Chuang-Tzu has written.) Such distribution of wealth is one of the near-universal features of traditional societies, and it is nothing other than the ability to acquire and distribute wealth, justly, which qualifies men for leadership—leadership into which they are thrust by the affection and respect of the people. This is true from early Medieval German monarchy to South Pacific chieftainship. It is a spurious and perverse mimickry of this Human principle which the Left uses to attract those dissatisfied with the Right—an example of what, as explained much earlier, is called "The Stealing of Virtue."

The Abkhasians, by the way, are called "the long-living people of the Caucasus": their health is extraordinarily good and they have extraordinary longevity. The Soviets, who claim to have liberated them from their "oppressive class system," have attempted and failed to account for their longevity. In light of the previous chapter and the self-interested-altruistic harmonious social functioning of this society, it is easy to come forth with a plausible explanation, not only of Abkhasian longevity, but of the reason that Soviet medical scientists cannot account—or fear to account—for that longevity.

The basis for such altruistic conditioning of self-interest as the Abkhasian version, which, in turn, through free enterprise preserves the natural system of rewards and maximizes the quality of goods produced, is no different in complex civilizations. Only the means differ somewhat. I use the best example of a complex civilization—traditional China.

Similarly to Abkhasian noble children, emperors' sons were raised not by their parents but by Confucian savants related to them as *shih-fus*:teaching-*fathers*, whose foremost task was to instill in them, through intense emotional instruction, Human-ness.

That is not a "myth." One of the results (historically documented) of such Human-izing was that it was illegal, by enactment of memorials from Confucian scholars, for landowners to form corporations, because when they do they can cooperate to fix land-rental rates, thus depriving tenant farmers (people from relatively unsuccessful families) of the natural justice-ensuring means of leaving the lands of exploiting landlords for the lands of just landlords (who, as a consequence, acquire more land and power and popular support than unjust ones and, at the highest level, become emperors who replace corrupt, exploiting ones). This moral-legal principle is reflected even in Taiwan—"even" because it suffered socially from Japanese colonization, because it is young, and because it is in many respects modern.* There Capitalism, the use of money, not honest labor, to make money, is considered immoral when practiced to excess. In accord with that principle, a few years ago President Chiang Ching-Kuo, whose popular nickname, because of his earlier actions against corruption among the powerful, is "The Tiger-Beater," called a group of Capitalist speculators who had quickly made millions by this inherently fraudulent and unnatural means into his office and asked them if they wished to donate their profits to farmers in the south of Taiwan who were in need of roads to transport their produce. Understanding that the request could turn into an order, those Capitalists quickly complied. Consequently they have become popular (and still amply wealthy) benefactors. This is "socialism" made Human and united with free enterprise, and it is "capitalism" made Human and united with altruism.

Another Chinese method is the ascription of the lowest prestige to the merchant class, a cultural feature true also of the European "Dark Age," which, as I said, was the West's most enlightened period. Traditional China had four classes, from top to bottom: Confucian scholar-officials (because they are responsible for the welfare of all classes), farmers (because they provide the most basic and therefore most sacred common need, food), artisans (lower than farmers because they satisfy secondary needs, such as home adornment), and merchants (lowest because their profession is the most self-interested one). There was a fifth, un-named, class, of prostitutes, gangsters, and the like, which was much lower, for reasons obvious to all save those who find the *Walking Tall* saga "objectionable." As a consequence of this lowest of named statuses, merchants were inclined to buy their way out of their embarrassment by using their money to pay the best teachers to educate their sons to become scholar-officials. The side-effect, quite intentional, was to prevent class stratification, specifically merchant dynasties—rightly a target of Marxism and wrongly asserted by Marxists to be curable through socialism, which has produced the

* Its successful Land-to-the-Tiller reform, the only one in the modern world, was executed through traditional, Confucian, not modern socialistic, principles.

most powerful entrenched class of property-owners in human history, the Communist elites.

Another Chinese method, used by many mainland landowners living in Taiwan today, was for a landowner to make his son perform the hardest of labors alongside his tenants, or to give him to be raised by his tenants for a few years. Such landowners' sons were dressed, as were their parents, no better than their tenants, and were given no more pocket money. The result is sympathy and respect for those who endure the rigors of manual labor, and just treatment of them.*

The majority of China-Specialists, especially today, when it is mandatory for one's prestige to have been "to the mainland," for which it is essential to agree or, minimally, not disagree, with its Communist rulers that Confucian China was unjust and "class oppressive," will deny all that has just preceded, calling it "naively idealistic," citing selected examples of social injustice and class oppression which, as I have said, have existed in China, but, all told, to a much lesser extent than in the West, to minimal possible degrees during healthy dynastic stages and widely during decadent dynastic stages. The only correlate is pioneer America—minus its betrayal of the Indians and slavery. Such apologists for their own careers and the Right-and-Left, with emphasis on the latter, should be asked: If these principles were never realized, why, then, did traditional China, unlike the "superior" European systems, last so long? Because the traditional system so oppressed its people? Or because Chinese have less Human-ness, spirit and intelligence than Europeans and Americans, so that they were unable to bring about a revolution of that "oppressive" system? The latter explanation would be preferred, of course, because it was only when Western interference was rampant in China that a revolution against the traditional system was able to be achieved. (The Chinese people were waiting, for 5000 years, for the spiritually and genetically superior Westerners to save them—praise The Lord.)

Another altruistic structural feature of traditional Chinese society was, and in Taiwan is, the clan. When clans get big enough they include both poor and rich families. Tied by kinship, the rich are naturally compelled to give or lend at low or no interest rates to the poor members of their clans. Bankers can hardly establish their parasitic mode of existence under such conditions, and what I have called

* I have met several, now poor, and some of their former tenants, who remain admiring and affectionate toward them.

"credit enslavement" is unthinkable. It is, by the way, illegal in Taiwan for there to be a poor person in any economically solvent extended family. No elderly people eat dog food in Taiwan, as many do in the United States.

It is also illegal in Taiwan for there to be people without food in the county of any wealthy family, and the heads of wealthy families are held legally responsible. When there have been crop failures, and after the Japanese were expelled from an impoverished Taiwan, wealthy people set up free restaurants in the afflicted areas.

Such politically institutionalized Human-ness may be contrasted with the dumping of unaffordable canned goods during the American depression and FDR's central-governmental, socialistic, pseudo-solution to the problem, which ignored and was implemented independently of the fact that such dumping of food symptomized a political-economic disease of catastrophic magnitude. It is *Jen*, Human-ness, and its lack that are in question.

Chinese-cultural incorporations of Human-ness into the political-economic system—"systems" which are the people themselves—naturally cause other desirable things to come about. For example, in half-traditional-half-modern Taiwan, taxes are extremely low by Western standards. And where a labor-intensive economy complements the personal responsibility of the rich for the poor, the distribution of income, as the former U.S. ambassador to the Republic of China pointed out, is more equitable even than in the United States. Instead of welfare, there is kinship.

It is unfortunate that the present epidemic of pro-Communist Chinese sentiment* among those who govern and those who intellectually influence our society has obscured the largely successful synthesis in Taiwan of Human culture and industrialization, and of the good, not the bad, features of the less-than-halves of the Right-and-Left. More than a mere "model for the Third World" (as it was held to be before said epidemic), it is also a model for the "First World." Its special virtue is the traditional Chinese, Confucian, social culture, adapted to modern social reality, which persists, and—if I may address myself as a doctor to said epidemic—it is because Taiwan is its last

* I cannot conscionably ascribe to the current use of "Peking" or "the mainland" instead of "Communist China," the purpose of which is to obscure the contradiction in a largely Christian and firmly democratic nation's acceptance of anti-familialism and totalitarianism.

bastion that Taiwan, not Communist "China," has the right to call itself "China."

What I have begun to do here is outline the basic structure of genuine civilization, and that requires book length, so I must restrict myself to "a few corners of the polygon." But I have shown, in common-sensical terms, that The Right and The Left are, indeed, two less-than-halves of a natural, civilized whole whose basis is *Jen*: Human-ness.

Let me add, in anticipation of the cynical, petty, and negative reactions of many social Knowledge-Specialists and Sinologists, that even if every self-interested-altruistic social function which I have exemplified never existed (and most anthropologists could think of numerous further examples of their kind), there is no reason whatever why they could not; and they should. As Winston Churchill once put it: "Perhaps there never was a King Arthur; but there should have been." Are we not supposed to be culturally Evolving? *Why, then, the heavy and pessimistic reliance on precedent when it comes to really doing so?*

Now here is the ultimate Confucian—Human—point. Each Human function relies for its existence and nurture, more than anything else, on healthy families, whose social functions, and therefore whose life-force, are leached out by the faults of the Right-and-Left. And it is in healthy families that we find the perfect expression of the Polar-Complete principle of oppositeness-identity, which Confucius called *hsiao* (pronounced *syao*). *Hsiao*, being the central specific concept of Confucianism, and Confucianism being the aspect of traditional Chinese science-art most distorted by Western China-Specialists, is probably the Chinese concept least understood in the West. So before clarifying it let me substantiate what I have just said by telling a little story.

When I returned from Taiwan to Yale in 1974 I attended a lecture on traditional and modern medicine, particularly medical ethics and governmental social policy, in traditional-and-then-Communist China by one of the "foremost" Sinologists, who had just returned from Communist China. The basic thrust of the lecture was to the effect that it was thanks to Buddhism, a moral system imported around the time of Christ from India, which established strong roots in China only several centuries later, and despite Confucian "elitism," that traditional Chinese medicine, first through Buddhist priests, was made available to non-wealthy Chinese. This reminded me, through an ironic twist, of my *shih-fu*, according to whose medical ethics, which predate Chinese Buddhism, Communism is the exact opposite of healthy government. By those ethics, it is his practice not to charge poor patients for his services, a result of which is that he and his family occupy Taiwan's lowest economic class. Said Sinologist added that under Mao Tse-Tung a similar democratization of Chinese medicine was now being implemented, thus comparing Mao Tse-Tung, murderer of fifty million Chinese, to the Buddha. This book, which prescribes the liberation of the Human-ness of peoples of all countries in which Human-ness is obscured (as in the Free World) or suppressed (as in the Communist World), is nothing more than a transformation, effected in the crucible of the Western scientific worldview, of the Human-ness of a tradition which survives in Taiwan and, *despite* Communism, mainland China—a tradition which it was Mao Tse-Tung's central objective to destroy. In view of that, it would be redundant to demonstrate that the preceding message from Mao through said Sinologist is an impeccable example of "The Big Lie." However I might cite the following quotation from a 2500-year-old text, of a following-son of Confucius in address to his teaching-father: "You have said: 'Leave the province that is upright and go to one which is chaotic, for there at the doctors' gates are found many diseased people.' "

The just summarized lecture of said "Confucian scholar" made my head spin, for I realized that its design was to sever Western consciousness and feeling from any and all the fruits of 5000 years of human civilization on the other, older and wiser, side of our planet. I raised my hand and asked him if it was really necessary to attribute altruism in traditional China to Western influences, for was altruism not the essence of the Confucian notion, *Jen* (Human-ness)? He replied that "It is impossible to say" to what extent Confucianism may have been responsible for altruism in traditional China, but that he believed the "high tide" of altruism occurred during the Buddhist influx, which was then carried over to Neo-Confucianism, the official Confucianism of the Sung dynasty—which, as I have explained, was the least Chinese and most Western, nearly Absolute-Fragmental, intellectual period in Chinese history. He added that in order to answer, the "kinds of Confucianism" in question would have to be specified, as though, as my *shih-fu* once put it in anticipation of my confrontations with such Sinological Confusion within our institutions of higher learning, there was a Confucianism based on *Jen*, the central aspect of Confucianism, and a Confucianism opposed to it. In this way, a way now common among Western Sinologists, this Learned

Being emphasized his opinion that if Confucianism had anything to do with being Human, it was the variety of Confucianism influenced by a Western (Indian, Buddhist) enlightenment, just as Mao-Tse-Tung's "humanism" was made possible by his Western (Marxist) enlightenment.

As I predicted several years ago, because Mao was and is hated by almost all Chinese, his statues and pictures are already being torn down, only a few years after his death, leaving Sinologists such as the just-quoted Learned Being to "puzzle this out," and, doubtless, to conclude that his hitherto unrecognized "closet Confucianism" is responsible for his failure to have used the Western Marxist Enlightenment to earn the undying love and admiration of his people, whose hate for him has nothing to do with the fact that he systematically violated every Human right, articulated by none other than Confucius, which they hold dear. In short, thanks to the efforts of "Chinese scholars" like the one just cited, Westerners not only misunderstand Confucianism, which is right under the noses of China specialists as the kinship-based social justices of Taiwan and the popular resistance against Communist totalitarian manipulations of mainland Chinese families and communities. This makes it easier for "progressive," culturally imperialistic, Western scholars and politicians to hold onto their religious hope that China will be totally overtaken by Western "Democracy" (of whatever Right or Left variety), and thus to affirm their own Evolutionary superiority.

Let us return to this notion of *hsiao*. It refers to the most powerful and Human relationship, the one between parents and children, from which springs, as I have shown through concrete examples, all other Human social features in all societies. Specifically, it refers to children's devotion, love and respect for their parents and, reciprocally, parents' devotion, love and respect for their children. Now, since in the West and the regions it has influenced—and only there—the prevailing Academic opinion about the parent:child relationship is that it may be culturally imposed, indeed, even psychological-disease-causing, as Freud proposed, and societal-disease-causing, as Marx proposed; it is necessary to discuss this matter. But first it must be understood that in Confucian texts it is the child's perspective on the parent:child relationship (*hsiao*) which is emphasized, because there is the understanding, as there is in all cultures save our own, that parents, being mature and responsible beings who have "found themselves" as humans, are in little or no need of being persuaded that unconditional devotion to their children is a good thing. In contrast with that fact of *hsiao* being *reciprocally* a parent:child relationship, prevailing Sinological opinion has it that *hsiao* refers to the total obligation of children to their parents, who, in turn, may treat them dictatorially and unfeelingly, for no requirements are set upon their behaviors toward their children. Examples of the selling of children by minorities of impoverished Chinese people during decadent dynastic periods are then trotted out to illustrate this "general feature of Chinese social culture." Such historical and sociological authorities should be asked if the epidemic run-away, prostitution, and suicide rates of American youth, and the rampant child-selling, or mere giving-away of their children, of young American women, most of whom are not in a state of poverty, must, therefore, be attributed to an American correlate of the—as they interpret it—traditional Chinese concept of *hsiao*, according to which parents can do whatever they want to their children. In other words, before one presumes to evaluate a foreign social culture one should, it seems to me, have established that one's own is more Human, and have made sure that one is not attributing to that foreign culture traits which are true, not of that culture, but of one's own, so as to preserve the illusion that one is in a position to evaluate it. Indeed, one should ask if it is not largely due to the prevailing Western Academic opinion that the parent:child relationship is over-valued and possibly psycho-social disease-causing that more and more of our children, who live in a world constructed according to the social-scientific paradigm of which that anti-Human opinion is a central component, are suffering so terribly.

Now, the only point in trying to show people that Confucianism is Human-ism and that the particularly attractive features of traditional Chinese society and its vestiges existed and exist, is to establish the world's largest set of evidence to the effect that the Confucian social paradigm is good for people and actually represents the interests dearest to their hearts. What is important is not what (if there were more than one) "kinds of Confucianism" there were, or whether they were natively Chinese or not. What is important is whether or not the social paradigm in question would have an influence superior to, healthier than, those which presently predominate. The endless and inconclusive arguments that are and can be waged about the preceding questions are nothing other than a distraction from the actually important consideration. It is for this reason that Sinologists such as the previously quoted Learned Being are 100-percent wrong, and are not, according to my (Human) definition, "gentlemen-and-scholars."

So, in order to preclude the onset of such interminable, inconclusive, and quickly boring arguments, I assert that Confucius never existed, or, if he did, was an imbecile who owned many slaves and hated women and children, and that anything good about traditional Chinese culture is purely accidental or a sly borrowing of unidentified Western social-scientific ideas, originally ideas diffused from Babylon or Atlantis.

According to Confucianism, all human empathy comes from parent-child love, and that love is natural. Since Western intellectuals are fond of comparing humans to animals and speculating about which animal traits have been retained by humans via the Evolutionary process, and to further discuss the parent:child relationship (which, again, is in all cultures other than our own not subject to discussion because it is sacred), let me tell another story. The retired farmer, Mr. Elmer Frizzell, who lives across the street from me had a bitch who was pregnant. The prospect of having to do away with some of her puppies reminded him of what had happened the last time he had done so. It had been in mid-winter, when the ground was hard, so that after drowning the pups the best he could do was bury them about eight inches under the soil in his woods, in a burlap bag. Later in the day, returning from work, he saw the bitch in his driveway pawing the burlap bag open. His voice went taut: "And, do you know, I walked up to her and there were tears streaming down her face." Has Evolution purged human females of instinctive emotions such as those of that mother of puppies? And if so, are we to assume (in consonance with the Academic understanding that is presently most chic) that the tautness in the voice of the old farmer is due to atavistic cultural conditioning which has no natural basis. Are we to believe that parent-love for children, as Marx and Engels concluded, is a mere justification for the Capitalistic ownership of one's children as low-cost laborers and extensions of one's land-owning powers. Are we to believe that mothers do not *naturally* hold their children more dear than anything on earth or heaven?

That matter cannot be decided "intellectually." "Scientific" experiments with animals and humans—including modern societies with their experimental "life styles," which are nothing other than laboratories testing the social-scientific hypotheses of people such as Keynes (father of modern Capitalism), Marx (father of Communism and modern socialism), Freud (father of institutionalized parent:child psychical conflict), and others like them—cannot determine what behaviors are "inherent" and what behaviors are "culturally conditioned," that is, what behaviors are acquired by social conditioning according to ideas about behavior dreamed up by inventors of social culture. The absurdity of a "scientific" approach to this matter—even aside from its being isolated from human feeling on the unfounded Greek Academic assumption that the mind is man's only decision-making capacity—should be obvious. Culture is made up of thoughts, and thoughts cannot exist without and are quite interdependent with the human body, which is inherent. Therefore the genetic inheritance of humans always has been, is, and always will be a conditioner of culture, and culture always has been, is, and always will be ultimately indistinguishable (Polar-Completely) from biological inheritance.* It follows that the human capacity to decide about anything, for example, this matter of whether or not total devotion to one's children is natural, naturally has two aspects, physical and mental, which are Polar-Complete aspects of a same thing, "linked" by a psychical-physical continuum of *ch'i*.

Despite the Western Mentalist "requirement" that such matters be decided intellectually, "educated" Westerners find themselves acting solely on the basis of the physical aspect of this continuum, or more exactly, on the basis of the more-physical-than-mental aspect of this continuum, emotion. For example, even the most ardent Marxist would experience unspeakable anguish and the desire for total revenge upon discovering that his mother, daughter, wife, or sister had been raped and beaten. What makes possible the obfuscation of this undeniable fact, and its implication of the existence and essence of Human Nature, is the queer mentality according to which the same Marxist regards as a negligible side-effect the rapes and beatings of thousands of Chinese women. Such was Mao's "Cultural Revolution," whose Marxist objective, communicated by inflammatory directives to not-yet-fully socialized children, was to destroy, through orchestrated violence and self-degradation, a culture based on Human

* There being a possibility that some academics will misconstrue this as suggesting a racist theory, I emphasize that here Human culture is distinguished from Mentalist ideas, and that there is such a thing as good or bad individual influence and intention.

Nature, specifically by fostering conditions for inter-sexual and intergenerational hate.* Their hearts frozen in the detached zone of the Mentalist paradigm, such Western social philosophers and the social planners they inspire are free to prescribe "medicines" which they could not tolerate themselves, and to promote society-chaotizing, anti-Human ideas, which often become social policies, which have, as one side-effect, a higher incidence of inter-sexual and intergenerational hate ("Class Struggle") and such attendant niceties as rape. For in a community which is self-governing, in which the men who live in it all ascribe to the right to protect their women and children and to punish those who injure them as they feel fit, and in which everyone knows everyone else, so that one treats others' wives and children as one treats one's own, few would dare molest the daughters of men from this community. Were one to do so, he would suffer consequences far more discouraging to other potential molesters of women than is the contemporary practice of sending them to psychotherapists, so they can share "being helped" with their victims, or the practice of sending them to prisons from which they emerge, punished but not rehabilitated, to again experience what our psychologists, in sympathy with rapists, not their victims, generously term "undeniable compulsions."

The Mentalist denial of Human Nature is the basic reason for the present, galling, epidemic of injustice and the consequent lack of social control (prevention of crimes against the person) in modern society. Where justice, after survival, is the chief object of society, it is worth explaining the connection I have just suggested between Human Nature, its denial or intensification, and injustice or justice. In the end it will become clear that the "false" in our hard-scientific sub-paradigm is *equivalent to* the "unjust" in our social-scientific sub-paradigm, and that the "true" in the Chinese hard-scientific sub-paradigm is *equivalent to* the "just" in the Chinese social-scientific one. It will also become clear that the ultimate product of the Mental-

* It is the popular Human reaction against the "Cultural Revolution" which has obliged Mao's successors to indict "The Gang of Four," and it is the fact that those indicters have used "The Gang" as scapegoats, while they themselves continue to suppress Human culture, which prevented them from executing Chiang Ch'ing: they do not have the right. Knowing that, she out-screamed them. (According to the Dialectic of Chinese Communism, Stalin's portraits have not been torn down and Mao's may be restored, once "the Masses" appear to be sufficiently recontrolled.)

ist paradigm is totalitarianism, and that the only genuine alternative to totalitarianism is Human self-government along Polar-Complete lines.

Here are two, opposite, examples of the adjudication of rape, one American-European and the other traditional Chinese, which symptomize the actual problem and its solution. I choose the crime of rape because it is one of the very few vilest of crimes and is a present national concern, since it, and fear of it, are epidemic. In the West, when a woman is raped she hesitates to report it, because she knows she will suffer from the judicial process. If, in hope of being avenged and protecting other women, she reports it and identifies the rapist, her person and her suffering are excluded from the body of "legal" evidence. Instead of *perceiving* that she has been raped, instead of responding in a Human manner to her agony, the officials who serve our supposedly Democratic system of justice take into account only circumstantial evidence—evidence around and other than the victim and her suffering. Her body is examined for evidence of violation, while what has been most violated, her soul, is ignored. Her testimony is compared with that of the rapist for logical consistency, and the influence that her shock and hysteria have on her "rational consistency" is ignored. And her past sexual behavior is evaluated to determine if she might have provoked the rape—as though being sexually provoked justifies raping, and as though rape, a brutal wounding of the soul, is a form of sexual relationship. She is, in effect, raped for being raped. The entire process implies the basic assumption that the victim's suffering is not real, and that she may have dishonestly identified the rapist. A priori, she is not trusted.

Further, instead of perceiving, in her suffering, the gravity of the crime, our system requires *not* fitting the punishment to the crime. The victim's suffering is personal, and therefore is not regarded as a legal matter. Consequently neither is the victim's natural desire for revenge satisfied, nor is there any serious deterrence to raping.

Finally, where her immediate relatives, for practical purposes her male ones in particular, feel they have the right to punish the rapist to the extent that he has caused his victim, and themselves, to suffer, neither are they "legally" permitted to do so, nor are they satisfied by the "punishment" of the rapist which the judicial system metes out. And in their stead are jurors who must *not* know either the victim or the accused, "because" her husband's, father's, son's, or brother's suffering is a personal, not also a legal, matter, and as though her suffering is not their business, but that of strangers. They also are raped, for the rape of one closest to them.

The traditional Chinese alternative will strike most Western readers as hard, but I ask the reader to suspend his or her judgment until the question has been fully covered, whereupon that "hard"-ness might be seen, rather, as humanity reappropriated.

In traditional Chinese society, according to customary law (democracy, being of and for the people, involves minimal written law), if a woman identifies her rapist—and normally she will—not only is the suffering of that woman and her immediate relatives taken into account; it is the very standard for all that ensues. The rapist has deeply wounded an entire family, and their corporate suffering is equivalent in gravity to the punishment by castration or execution of the individual rapist. This, of course, is exactly how the American common people feel about it—at least until a second thought about the dangerous illegality of such a sentiment. It is Human-Natural. Since it is the family which has suffered, it is the family which has the right to determine the punishment, on behalf not only of themselves but of all other families which might, otherwise, be victims of rape.

But unlike societies which have natural justice without ennobling culture, such as backwater Western exceptions where there ensues the endless vendetta, traditional Chinese society took the matter a crucial step higher. It institutionalized the example of taking another's suffering to be one's own, and the conception of society as a multiplicity of families. In reciprocal, exemplary, place of the victim's family, *the rapist's* family, or in their absence, those closest to him, was expected to enact the punishment. Ideally, at the order of his closest senior male relative, the rapist should execute or castrate himself before the altar of his ancestors, whom he has shamed. If he failed, the senior relative should do it for him. If his family failed to thus affirm its desire to rectify the situation, to affirm its compassion for the victim and her family, and thus to remain qualified to belong to society, then the family of the rapist had the right to execute the punishment. The case would fall into the hands of a public legal system only when neither of these alternatives had been realized, or when the just limits of revenge were exceeded (which, after such just punishment, would be exceptional). The public legal system acted on the same, Human, standards as the families normally would. In this way, deterrence of crimes such as rape was maximized so that the people, *democratically*, were protected; and the people were paid the due respect of being assumed normally capable of justice, and of being accorded justice by common human standards. The low rate of violent crime in traditional Chinese society was, of course, a well-known fact, although ironically it has been attributed chiefly to the supposed "passive nature of the race." The relative freedom from violent crime, which to a decreasing extent remains typical of Chinese society, was and is due to nothing other than the Chinese assumption of full, Human responsibility for justice—the most *active* act that a human can perform. A truly spiritual act.

Note how, through the empathetic identification of the criminal's family with the victim's family, a clear distinction between criminal and victim, punishment and the alleviation of pain, is drawn and enforced. This is the same pattern by which, much earlier, I showed that matter and energy are mutually identified to make a clear distinction between them. The same, Polar-Complete, paradigm underlies both cases. Now let us see how our Mentalist one underlies the injustice and lack of social control we increasingly suffer.

As human beings, Westerners know what is just punishment for and effective deterrence of rape, but we find it hard to accept a system which is just in that regard. Obviously, there are underlying factors in our social paradigm which interfere with justice, although they are obvious only from outside the system. They are symptomized by the preceding Western example. One factor is that Westerners have been conditioned for 2500 years or more to regard themselves—all humans —as incapable of administering justice. Indeed, there is *fear* of such responsibility. There has always been a supposed need for a "higher," therefore non-human, Authority which might shoulder part of the responsibility. In the beginning it was God, and now it is a set of rules for bureaucrats, which constitute the State. Of course, humans are then required to interpret the will of God or the State's laws, so the result is human decisions which attempt to be non-human: rational but not also Human-Natural. This is a justice of mind without heart. And that is a contradiction in terms. Man, who has heart, is the measure of all things, whether he has the strength to measure them or not. When he does not, there is no measure and no genuine justice at all.

Now we are close to the central factor: the Mentalist split of mind from heart, based on a fear and mistrust of Human Nature, and the elevation of the mind over the heart. The evidence recognized with respect to the Western rape case, the reader will recall, was all "intellectual." Emotions, the very souls of the victim and her relatives, were excluded. The social-scientific institution of this unnatural split originated in classical Greece, where a distinction was drawn between society, a "higher," rational sphere, and the totality of families which constitute it, a "lower" sphere of human feeling identified with "animal" life. According to this Binarily-Conning view of the social world, justice, as an aspect of the "higher" sphere, must be an intellectual process, and not also a matter of human feeling or common sense. With the decline of families' informal but real power, the situation has been gravely aggravated since classical Greece, and again, since Rome. The legal system has radically alienated itself from Human reality because of the loss of natural, "lower," controls. The law now protects the public, "higher," sphere exclusively, *as opposed to* the "lower" sphere of the families which constitute society. What is protected is a semi-imaginary, Mentalist "society"—ultimately, the State and its bureaucrats' positions within it. The Mentalist split increases the power and reality of the semi-imaginary society-State through a confusion of criminals with their victims, and of their victims with criminals; and reciprocally, it crushes Human-ness.

Criminals are confused with their victims when sympathy for the rape victim and her family is transferred, through a perverse sleight-of-hand, to the rapist. The "Humane" justification for this is, "If one person hurts another, and that is wrong, then how can hurting yet another person (the criminal) be right?" This confusion, which is purported to be the ultimate in "Humanism," is a direct result of the creation of and elevation over the "lower sphere," of "Society." In it, each individual may be confusedly regarded as "equal before the Law," *even after guilt has been established.* Where this split is not made, as soon as one has raped, one is no longer a member of society and no longer has normal rights. One is no longer "equal," for one is *socially dead.* For in this view no individual acts independently of families. The criminal has, rather than societal rights, duties toward the victims of his crime, which involve physically affirming his social suicide. Society as a higher sphere of family-less individuals is an ozone in which the actually "dead" are permitted to survive to haunt and torment the living, cavorting as it were with the Authorities whose ecological niche is this zone. But this becomes obvious only when the arbitrary nature of the society/families, mind/heart, split is recognized.

In turn, victims are identified with criminals when, as is the case in present America and Europe, the blame for rape is placed on men-in-general, precisely because of a fear or disinclination to recognize that the problem is the political aspect of the Mentalist paradigm. This is the most vicious dimension of the problem. According to the prevailing school of Feminism, which rightly seeks to prevent crimes such as rape but is social-scientifically confused, it is the male subjugation and devaluation of women (supposedly typical of all societies) which is responsible for the high frequency of rape and the injustice of the legal system in its respect. Since this is close to the truth, it obscures the truth and leads to its opposite, the total obfuscation of truth and justice.

In the West, as everywhere, the law, be it formal or customary, is chiefly in the hands of men, just as the family is chiefly in the hands of women. But this becomes a problem only when, in addition, "The Law" is assigned to a "higher sphere" as opposed to the "lower" sphere of family and kinship. Then, by virtue of the "higher" sphere being thought of as masculine, the family is identified with women only, and women, by association with it, are conceived as "lower." It is discrimination against *the family and human feeling* that is responsible for the Western epidemic of rape-injustice, not discrimination against women. The split comes first; the devaluation of women is its side-effect. This is too obvious to warrant discussion as soon as one awakens from the hypnotic fixation on sexual conflict in modern society. One then realizes that the male relatives and husbands of raped women are raped when they are, that they are again raped when the direct victims are again raped by our "legal" system, and that all males, with the exception of a minority of alienated individuals, suffer, *impotently,* with their women in anticipation of the possibility of such double-rape becoming a reality in their own lives. Obfuscating the Mentalist split between heart and mind, such Feminist agents of justice are creating another, equally vicious, split between the sexes, thus contributing to the subjugation and breakdown of the family which is the root of the whole problem. Just as the right to have sexual intercourse in the context of love is the most cherished sexual aspect of a woman, the right to protect the women and children closest to him is the most cherished sexual aspect of a (civilized) man. Present "wisdom" on rape psychologically castrates men in the name of protecting women, and thus further alienates women from male protection, so that they can be raped more frequently.

The mis-identification of all men with the "Oppressing Class" and of women with the "Oppressed Class" is the ideology which underlies

Feminist wisdom, thanks to deliberate agents of Marxism within their chiefly innocent midst, and thanks to the increasingly Marxist character of so-called "higher education." That mis-identification is Marxism's most effective strategy for preparing non-Communist societies for totalitarianism: the ultimate in "Law and Order" of the "higher sphere." It is also the ultimate strategy for maintaining totalitarian control where it is already established, as by the "Cultural Revolution" of Mao Tse-Tung and of his wife, Chiang Ch'ing, for whose release from tentative death-sentence some well-known European Feminists are now (February, 1981) internationally petitioning.

Fortunately, despite the superlatively conservative, classically Greek, Marxist ideology to the contrary, Human Nature does exist, and what is natural has limits, so that a popular, natural reaction is taking place against this spiral of galling injustice and its "higher" objective, the coercive State and its invasion and misdirection of private, familial, life and Human Nature. Recently (Winter, 1981), a Californian rape-victim—who in my eyes is a folk-hero for whom a national holiday should be established—refused to undergo the foregone humiliation and frustration of our system of justice. She shot each of her rapists to death as they ascended the steps of the court-house. What's more, she was acquitted of manslaughter, albeit on the Human-ness-denying ground of "temporary insanity." She was acquitted because of the Human *force*, the Human *ch'i*, of her exercise of human right. Such Feminist-Masculinist, *truly* revolutionary, acts, such temporary *lucidity*, should increase, as we Westerners, by virtue of the limits set by our Human Nature, become more and more "Taoistic and Confucian." There is a current American proverb: "The law exists to protect criminals, not their victims." Its Taoist correlate, 2500 years old, is Lao-Tzu's "When laws rise up, the Natural Way (the Way of Justice) has declined." Likewise, people such as this Californian agent of the Tao might well quote Confucius, who said, "If I requite evil with charity, then how shall I requite good?" *This* is the common people's answer to the "Humane" pseudo-argument designed to fixate all human concern on the fact that prison is unkind to and does not rehabilitate rapists: "If hurting another is wrong, then it is wrong to hurt one who has hurt another." Of course, "Two wrongs *don't* make a right." Especially when human justice is regarded as a wrong, and the subjugation of families and human feeling to the State and unfeeling "rationality" is regarded as a right.

The whole matter of justice rests on a choice between two scientific paradigms. "Hard" Western science mistrusts the senses and splits the mind from them, to conclude that a "black body" is Absolute-Fragmentally black (the error being "negligible"). Identically, Western social-political science splits the mind from the common people's sense of justice, mistrusting common sense, its standard, human feeling, and its vehicle, human empathy (*Jen*). And any side-effects, such as epidemic rape, of the resulting system of justice are, supposedly, "negligible and inevitable." Just as the Polar-Complete pattern of all concrete phenomena is obfuscated to produce physical side-effects, the Polar-Complete pattern of all societal phenomena is obfuscated to produce societal and psychical ones. But just as light inexorably escapes a "black body," so does the at-once-rational-and-emotional lucidity of victims of our social-scientific sub-paradigm escape the obscurity of its Mentalist "justice." In turn, just as any two opposed physical properties are ultimately identical, so are any two opposed features of human society: men and women; society and the families which constitute it; rational justice and emotional justice. One can *obscure* this reality, but one cannot *destroy* it. The Tao-paradigm is not merely a theory; it is also a natural property, a world-force that exists everywhere, including the steps of a Californian court-house. As my *shih-fu* said after explaining the differences between the Western and Chinese paradigms:

> Now that you understand the differences, let me say this, and hear it well. The West, also, has the Tao. It is just that in the West and the countries which have Westernized, the Tao is obscured, whereas in healthy society it is illuminated.

How is it "illuminated"? With respect to justice, it is illuminated by moral education, by which is meant, not preaching, but reaching the heart. With respect to society as a whole, it is illuminated by expanding instead of suppressing the root expression of Human Nature, the parent:child relationship, in which love and empathy exist at their greatest intensity, and by which men and women are *sacredly* bonded as actual or potential coparents. That question of moral education concerns the only legitimate compunction one might have about a system of natural justice based on Human Nature, so I discuss it first.

Is not the tendency to convict and punish innocents because they are not of one's own "kind" as Human-Natural as is the *undeniable* compulsion to kill the rapist of one's mother, wife, daughter, sister, or self? This "xenophobic scapegoat syndrome" *is* natural, but unlike natural justice, it arises out of the weakest, not the strongest, aspect of

Human Nature. Therefore it can be prevented by cultivating the strong aspect of Human Nature. And it happens that it has been strengthened, in the West and many other societies, where the weakness of the architects of such societies has prevailed. This institutionalization of lower Human Nature is nothing other than the concept of a Chosen People. Yoruba tribesmen used it, ethnically defined, to justify starving Ibo tribesmen to death. Vietnamese used it, ethnically defined, to justify starving Khmers to death in Viet-Nam—after which the Khmers migrated on elephant back to Cambodia, their homeland, where the genocide has been accelerated by Vietnamese and racist Chinese minions of Communism. (The latter's sin is the greatest, for as I'll show, it was Chinese civilization that institutionalized the prevention of such ugliness.) The Crusaders used it, religiously defined, to justify pillaging the Mideast. North Americans (originally Europeans) used it, religiously and racially defined, to justify enslaving brown people* and exterminating American Indians. Pakistanis used it, religiously defined, to justify the rape and destruction of Bangladesh. What Christians have done to Jews need not be repeated; it is unspeakable in any case. It is precisely this evil, in the West originally religious, although misinterpreted, concept, which is translated at the level of individual interactions into the xenophobic scapegoat syndrome, by which a theft in a Russian village automatically results in the mutilation of a Jew, and a rape in redneck America automatically results in the murder of a brown man. And predictably, as everything worsens in Europe and North America, Christian whites are beginning to blame Jews and "Blacks" for the problems they share with them. The "logic" of the Chosen People concept, as it has come down to us, is: "Our kind is chosen by God; therefore you are less than we; we are human; therefore you are less than human and do not have human rights."** (Unless you become like us, skin color permitting.)

The assumption, then, is that our non-natural system of justice prevents xenophobic scapegoating, which the *common* people will naturally commit if left on their own. But our *un*common system does not,

* As my former teacher, Cornelius Osgood, once observed, "I have never seen a black human being and therefore find it epistemologically difficult to call any people 'black.'"

** As I explained earlier, how a people use this concept is a function of their popular culture. I speak here of that concept's potential. The Christians, for example, generated this "logic" from a prior Jewish concept whose intention was quite different.

in fact, prevent this; rather, it is consistently an agent of it, as everyone knows. American judges tend to sentence Blacks more severely than whites convicted of the same crimes.* So this justification for our non-natural system of justice is quite false. And the proper objective is to neutralize the Chosen People concept with its opposite, as part of a system of natural justice and self-government which in every other respect is plainly superior to the Mentalist one.

As all anthropologists know, the chief mode of moral education is not church sermons or its effete surrogate, "values-clarification," but personal example and stories told to children by their parents and grandparents. In the West, the stories told to children are about child-eating trolls and witches, whose modern version in adolescent folklore is the baby-sitter who, stoned on pot and caring for the baby for money, not kinship-love, carelessly cooks it instead of the chicken; about malevolent sorcerers and kings from foreign countries; and in TV kiddie-program form, villains who are almost unexceptionally the eldest adult male in the cast, and stereotypically foreign- (in fact, usually Chinese-)** looking ones. The net effect is first to establish that the (Western) world is extremely dangerous for children (which is true), and second to establish that evil in the world is due to people not of one's own "kind." In diametric contrast to this blot on the educative folklore of the world, every Chinese child (modernized Chinese not being culturally Chinese) was and is told stories such as the following one, which, thanks to Western technology, has also been made into a movie. To make this powerful story short, I summarize it. It is called *The White Snake*, and it is probably the most popular children's story in the Orient, from Korea to Singapore. A man meets, falls in love with, and has children by a mysterious woman from he knows not where. She is beautiful and is an excellent wife and mother of his children. One day, to his horror, he awakens in the middle of the night to witness her transformation into a white snake. A white snake with Human potential had fallen so deeply in love with this man that she had been enabled to transform herself into human form. There follows a series of heart-wrenching developments, the moral of which is that regardless of her origins, he should accept her as a Human wife. And the moral in that moral is: Even if a being is genetically non-human, if it acts like a human, treat it like one. There is little room,

* See Robertson, Chapter 12, cited in the *Bibliography*.
** Ming the Merciless, Flash Gordon's nemesis, is the archetype.

then, for racism (or for the self-image of being Chosen People and its side-effect, treating those outside one's group as not-fully-human) in the minds of children impressed by this story. And that was precisely the high-medical intention of the Chinese savant who invented it.

Note that the Human love which this story energizes is focused on the parent:child relationship and its source, the husband-wife relationship. The relative lack of racism in traditional China* and the resulting interracial marriage was one of the principal reasons numerous physically and culturally distinct groups** long ago blended into a peaceful whole which was still expanding until the Chinese Communists singled out "minority ethnic groups" to show the world how well *they* treat *them*—in the same spirit that Western Whites proudly show off their Indian reservations.

Such stories, by transcending even differences between living species, transcend the difference between one's family and one's community, one's community and one's nation, one's nation and another's, thus extending the kinship-feeling within the family to all other humans, and indeed to all other life-forms. Such moral education is an agent of the expansion of kinship-love. Human society *is* a family—a family united not through a culturally predatory God but through humans, specifically through human altruism. It follows that for altruism and justice to be maximally intense and far-reaching, the family must be maximally strong and valued highest. Its core is the parent:child relationship, which is Polar-Complete. The child is the parent, the parent is the child. Extended, the wife is the husband, the husband is the wife; the family is the nation; and the ruler is the common people. Polar-Completely. The love that is extended in this Polar-Complete manner is what is meant by *hsiao*. Its hard edge is justice; its source is love of children.

That leads to my concluding high-medical remarks. The parent:child relationship is naturally sacred because it is in fact the only possible general, common, source of Human-ness and because it is the most powerful one. This is because only that bio-social relationship is

* Unfortunately, the same cannot be said of China since Western contact and its own, simultaneous, decadence, which mixed with Western culture-shock to produce anti-white racism.

** Indeed, the earliest Chinese history identifies a *foreign* people as "The Black-Haired People": plainly the writer's people did not have black, or the blackest of, hair. Likewise, the literature has "curly-red-bearded" and "black-faced" heroes.

available to almost everyone, is general, common, and because only in that relationship or in relationships modeled after it, and in the husband-wife relationship which is based on it, is there unlimited and unconditional love. That love, or empathy, is the basis of altruism, or *Jen*, the Chinese character for which is the figure of a human and the figure two: two interacting, empathetic, humans. There are limits to the devotion between friends, unless they are so close that they regard each other as *brother* or *sister*, and there are narrower limits to the bond between acquaintances, and there is very little feeling between non-acquaintances. But there is unlimited feeling—under healthy social-cultural conditions—between members of same families. And when this feeling is nourished by Human culture, it then overflows its familial boundaries and Human-izes relations even between non-acquaintances, so that even those without parents are treated by others as though they were their children.

A genuine parent never gives up on the most refractory child. And a child may arrive at a point where he or she hates his/her parents, but the intensity of that hate derives from nothing other than the love that is between them. And the expression of that hate is likely to be characterized by *parent*-seeking behavior, a fact delightedly exploited by the "chicken-hawks" (adult homosexuals) who play "parent" to young runaways, exploiting their emotional needs to seduce them into homosexually satisfying their desires. It was along these lines that a dramatization of the insufficient love for children, which is the axis of our social paradigm, was recently effected, in San Francisco, where, as one of the high points of the Parade of Gay People, a nearly naked young boy "walked" down the street on all fours, like a dog, a "chicken-hawk" behind him holding a leash. From the Western historical-sociological perspective, I am sure that the relevant question to ask is if similar dramatizations were performed by the usually homosexual Greek Academics, the architects of our scientific paradigm, and their students; and whether or not, if so, long-term cultural continuity might be operative. This hypothesis, I am sure, would elicit a generous grant from one of the American academic granting institutions.

The unlimited extent of parent:child love stands to reason, because children are in fact nothing other than psychical-physical extensions of their parents, Polar-Completely *are* their parents. And the energetic viciousness of the ultimate misdirection of that love stands to reason, because that love is so powerful.

There is also a vigorously scientific basis for parent:child love. It is extremely Polar-Complete, the ultimate human expression of Nature's universal pattern. I explained with reference to the nature of life itself that unlike the machines after which we model our life- and social sciences, life-forms do not get started. There is no point at which life begins, because there is no death before it. (That is why our doctors will never be able to provide a certain decision with respect to the morality of abortion.) Parents extend themselves as their children: the fertilized egg is simply their genetic energy-matter relocated. Parents and children are actually one living being which at first has two and then has more than two physical identities, just as the color spectrum may be seen to have any number of primary colors. Life does not exist in discrete units; it unfolds and, with help from savants, flowers. The more primitive, that is, natural, a culture, the more directly it tends to express this. The more civilized a culture, the more it defends and perpetuates it as kinship-and-descent-based society. Of the former, the Walbiri aborigenes of Australia are a superior example. As Meggitt explains, they have what is called "dream-time," in which the present generations, through a suspension of time, are regarded as the same people as the first humans and creators of reality. The Chinese ancestral temple, which was Mao's "Cultural Revolution" 's central target, is a civilizational version of the same thing. The small-town American graveyard is a casual expression of it, overshadowed by a church which takes one's ancestors into another world, and a science which dreams of making individuals, *as such*, immortal in this world.

The (unobstructed) parent:child relationship being the most powerful, and a natural, generator of Human-ness, it is only logical to provide for society a model of society based on that relationship. Human-ness is thus able to extend itself in the most natural manner to relationships between relatives beyond the family sphere, and between non-relatives who, as a consequence of this model, treat each other as though they were relatives. A man who thinks of all elderly women as like his grandmother, whom he loves, cannot tolerate the sight of a mugger beating an elderly woman up—and being surrounded by such men, few if any people would dare mug elderly women. In turn, a punk whose grandmother deserted his mother has no feeling whatever, or sadistic feeling, toward elderly women. Nothing other than the breakdown of our familial system, made possible, basically, by the other-worldly and politically centralized West never having had a social culture which would prevent that possibility, is responsible for the epidemic of sadistic crimes against elderly people in our cities and the ease with which their children leave them alone in their own apartments susceptible to such atrocities, instead of living with and protecting them.

In contrast to modern society, traditional Chinese society called itself not a "state" or a "Democracy" but a *kuo-chia,* literally, "large-territorial family."

Any idea which denies or jeopardizes or distracts human vital energy from the most basic Human relationship, the parent:child relationship and its source, the human couple, is anti-Human, because, obviously, the generator of Human-ness must be operating at its maximal capacity at all times if human injustice and disorganization and the societal, psychical, and physical diseases it causes are to be kept to a minimum. That is why I say that, even granting the fact that religious morality is Human (for it has to do chiefly with family-supporting morals), ultimately it is anti-Human, because the source of these rules, according to religious doctrine, is not human beings, not the human heart, but a non-human agency, God. And the history of the church is a history of its expropriations from fathers and mothers of the most basic rights over their children: their right to ultimate authority about what is right or wrong and their right to legitimately marry (which has been expropriated, in turn, from the church by the state).*

I have said that if our scientists do not realize that their paradigm is defective, the common people will arrange the matter for them. Now that I have shown, as best I could, that the basis of a scientific paradigm itself is, in turn, its Humanism or lack of it, it is clear that the reason that the common people may ultimately arrange things for our scientists is that the world they have constructed for the common people attacks the common people's interest at its heart: the parent:child relationship. Our physical sciences chiefly are agents of industrial pollution. Our social sciences chiefly are agents of social-cultural pollution in the form of Right Capitalism or Left Socialism, agents of a future of one's children over which one has no control, but which promises to involve more rape, more drug addiction, more divorce, more life-flow obstruction, more cancer—and more abuse of these very parents as elderly people.

* Until the advent on their social territory of Christianity and Communism, Chinese families married independently of any external agency; so did the pre-Christian Europeans. Cross-culturally, this is the normal case.

There is a limit to human tolerance of anti-Human, disease-causing, conditions. If the common people whom our Knowledge-Specialists guide never come to intellectually understand the disease-causing nature of that "guidance," they will "understand" it "at the gut level." In turn, as shown, the potential for a Human paradigm, because it is natural, is constantly present. So, as our social-"scientific" paradigm and reality erode from the friction of stress, the potential for a Human scientific paradigm increases; and if the social reality constructed according to our mind-and-heart-divorcing paradigm is not healed, present social reality, which politically is an interdependent Left-and-Right, will break down. The alternative is that our Knowledge-Specialists will Human-ize themselves and the reality they influence. Hence, either way, there is every reason for the highest optimism.

The ultimate prescription, then, is self-government—"using Human ideas to redirect the *ch'i*" in a natural flow, to use a Chinese high-medical expression. Self-government's basic unit is the individual. Because the individual is a Polar-Complete aspect of many human relationships, as a result of self-government those relationships are changed. Hence, every individual's behavior influences every other individual's behavior and therefore actually *is* everyone else's behavior, in one Polar-Complete respect, as well as itself, in the other Polar-Complete respect. That is, social behavior, like the physiology of the human body and every other phenomenon investigated in this book, is a continuous field of which each node (here, not electrons, atoms, colors, brain cells, or organs, but human individuals) is a determining factor. And that factor is (as in all the other cases) demi-physical as well as physical. It follows that the intensity of the spirit of each individual determines his or her therapeutic efficacy. And the spirit is most intense when there is an object which most intensely attracts it. So we are returned to the parent:child relationship, which is the relationship of maximal attraction.* These relatively scientific, or "Taoistic-medical," understandings understood, and their understanding necessary to understand Confucius, I now quote from the Confucian *Ta Hsueh:Great Learning*, the first text a student of traditional Chinese, or Human, science must study:

> In order to bring human affairs as close as possible to the per-

* We tend to forget the *time* dimension because of our intergenerational discontinuity. *Sexual* attraction not only fluctuates but is ultimately about the parent:child relationship, which does not.

> fection of the Tao, the ancients had to make explicit and obvious to everyone the luminous virtues inhering in human beings. To do so it was necessary to regulate in a natural and harmonious manner the territories in which they lived. To do that it was necessary to arrange in a natural and harmonious manner their own families. To do that it was necessary to cultivate their own persons. To do that it was necessary to rectify their own hearts. To do that it was necessary to have sincere intentions. To have sincere intentions it was necessary to attain genuine understanding. Attaining genuine understanding is accomplished by investigating the nature of physical things. Physical things investigated, the desired understanding was attained. . . . When the root [genuine understanding] is neglected, what arises from it cannot have a natural and harmonious configuration. It has never been the case that what was of great importance [genuine understanding, ultimately of the common luminous virtues, namely the Human-ness which is generated by healthy parent:child relationships] has been slightly cared for, and, at the same time, what was of slight importance [for example, one's Academic prestige, one's political power, one's great material wealth] has been greatly cared for.

But that was true of another reality constructed, in harmony with Nature, including Human Nature, according to a genuinely scientific, Human, paradigm. Nevertheless, all the realities constructed by human beings overlap, do not have absolute boundaries, so the movements of one resonate in all others. Those which resonate in harmony with Nature are relatively animate and therefore are self-regenerating and enduring. Those which do not are relatively inanimate, anti-animate, and therefore do not endure.

APPENDIX:

THE POLAR-COMPLETE WORLDVIEW IN THE WEST AND SOME DETAILS OF THE WESTERN BATTLE FOR WORLDS

There are some exceptional worldviews of Western origin which, looked at only in one or another aspect, appear to be Polar-Complete, or, at least, much the same as the traditional Chinese one, and some writers, accordingly, have suggested that they are all of a same type. As I see it, some are and some are not basically the same as the Chinese paradigm. Northrop, in his *The Meeting of East and West,* speaks of an "Oriental" system which includes the Hindu and Chinese ones, and which is more faithfully empirical than ours. Capra observes that Hinduism, Buddhism, the system of Heraclitus, and Taoism (a Western term for an aspect of the Chinese system) all have certain basic concepts which are consistent with the most advanced findings of modern physics. Northrop's and Capra's perceptions of these similarities, in the understanding of this student of Chinese science, are accurate and very valuable. However, if one's objective is to understand the Polar-Complete alternative to the Western Absolute-Fragmental paradigm, it is as important to understand the differences among all these worldviews, and which features are their *centers of gravity*. This is possible only when these systems are compared as *wholes*—an enterprise to which neither of the preceding writers was devoted. With the understanding that there are two basic types of worldview under consideration, this can be simply done.

Again, there are three basic features of the Absolute-Fragmental paradigm: its basic concepts, and, therefore, all its derived ones, are Absolutes, its perspective is Fragmental (one-sided, partial), and its attitude is socially sterile, involving a dwarfed image of human beings and a disdain for the ordinary. In contrast, the three basic features of the Polar-Complete worldview are non-Absolute concepts, an all-sided, wholistic, balanced, perspective, and, most important, a consummately Humane attitude, wherein humanity is central.

India has produced a magnificent, plural worldview and world in which parallels with the traditional Chinese scientific paradigm as well as with classical Greek, Absolute-Fragmental, worldviews are to be found. However, as a whole, the Hindu paradigm is Western, Absolute-Fragmental. Indeed, India is the Mother, via westward diffusion, of much that is basic to the Western worldview, such as the European languages, with their grammatical reflections of time, person, and sex, and the European writing systems, which are based on the choice to represent sound, as opposed to meaning as in the Chinese, Egyptian, and Central American Indian systems. These Indic-European cultural traits are classically Absolute-Fragmental, and, since they concern the word (of which mathematics is a variety), have fundamentally influenced all of Western thought.

To illustrate the Indic-European trait of grammatical inflection, the English verb *to walk* has one form, "*am* walking," when the subject is *I*, and has another form, "*are* walking," when the subject is *you*. It has one form, "*am* walk*ing*," when time is present, and another form, "*will walk*__," when the time is future. When an object belongs to a male, we use the form "his"; when it belongs to a female, we use the form "hers." In this way, our language marks, forces us to remark, the differences *I* and *you, present* and *future, male* and *female*. Chinese is free of such formal obligations. The verb form for *walk* does not change from one tense to another and does not change according to person of the subject; the pronoun forms do not change according to the person's sex. This relative formality of our languages has three distinctively Absolute-Fragmental effects. One is that we are conditioned to obey many more rules when we speak than are the Chinese, and, by extension, we are conditioned to obey many rules, to be subordinate to an abstract system. (Our languages, of course, are but one aspect of a greater Western cultural pattern.) The Western mind is not as free to express itself as is the Chinese one: the abstract language system takes up human space, diminishes, and restricts the expressive spirit of the speaker. This is consistent with the basic feature of our Absolute-Fragmental worldview, in which humanity is subordinated to God or The Environment, to an abstract system of "Divine Will," "Evolutionary Law."

Another effect of the originally Indian formality of our languages is, simply, the tendency to be preoccupied with form as opposed to content, with words and speaking, more than with what words and speech refer to—the human and natural concerns and realities that are the reason for speaking in the first place. Our highly formal languages obstruct perception, are a film between us and what we talk about. (Before we can refer to a person by other than *his or her* name we must compute the form for sex: *he/she;* and so on.) This preoccupation with form, this abstract perspective, is consistent with Idealism and with the more general detachment from human reality that is true of both sides of the Binary Con. (The freedom from such formal constraint is one of the reasons that Chinese poetry strikes Westerners who can read it as relatively pure, natural, directly reflective of Nature and human feelings.)

A third effect is that our basically Indic languages encourage us to emphasize differences—among present, past and future, among I, you, and third person, among male, female, and neuter (and thus, not only between sexes, but between animate and inanimate). Emphasizing differences, obliged by grammar to do so, one is inclined to think in Absolute-Fragmental terms: time is not a continuum; you and I are Absolutely different, are totally separate; the sexes are Absolutely different, opposed; Spirit (life) and Matter (the inanimate) are Absolutely different, discrete, if intermixed, things.

The difference in the effects of writing based on sound and writing based on meaning is more straightforward. The Western writing system represents only the sound parts of words, in no way represents the meanings of words. So, everything that is written in the past is in no way insured against being interpreted in the present differently from what its writer originally intended. Accordingly, words in our language are relatively free to, and do, change their meanings over time. Western meaning is entropic. A science entirely devoted to *guessing,* from overall context, what these changes have been, historical linguistics, has arisen to attempt to remedy the problem. In short, our originally Indic sound-based writing system—and it may be added that Arabic and Hebrew are also sound-based—disengages present from past knowledge and consciousness, separates the successive generations one from another, makes knowledge and consciousness unstable over time, and minimizes the chances of knowledge accumulating—a Fragmental effect.

In contrast, the Chinese, meaning-based writing system, basically a system of idea-pictures, helps to preserve original meanings and thus maximizes the accumulation of knowledge, overlapping the present generation with all the preceding ones—a Complete effect. Second, the sound-based system fails to provide a common medium of communication for people whose languages *sound* different and are mutually unintelligible, and thus minimizes communication among human groups and societies and encourages each to distinguish itself from each other on the basis of its language. Accordingly (but, of course, not only for this reason), India, the Mideast, and Europe are loosely associated aggregates of little provinces or countries, each of which is fiercely chauvinistic; to America is chiefly extended the English/French/Spanish division. India has about two hundred different languages with separate groups attached to them, is the social-cultural archetype of this Fragmentation. In contrast, the meaning-based system does provide a common medium of communication for peoples whose languages sound different, and accordingly (but, of course, not only for this reason), China is a huge social-culturally united whole despite the many mutually unintelligible spoken languages and dialects of its peoples.

At the level of worldview itself, all of the Hindu variants, which are arranged on a continuum from folk religion to sublime and brilliant esoteric philosophy, have the Idealistic, "other-worldly," penchant. It is this central characteristic which makes the correlation between the Chinese paradigm and Hinduism's highest variant, Advaita Vedanta, misleading when taken alone. For example, according to Advaita Vedanta all dualities are ultimately identical pairs, which is an aspect of the Polar feature of the Chinese system, grounded in seriously scientific empiricism. But, in turn, according to Advaita Vedanta, Matter is mental: the Idealist side of each pair is given primacy, and the standard, Idealistic belief that the "world of the senses" is an illusion is firmly set forth. Like its purely-Absolute Western religious-Idealist correlates, at its peak the Hindu worldview is a vehicle for the savant's escape from and absconding from responsibility for human reality—the living social body is cut into writhing castes; and Hindu mystics, their backs to it, gaze through the sun.*

The penchant is dramatically realized in extreme Absolute-Fragmental form. All of Hinduism has, in one form or another, the objective of "getting off the Wheel of Life" to become one with God (Brahman). This Wheel of Life, probably under the pressure of the ancient Persian colonization of India, was straightened out into a ladder, a caste system up through which one migrates by reincarnation and accumulated merit, to the highest caste and, finally, The Exit. In this way, Hinduism supports a terminally stratified social structure, a dissected body, in which one's birth is one's social and spiritual destiny regardless of one's merit, until death. The people are Fragmented into "genetic" groups separated by Absolute walls. (The European sacred royalty-freeman-serf-slave ladder is a reflection of this.) The inevitable adaptive response to this Absolute-Fragmental tyranny was the creation of *sub*-castes, whose relative "heights," unlike those of the castes, are negotiable through incessant competition. All told, the objective is to use an upwardly mobile group to get ahead of everyone in other groups and leave society behind. This diametrically contrasts with traditional Chinese society and worldview, with its remarkable lack of stratification and its democratic provisions for acquiring social status through individual achievement regardless of one's social class by familial or personal profession. In turn, the similarity between Hindu and modern American society is that of seed and fruit: American individualistic competitiveness replaces Hindu group competitiveness, and the hierarchy of clubs through which the individual moves, leaving his former clubfriends as far behind as possible, replaces the juggling hierarchy of Hindu sub-castes. (The anthropologist F.L.K. Hsu has pointed out some of the preceding connections in his book *Caste, Clan and Club.*) The Hindu and American social systems are direct reflections of historically connected world-and-humanity-disdaining Idealist worldviews propagated by "detached" intellectual elites, their faces glowing with the light of a supernatural, superhuman God. Despite some of its variants' Polar features, the Hindu worldview is classically Absolute-Fragmental and Western.

The history of Buddhism further illustrates this fundamental difference. Although the Buddha was a Hindu, of the Warrior caste, and

*Of course, there have been great, individual, exceptions, such as Gandhi. I speak here of usual, cultural, generally-influencing traits—worldview.

much of Buddhism's metaphor is Hindu in form, Buddhism flourished only briefly in India, under the only (native) emperor who ever unified it, the Buddhist Asoka, and took root not in India but in China, Korea, Japan, and Indo-China (in that order of success). This is because Buddhism, unlike its Hindu antecedent, says nothing of a God and has as its final objective not escape from, but selfless immersion in, the suffering in this world—and with the full knowledge of its causes that makes that suffering immeasurably sharper, in order to maximize Human-ness in this world by enlightening all others, each according to his or her capacities. Buddhism is objective compassion —for humans, by humans, and not for or by God. This is what made it consistent with the already established Chinese system which accepted it and which it, in turn, embellished. Buddhism, then, is most like the Taoist-Confucian paradigm which arose independently of it in China, in its Human-ism. There is nothing about it which has particularly to do with science, as what our Sinologists call "Taoism" does. But it agrees with the Polar-Complete paradigm (and Advaita Vedantism) in neutralizing the basic subject:object opposition, of which I speak in very specific scientific terms in Chapters Four and Seven.

The atypical Western paradigm most immediately thought of by most as Western, and which is Polar-Complete, is Heraclitus'. As best I can make out from the fragments of it which remain available to us, there is no difference at the level of basic ideas and attitudes between it and the Chinese one. And, therefore, it was this paradigm which the architects of the one we have inherited devoted much energy to suppressing (for reasons made clear in the text, they suppressed, insulted and obscured, not refuted, it, for they could not refute it). Since only fragments remain, I summarily interpret some of their central concepts to show their consonances with the Chinese ones.

Like all originators of scientific paradigms, Heraclitus used metaphors; he used words with established meanings to signify new meanings that are similar to the established ones. Metaphor is a way to "use what one knows to get at what one does not know," as Chuang-Tzu put it. Even our physicists do this. They call the universe in the (again posited) pre-"Big Bang" state the "Cosmic Egg" (a metaphor borrowed from pre-Aristotelian Greeks); Einstein in overthrowing Newton showed that gravity is not a relation but a "field"; loci in space that refuse to influence our measuring instruments are called "black holes," "kernels" of matter, and "bottomless pits." The terms of modern chemical structure are entirely metaphoric. Stephen C. Pepper, in his *World Hypotheses*, holds that certain types of worldview, or scientific paradigm, have "*root* metaphors"; actually, if one looks deeply enough one sees that *all* do.

Heraclitus used two root metaphors: *fire* as all substance-and-change and the relations inhering in substance-and-change, themselves. Fire, by which he meant "a fire," the fuel and the flame, is two opposite and interdependent things as one thing. The fuel, for example, wood, is (in itself) chiefly cold, solid, and static; but to a lesser extent, the wood is hot, subtle, and dynamic, for it in part makes up the flame. Thus, the wood and the flame are actually one thing with two Polar, ultimately indistinguishable, aspects, neither of which can be *accurately described*, as a set of properties, without reference to properties of the other. Therefore, neither of them could *exist* without the other. This logical deduction is empirically confirmed. There can be no (wood-) flame without wood, and, where flame includes everything that transforms other things by consuming them, as in decay, the digestive refining of soil, and the melting of ice, there could be no wood without flame.

Further, fuel and flame mutually increase and mutually decrease each other, so that sometimes there is more burning than increasing of fuel (as in Autumn), sometimes the opposite (as in Spring), and ultimately constant balance (as in a year).

By analogy to a fire, half the universe is like the flame and half the universe is like the fuel: the universe consists of Polar opposites in a continuous, cyclic relation of "want and surfeit" to each other. And the Polar opposites, being Polar, are *opposite yet ultimately indistinguishable*, ultimately one and the same thing. As Heraclitus put it, "Men do not know how what is at variance agrees with itself." (Compare Chuang-Tzu's contemporary statement to the same effect, above.) To use the standard terms of the Absolute-Fragmental worldview, the flame is Spirit and the fuel (and all other relatively earthly substances) is Matter. So, by analogy to the fire:fuel relation, Spirit is in minor "part" Matter and Matter is in minor "part" Spirit; and Idea (or Form) and Energy are aspects of the polar Unity, Spirit-Matter. In the text I show how this basic ontological feature of the Polar-Complete worldview leads to universal, wholly consistent, non-"side-affecting" (Humane) science.

The idea of Polar Unity is the axis of the Polar-Complete worldview and must be clearly distinguished from any notions of unity forwarded by proponents of the Absolute-Fragmental one. As Heraclitus stressed, the Unit *is* differences, *is* Polar opposites. Unity cannot exist without Polar opposition. And conversely Unity and difference themselves, after all, are Polar opposites, are interdependent in the way that flame and fuel are.

In fundamental contrast, the unity of Idealism (as in Hinduism and mystical Christianity) is God, which (or who) exists independently of all that proceeds from it, and the unity of Materialism is Matter, which exists independently of all that proceeds from it. The unity of Realism, in turn, is twofold. The ultimate unity is God, as in Idealism, and the secondary unity is that of Form and Matter, which, it must be emphasized, are not *Polar* opposites, are not ultimately indistinguishable, but, rather, are Absolutes, are Absolutely different. Hence, there is a "line" between them. It may be posited that they are conjoined, are thoroughly and inexorably intermixed, as did Aristotle. But since they are not ultimately identical, they are not truly united, one. In the place of this true Unity, this "variance that agrees with itself," there is an implicit "glue" (energy) that holds the absolutely different halves of the universe together. God, the one source of both, is the "gluer." Accordingly, Form and Matter are Absolutely discrete along an Absolute actual/potential "line." This Realist statement of both Idealism and Materialism, this fractured "unity," is the basic formula for all the contradictions, and all the side-effects, of science as we know it.

In the place of a primary and independent unity, God or Matter, Heraclitus posited the "Word" (*Logos*), the principle of Polar unity, of opposition-identity, which is *in* all things—including human beings, in whom it takes the form of thought. Thought (with speech) is the focal symptom of what makes us human—our Human Nature. It is our chief survival tool: our tiger's claw, our fish's fin, our monkey's agility. The *Logos* is inherent in humans, but it atrophies unless it is cultivated, by a truly scientific means of learning. Otherwise put, as our unstable and conflictful scientific and social history testifies, it is possible for human beings to have a false worldview whereby they unscientifically "know" the universe, are disharmonious with it, out of sync with it, only pretend to know it while actually they project a primitive, Absolute-Fragmental, form of thought (also inherent in humans) onto the universe. Heraclitus was fully aware of this, for such was the case during his time. He therefore stressed that men must be content with the necessary (which means natural) process of the universe and not rebel against it. In terms of knowing, this means that they must *reflect* Nature, not, as did Thales, Pythagoras, Parmenides, Democritos, Plato, Aristotle, and their successors of 2500 years, *forcibly project* an aspect of themselves upon Nature, dividing it up into Absolute Fragments for which there is no evidence. In sum, there are two kinds of word, true and false, and it is according to kind that a true or false reality is created, through a worldview based on a kind of word.

I turn from the general picture of things (space) to causation (time). Where Idealists and Realists posit God as ultimate Cause, and Materialists, as shown, posit (non-Natural) "laws" (and thus reveal themselves to be "closet Idealists"), Heraclitus posited what he metaphorically called *exhalations,* out-breathings. The notion of "breath" as cause, especially of life, is very widespread on our planet, and is familiar to us: *Genesis* has God "breathe" life into the "clay" that Adam was made of. Heraclitus' "breaths," however, are not the means of and substances of a God, a "first" cause, but are general phases in the cycle of increase and decrease in the powers of different forms of Spirit-Matter, specific aspects of the *Logos,* the natural principle-force. They are *natural* causes of, or more precisely, *conditions for,* phenomena that in relation to them are yet more specific, are their effects, or more precisely, *unfoldings, fruits.* (They are *ch'is.*) Night and day, for example, are specific effects of the dark "breath" and the bright "breath"; others are Winter and Summer, midnight and noon, new moon and full moon, eclipsed sun and Summer-solstice sun, *death and life.* In accord with the Polar-Complete *Logos,* none of these "breaths" is Absolute or independent of its Polar opposite, nor are their effects: night and day, Winter and Summer (and Fall and Spring), midnight and noon, new moon and full moon, eclipsed sun and Summer-solstice sun, *death and life* are all continuous, ultimately indistinguishable pairs, each member of which is never purely, Absolutely, dark or purely, Absolutely, bright. Likewise, each Polar phenomenon is in fact "instantaneous," of "vanishingly short" duration, in fact *non-existent* as a thing-in-itself. Each is a function of each other, is "relatively there" in relation to its opposite, which is "relatively not there," in time. This relative "there-ness," or *intensity of presence,* has the position in the Polar-Complete worldview that Aristotle's absolute Potential-*vs.*-Actual has in his. So, like properties of things, time is Polar, is continuous, has opposite terms that are ultimately identical, has no basic units, no Absolutes. This logical-empirical truth Heraclitus emphasized in saying "All things are in a state of flux."

That understanding has been distorted through interpretations configurated by the *mystical* Absolute-Fragmental worldview that supplanted it. Aristotle, for example, objected to this notion because it violated his deduction that there is Being as well as Becoming, understanding Heraclitus to mean that there is only Becoming: that basically, all that exists is Becoming, itself. Heraclitus was simply summarizing the empirical evidence (which I will fully provide later on), and, unlike those who supplanted him for 2500 years, never fell into the trap of reification, of "making into things" abstractions such as Being and Becoming. It was not that Heraclitus differed from his supplanters in positing Becoming as the basic state of all things, but that he was operating scientifically, instead of reifying abstractions from a position totally alienated from Nature. He meant that change is constant, and that there is no such thing as a discrete, individual thing or event. No discrete-individual night, no discrete-individual day and no discrete "moment" during which such a thing might exist. Rather, there is *relative*-night and *relative*-day, increasing-decreasing aspects of a whole, day, which, in turn, is an aspect of the Polar bright-dark "breath."

True Humanism is implicit in Heraclitus' worldview, because it lacks the Idealist and Realist placing of God over man, "other" world over "this" world, and location of source of life outside Nature; *and* because it lacks the Materialist and Realist view of man as machine and society as "mass." It follows that human beings, being capable of achieving *Logos*-thought (what has been perverted into the notion of "being logical, rational"—something quite different), can control themselves and their relation to the rest of the universe, and have the full responsibility to do so.

Natural human behavior comes to full flower when it is not obstructed. The pincer movement of Idealist impotence under "God" and Materialist impotence under (human manipulators pretending to represent) "Natural Law" is precluded; the human being is not obstructed in his task of becoming fully adult, fully human. Heraclitus, who had a dry and ironic sense of humor, once quipped to some scholars hesitant at his door, "Fear not, *the gods* are present even here." He meant, "Cast away your 'God'-inspired fears of *fully human* being, and do come in."

Finally, I mention Sufism, which appears to have originated in Persia and, as suggested by Gurdjieff, may derive from the Chinese scientific paradigm. Like true Taoism,* true Sufism has been misrepresented by Western scholars, and there is a religious derivative of it, with cults, to provide such "scholars" with ample, selected, indication that Sufism, entire, has nothing to do with science, or genuine understanding. I have quoted Gurdjieff above some chapter headings to show that, as the Chinese saying goes, "Truth does not vary from one place to another." Insofar as translations of Sufi writings indicate, there is little difference between the Chinese and the Sufi paradigm, so I will say no more about that. That Sufis were never able, as were the vehicles of the Chinese paradigm, to set up an enduring and large civilization I attribute to Sufism having arisen in the West at a time when the roots of the Absolute-Fragmental paradigm and the reality constructed through it had already been firmly established.

SOCRATES ENCIRCLED

1. *Socrates' "proof" that sense-data (colors and so on) are not real, do not exist.*

The color white, for example, exists neither in the thing nor "in the eye"; rather, it "passes between them." (Modern science agrees.) Therefore it has no position of its own, has no rest, has no Being. It is neither active (energy from the object) nor passive (sensing on the part of the eyes); hence it is between them, therefore has no Being. (Modern science agrees.)

Both his "therefores," obviously, follow from nothing stated in his argument. Rather, both presuppose that anything which truly exists, which is real, is perfectly homogeneous: is *all* object or *all* subject, *all* active or *all* passive. The argument is implicitly based on the proposition it purports to prove, is a vicious circle. (Vicious, of course, only for those who fall for it.) Not making that prior assumption, one may use Socrates' own argument as a little proof to the effect that, as our senses and simple inference

* As explained, there is no Tao-*ism,* no Taoist "school."

tell us, color is both object-energy and perceiver-energy at once, and that, furthermore, the two cannot be Absolutely distinguished. Put in abstract terms, color Exists, not-Exists, and Becomes, all at once. Socrates simply arbitrarily defines, at the start, anything "that exists" as an Absolute-Fragment. If human experience is more subtle and complex than that, well, demean it and ignore it. He also makes the prior assumption that what "is real" is "self-existent," cannot depend for its existence on anything else. Again, this is simply to insist that "what exists" must be an Absolute-Fragment—to insist that the real world be a dead mosaic of discrete homogeneous fragments and to preclude all data which disobeys him: to preclude from intelligent attention the whole universe minus his small, quasi-logical mind. (I've shown that modern science is essentially no different.)

2. *Socrates' "proof" that perception is not a valid medium of truth.*

(a) Different men have different perceptions; dogs and men have different perceptions, of same things. Therefore perception is not a valid medium. This makes the prior, unstated, assumptions that (1) there are "same things" out there to be differently perceived, (2) that difference implies falseness. Again, a priori the argument is based on the formula it claims to prove. The things "out there" are Absolutely separate from the perceptions of them (object and subject are Absolute-Fragments); and perceptions must be perfectly homogeneous, Absolute-Fragments, to qualify as truths.

The *lifeless poverty* of those assumptions, which are fundaments of our scientific paradigm, is revealed in light of the Chinese alternative. The real world of objects and perceptions, of all different living-perceiving things, is a Harmonious Manifold, *T'ai-P'ing.* Recognitions of it are truths. The truth is that whole—with its regularities within irregularities, its almost-Absolutes within its non-Absolutes—and its *life,* which is a function of that manifold harmony. Each part of that truth, each particular perception and thought about it, is non-Absolute-Fragmental, Polar-Complete (energy is relatively energetic energy-matter, matter is relatively material energy-matter; human and dog perception are similar; and so on).

There is more to that. If all men agree that they and dogs perceive differently, all men have a same perception in this case. The Western argument is in this respect based on what it seeks to disprove.*

(b) Perception is involved in dreams and such perception is false. Therefore at best perception is either true or false. To that I can only say, dream-perception is not necessarily *false,* it is necessarily *dream,* perception. Why exclude dream data from reality, unless it be that one's own dreams, because of one's doubts about one's own human validity, are exceptionally unpleasant? As Chuang-Tzu, in the simultaneous Chinese sector of the Worlds-Battle, said: "I dreamed I was a butterfly; but can I be sure that I then awoke, that it is not that I then slept and dream I am a man?" Surely Socrates' own "certainty" about this matter is based on his *perception* of a difference between dreaming and waking, a perception shared by all other humans.

3. *Man, contrary to what Protagoras asserted, is not "the measure of all things."*

If man, with his perception, is the measure of all things, then a tadpole, which also perceives, is the measure of all things, and man, in turn, can judge the gods. To that I say, whatever a tadpole perceives, all we can know about it is through human perception of tadpoles and the objects they perceive—*we,* who are humans. The intention of his argument, to demean human perception by comparing it to a tadpole's, is based on a *human perception* of the difference between human and tadpole perception: the argument is circular. Rather than demeaning human perception, objectively seen Socrates' argument elevates it. The similarity between the simultaneous Greek and Chinese Battles for Worlds is remarkable: when Chuang-Tzu's opponent challenged the knowing behind his assertion that the goldfish they were ob-

* The same is true of its modern version, the Cultural-Relativist position that, since each language and culture has a unique structure—and each part is a function of the overall structure—no language-and-culture can be translated into another even if the elements of that other are reorganized (as I am doing in this book). Hence, there are no cultural universals and therefore no cultural truths (truth *is* universal). But, as I pointed out to my teachers at Yale, if one has determined that one's own and another people's language-structures are different and has expressed those differences in one's own language, one must already have translated the other language into one's own. Hence there are language-universals, based on human nature, and, possibly, cultural truths.

serving were happy, Chuang-Tzu said: "How do *you* know that *I* don't know they are happy?" There is accurate and inaccurate perception, fine and gross perception, heightened and suppressed perception. But that is a matter which Western science can say nothing about since it has denied perception as a medium of truth. The argument that man is not the measure of all things because if he were he could judge the gods is not an argument but an attempt to scare Theaetetus away from asserting his humanity—an old and still relied-upon religious trick. If successful, it permits the trickster to mediate for the gods at the expense of his victim's self-development as a human. (With that, I judge the gods.) As Heraclitus double-ironically said to two "scholars" hesitating at his door: "Fear not and enter. The gods are present even here."

BIBLIOGRAPHY

Note. This is not a bibliography exhaustive of relevant sources; since the Western and Chinese sciences are covered as wholes, such a bibliography is impracticable. It is a bibliography of sources cited. However, it points in all the directions of interest to a reader who might wish to consider further the questions raised and answers given in this book.

Adler, Jerry, and Marianna Gosnell. "Stress: How It Can Hurt." *Newsweek,* April 21, 1980.

Aristotle. "The Metaphysics, Natural Science, Zoology." In *Wheelwright's Aristotle,* trans. P. Wheelwright. Odyssey Press, 1951.

Barret, William. *The Illusion of Technique.* Garden City, N.Y.: Anchor Press (Doubleday), 1978.

Becker, Robert O. "Boosting Our Healing Potential." In *Science Year: The World Book Science Annual 1975,* eds. Harrison Brown, et al. Chicago: Field Enterprises Educational Corp., 1975, pp. 40–55.

Benet, Sula. *Abkhasians: The Long-Living People of the Caucasus.* New York: Holt, Rinehart & Winston, 1974.

Berger, Peter L. *The Sacred Canopy.* Garden City, N.Y.: Doubleday, 1967.

Bethell, Tom. "Burning Darwin to Save Marx." *Harper's,* December, 1979.

Beveridge, W.I.B. *Influenza: The Last Great Plague.* New York: Prodist, 1977.

Birren, Faber. *A Grammar of Color: A Basic Treatise on the Color System of Albert H. Munsell.* New York: Van Nostrand Reinhold, 1969.

Bishop, J. Michael. "Enemies Within: The Genesis of Retrovirus Oncogenes." *Cell,* Vol. 23:1, January, 1981, pp. 5–7.

Boeth, Richard. Review of William Barret's *The Illusion of Technique.* Under "Books" in *Newsweek,* October 2, 1978.

Bragg, Sir William Henry. *The Universe of Light.* London: Bell, 1923.

Bricklin, Mark. *Natural Healing*. Emmaus, Pa.: Rodale Press, 1976.

Capra, Fritjof. *The Tao of Physics*. Denver: Shambhala; New York: Random House, 1975.

Castaneda, Carlos. *The Teachings of Don Juan*. New York: Ballantine Books, 1968.

Chang Chung-Ching. *Shang-Han Lun*. Second century B.C.

Ch'eng I and Ch'eng Hao. *Erh-Ch'eng I-Shu and Erh-Ch'eng Ts'ui-Yen*. A.D. eleventh–twelfth centuries.

Ch'i Po. *Huang-Ti Nei-Ching (The Yellow Emperor's Internal Classic)*. Traditional fourth millennium B.C. Reconstructed and elaborated by Imperial Physicians of the Han dynasty, third century B.C.

Chuang-Tzu. *Nan-Hua Ching (The Classic on Southward Florescence)*. Fifth century B.C.

Conklin, Harold. "The Relation of Hanunoo Culture to the Plant World." Doctoral dissertation, Yale University, 1954.

Copleston, Frederick. *A History of Philosophy*. Garden City, N.Y.: Image Books (Doubleday), 1959.

De Riencourt, Amaury. *Sex and Power in History*. New York: Delta, 1974.

Dewar, Stephen. "A Second Nervous System?" *The Canadian*, n.d., entire article reproduced within "Beyond Bionic Man," *Reader's Digest*, November, 1979.

Douglas, Mary. *Purity and Danger*. New York: Praeger, 1966.

Dubos, Réné. "Second Thoughts on Germ-Theory." In *Subversive Science*, ed. Paul Shephard. Boston: Houghton Mifflin, 1969, pp. 223–29.

Dudley, Horace Chester. *The Morality of Nuclear Planning*. Glassboro, N.J.: Kronos Press; Hinsdale, Ill.: Redsafety Associates, 1976.

Duke, Marc. *Acupuncture*. New York: Pyramid Books (Harcourt Brace Jovanovich), 1972.

Durkheim, Emile. *Suicide*. Glencoe, N.J.: Free Press, 1951.

Etzioni, Amitai, and Clyde Nunn. "The Public Appreciation of Science in Contemporary America." *Daedalus*, No. 103, 1974, pp. 191–205.

Francis, Diane. "Sex, Cancer and the Perils of Promiscuity." *MacLean's*, October 6, 1980.

Gilder, George. *Sexual Suicide*. New York: Bantam Books, 1975.

Goffman, Erving. *Asylums*. Chicago: Aldine, 1962.

Goodman, Paul. *Growing Up Absurd*. New York: Random House, 1956.

Graham, A. C. "Being in Classical Chinese." *Asia Major*, No. 7, 1959, pp. 79–112.

Gris, Henry, and Dick William. *The New Soviet Psychic Discoveries*. New York: Warner Books, 1978.
Gurdjieff, G. *All and Everything*. London: Routledge & Kegan Paul, 1950.

Hansch, Schawlow and Series. "The Spectrum of Atomic Hydrogen." *Scientific American*, March, 1979, pp. 94–110.
"A Haunt of Flies." In "In the News," *Discovery: The Magazine of Science*. New York: Time-Life, Inc., October, 1980, p. 11.
Hsu, F.L.K. *Clan, Caste and Club*. Princeton, N.J.: D. Van Nostrand, 1963.
Huai-Nan-Tzu. *Huai-Nan-Tzu*. Third century B.C.

I Ching (The Classic on Naturally-Smooth Transformations). Second-first millennia B.C.
Illich, Ivan. *Medical Nemesis*. London: Calder & Boyars, 1975.
Imperial Physicians of the Former Han Dynasty. *Su-Nü Ching (The Pure Maiden's Classic on Societal-Psychological Health)*. Second century B.C.
Inglis, Brian. "The Epidemic Trigger." *Omni*, November, 1980.

Jastrow, Robert. *Red Giants and White Dwarfs*. New York: W. W. Norton, 1979.

Key, Wilson Bryan. *Subliminal Seduction*. New York: New American Library, 1974.
Koestler, Arthur. *The Roots of Coincidence*. London: Hutchinson, 1972.
Koestler, Arthur. *The Sleepwalkers*. London: Hutchinson, 1959.
Korzybski, Count Alfred. *Science and Sanity*. Lakeville, Conn.: International Non-Aristotelian Library Publishing Co., 1958.
Kuhn, Thomas S. *The Structure of Scientific Revolutions*, 2nd ed. Chicago: University of Chicago Press, 1956.
K'ung-Tzu. (Confucius). *Hsiao-Ching (The Classic on Hsiao)*. (Recorded by his following-son, Tseng-Tzu, fifth century B.C.)
K'ung-Tzu. (Confucius). *Lun-Yü (Analects)*. Sixth century B.C.
K'ung-Tzu. (Confucius). *Ta Hsüeh (The Great Learning)*. Sixth century B.C.

Lao-Tzu. *Tao-Te Ching (Classic on the Virtues of the Natural Way)*.
Lane, Earle. *Electrophotography*. San Francisco: And/Or Press, 1975.
Levine, Rick. "Cancer Town." *New Times*, August, 1978, pp. 21–32.
Lock, Margaret. *East Asian Medicine in Urban Japan: Varieties of Medical Experience*. Berkeley: University of California Press, 1980.

Lynch, James J. *The Broken Heart: The Medical Consequences of Loneliness*. New York: Basic Books, 1977.

Lyons, John. *Semantics*. New York: Cambridge University Press, 1977.

Maruyama, Magoroh. "Mindscapes and Science Theories," *Current Anthropology*. Chicago: University of Chicago Press, 1956, pp. 589–99.

Maruyama, Magoroh and Arthur M. Harkins. *Cultures of the Future*. The Hague, Netherlands: Mouton, 1978.

Marx, Karl. *Das Kapital*. London: Lawrence & Wishart, 1970.

Marx, Karl. *The Communist Manifesto*. New York: Penguin Books, 1968.

Meggitt, Mervyn. "Understanding Australian Aboriginal Society: Kinship Systems or Cultural Categories?" In *Kinship Studies in the Morgan Centennial Year*, ed. Priscilla Reining. Washington, D.C.: Anthropological Society of Washington, 1972, pp. 64–87.

Melzack, Ronald. "Shutting the Gate on Pain." Chicago: Field Enterprises Educational Corp., 1975, pp. 56–67.

Melzack, Ronald. "Stimulus/Response: How Acupuncture Works." *Psychology Today*, 1973.

Mencius (Meng-Tzu). *The Works of Mencius*, Books III, VI, and VII. Fourth century B.C.

Mendelson, Robert S. *Confessions of a Medical Heretic*. Chicago: Contemporary Books, Inc., 1979.

Miller, Clyde. *The Process of Persuasion*. Cited in *The Hidden Persuaders*, Vance Packard. New York: McKay, 1957.

Morris, Henry M. *Scientific Creationism*. San Diego: Creation-Life Publishers, 1974.

Mueller, Conrad G., Mae Rudolph, and the editors of *Life, Time*, Inc. *Light and Vision*. Chicago: Time-Life Books, 1972.

Nan-Ching (Classic on What Is Difficult). Second century B.C.

National Research Council. *Science and Technology: A Five-Year Outlook*. San Francisco: W. H. Freeman, 1979.

Nauta, Walle J. N., and Michael Feirtag. "Organization of the Brain." *Scientific American*. September, 1979, pp. 88–111.

Needham, Joseph. *Science and Civilization in China*. New York: Cambridge University Press, 1954.

Needham, Joseph. *Three Masks of the Tao*. London: Teilhard Centre for the Future of Man, 1979.

Needleman, Jacob. *A Sense of the Cosmos*. Garden City, N.Y.: Doubleday, 1975.

Northrop, F.S.C. *The Meeting of East and West*. New York: Collier Books, 1946.

Packard, Vance. *The Hidden Persuaders*. New York: McKay, 1957.

Pepper, Stephen C. *World Hypnotheses: A Study in Evidence*. Berkeley: University of California Press, 1970.

Plato. *The Theaetetus*. In *The Works of Plato, The Jowett Translation*, ed. Irving Edman. New York: Modern Library, 1956.

Porkert, Manfred. *The Theoretical Foundations of Chinese Medicine: Systems of Correspondence*. Cambridge, Mass.: M.I.T. Press, 1974.

Porkert, Manfred. "The Dilemma of Present-Day Interpretations of Chinese Medicine." In *Medicine in Chinese Cultures: Comparative Studies of Health-Care in Chinese and Other Societies*, eds. A. Kleinman et al. Washington, D.C.: N.I.H.D.H.E.W. Publications, 1975.

Pribham, Karl. *Languages of the Brain*. Englewood Cliffs, N.J.: Prentice-Hall, 1971.

Prirogene, Ilya. Personal communications quoted by Marilyn Ferguson in *The Aquarian Conspiracy*. Los Angeles: J. P. Tarcher, 1980.

Reich, Wilhelm. *Selected Writings*. New York: Noonday Press, 1969.

Robertson, Ian. *Sociology*. New York: Worth Publishers, 1977.

Ross, Val. "The Great Psychiatric Betrayal." *Saturday Night*, June, 1979, pp. 17–23.

Roszak, Theodore. *The Making of a Counter Culture*. Garden City, N.Y.: Doubleday, 1969.

Ryckmans, Pierre (Simon Leys). *Chinese Shadows*. New York: Viking Press, 1977.

Schumacher, E. F. *Small Is Beautiful*. London: Blond & Briggs, 1973.

Shao Yung. *Huang-Chi Ching-Shih Shu*. A.D. eleventh century.

Silverman, Harold, and Gilbert I. Simon. *The Pill Book: The Illustrated Guide to the Most Prescribed Drugs in the United States*. New York: Bantam, 1979.

Simpson, George Gaylord. *The Meaning of Evolution*. New Haven, Conn.: Yale University Press, 1967.

Slater, Philip Elliot. *Earthwalk*. Garden City, N.Y.: Anchor Press (Doubleday), 1974.

Solzhenitsyn, Aleksandr Isevich. *The Cancer Ward*. New York: Dial Press, 1968.

Sun Sze-Mo. *Ch'ien-Chin Yao-Fang*. A.D. eleventh century.

Thomas, Lewis. *The Medusa and the Snail*. New York: Viking Press, 1979.

Totman, Richard. *Social Causes of Illness*. New York: Pantheon, 1979.

Von Bertalanffy, Ludwig. *General System Theory*. New York: G. Braziller, 1968.

Watson, Burton. *Chuang-Tzu: Basic Writings*. New York: Columbia University Press, 1964.

Whorf, Benjamin Lee. *Language, Thought and Reality*. Cambridge, Mass.: M.I.T. Technology Press, 1956.

Wintrope, Maxwell M., et al. *Harrison's Principles of Internal Medicine*, 6th ed. New York: McGraw-Hill Book Co., 1970.

Wright, William David. *The Measurement of Colour*. London: Hilger, 1969.

Wu Cha-Jen. *Chung-Kuo Shih-Pao (The Chinese Times)*. Taiwan, Republic of China, August 19, 1978.

Wu Ch'eng-En. *Hsi-Yu Chi (Record of the Westerward Journey)*. A.D. sixteen century.

Zukav, Gary. *The Dancing Wu-Li Masters*. New York: Morrow, 1979.

INDEX